Handbook of Usability and User Experience

Handbook of Usability and User Experience

Research and Case Studies

Edited by
Marcelo M. Soares, Francisco Rebelo
and Tareq Z. Ahram

CRC Press is an imprint of the
Taylor & Francis Group, an **informa** business

First edition published 2022
by CRC Press
6000 Broken Sound Parkway NW, Suite 300, Boca Raton, FL 33487-2742

and by CRC Press
2 Park Square, Milton Park, Abingdon, Oxon, OX14 4RN

© 2022 Taylor & Francis Group, LLC

CRC Press is an imprint of Taylor & Francis Group, LLC

Reasonable efforts have been made to publish reliable data and information, but the author and publisher cannot assume responsibility for the validity of all materials or the consequences of their use. The authors and publishers have attempted to trace the copyright holders of all material reproduced in this publication and apologize to copyright holders if permission to publish in this form has not been obtained. If any copyright material has not been acknowledged please write and let us know so we may rectify in any future reprint.

Except as permitted under U.S. Copyright Law, no part of this book may be reprinted, reproduced, transmitted, or utilized in any form by any electronic, mechanical, or other means, now known or hereafter invented, including photocopying, microfilming, and recording, or in any information storage or retrieval system, without written permission from the publishers.

For permission to photocopy or use material electronically from this work, access www.copyright.com or contact the Copyright Clearance Center, Inc. (CCC), 222 Rosewood Drive, Danvers, MA 01923, 978-750-8400. For works that are not available on CCC please contact mpkbookspermissions@tandf.co.uk

Trademark notice: Product or corporate names may be trademarks or registered trademarks and are used only for identification and explanation without intent to infringe.

Library of Congress Cataloging–in–Publication Data

Names: Soares, Marcelo Marcio, editor. | Rebelo, Francisco, 1962- editor. | Ahram, Tareq Z., editor.
Title: Handbook of usability and user experience / edited by Marcelo M. Soares, Francisco Rebelo, and Tareq Z. Ahram.
Description: Boca Raton : CRC Press, 2022. | Includes bibliographical references and index. | Contents: v. 1. Methods and techniques -- v. 2. Research and case studies.
Identifiers: LCCN 2021024350 | ISBN 9780367357702 (v. 1 ; hardback) | ISBN 9781032070292 (v. 1 ; paperback) | ISBN 9780429343490 (v. 1 ; ebook) | ISBN 9780367357719 (v. 2 ; hardback) | ISBN 9781032070315 (v. 2 ; paperback) | ISBN 9780429343513 (v. 2 ; ebook)
Subjects: LCSH: Human-machine systems. | User-centered system design.
Classification: LCC TA167 .H365 2022 | DDC 620.8/2--dc23
LC record available at https://lccn.loc.gov/2021024350

ISBN: 9780367357719 (hbk)
ISBN: 9781032070292 (pbk)
ISBN: 9780429343513 (ebk)

DOI: 10.1201/9780429343513

Typeset in Times
by Deanta Global Publishing Services, Chennai, India

Contents

Preface ix
Editor Biographies xi
Contributors xiii

SECTION 1 Usability and UX Concepts and Implementations

Chapter 1 UX Concepts and Perspectives – From Usability to
User-Experience Design 3

Manuela Quaresma, Marcelo M. Soares and Matheus Correia

Chapter 2 User Requirements Analysis 17

Martin Maguire

Chapter 3 Designing the Interactor – From User-Experience to Interaction
Experience: Conceptual Foundations and Contributions towards
a New Paradigm 27

Sónia Rafael and Victor M. Almeida

SECTION 2 Usability and UX in the Automotive Industry

Chapter 4 Usability Evaluation of Exoskeleton Systems in Automotive
Industry 47

*Maria Victoria Cabrera Aguilera, Bernardo Bastos da
Fonseca, Marcello Silva e Santos, Nelson Tavares Matias and
Nilo Antonio de Souza Sampaio*

Chapter 5 Proposals for the Usability of Automated Vehicles' HMI 59

Manuela Quaresma, Isabela Motta and Rafael Gonçalves

Chapter 6 Is the Driver Ready to Receive Just Car Information in the
Windshield during Manual and Autonomous Driving? 81

Élson Marques, Paulo Noriega, and Francisco Rebelo

v

SECTION 3 Usability and UX in Digital Interface, Game Design, and Digital Media

Chapter 7 Interface Design and Usability Evaluation of Voice-Based User Interfaces 117

Martin Maguire

Chapter 8 Accessibility Features in Digital Games that Provide a Better User-Experience for Deaf Players: A Proposal for Analysis Methodology 135

Sheisa Bittencourt, Alan Bittencourt and Regina de Oliveira Heidrich

Chapter 9 Game On: Using Virtual Reality to Explore the User-Experience in Sports Media 157

Ragan Wilson, Nina Ferreri and Christopher B. Mayhorn

SECTION 4 Usability and UX in Fashion Design

Chapter 10 Expropriating Bodies: Immateriality and Emotion in Costume Design 173

Alexandra Cabral

Chapter 11 UX, Design, Sustainable Development and Online Selling and Buying of Women's Clothes 191

Carolina Bozzi, Marco Neves and Claudia Mont'Alvão

SECTION 5 Case studies in Usability and User Experience

Chapter 12 User Testing in an Agile Startup Environment: A Real-World Case Study 221

Carlos Diaz-de-Leon, Sarah Ventura-Basto, Juliana Avila-Vargas, Zuli Galindo-Estupiñan and Carlos Aceves-Gonzalez

Contents

Chapter 13 User-Experience and Usability Review of a Smartphone Application: Case Study of an HSE Management Mobile Tool 241

Marcello Silva e Santos, Sebastian Graubner, Linda Lemegne, and Bernardo Bastos da Fonseca

Chapter 14 Encounters and Difficulties when Gathering User Experience Data 259

Arminda Guerra Lopes

Chapter 15 Parametric Design Method for Personalized Bras 279

Yuanqing Tian and Roger Ball

Chapter 16 Dimensional Aspects of Usability of the Beds 293

Aleksandar Zunjic

Chapter 17 Usability of the Back Seat of Wagon Cars – Recommendations for Design 319

Aleksandar Zunjic and Vladimir Lesnikov

Chapter 18 System Perspective in Usability and UX Design: A Case Study of an Indian Cooking Spatula 335

Somnath Gangopadhyay and Sourav Banerjee

Index 347

Preface

Attending to the needs of users is one of the main premises of design. Designing products that meet this need in a safe, efficient and comfortable way is fundamental to promote user satisfaction. Usability is the way to meet the needs of users efficiently and effectively. Good usability guarantees great satisfaction in the use and consequent satisfaction to the user.

Global challenges altered social and working relationships and the ways of interaction with products and systems. In order to attend to these new requirements, innovative methods of interactions must be thought of to promote a better user-experience.

This volume of the *Handbook of Usability and User Experience* presents research and case studies to be used for designing products, systems and environments with good usability and consequent acceptance, pleasure in use and good user-experience.

This second volume of the *Handbook of Usability and User Experience: Research and Case Studies* comprises 18 chapters, divided into five sections. The first section, with three chapters, discusses "Usability and UX Concepts and Implementations." The first chapter presents concepts and perspectives of user-experience; the second discusses methods that use requirements analysis activity elicitation, recording and analysis to guarantee the user-experience; the third introduces a model that aims to overcome and challenge the psychosomatic and sensorimotor limits of users and computational systems, and prevent the primacy role of either of the interaction agents involved.

The second section focuses on "Usability and UX in the Automotive Industry." It contains three chapters. In these chapters, the following are presented: a usability evaluation for exoskeleton in the automotive industry, describing methods and techniques for user interface test; research for automated vehicles' human–machine interface (HMI) focusing on system information, generated from design workshops using a co-creation method; and a study which investigates whether augmented reality information can positively influence the user-experience during manual and autonomous driving is also presented.

The third section, comprising three chapters and introduces "Usability and UX in Digital Interface, Game Design and Digital Media." The first chapter of this section reports on case studies of the use of voice-based assistants in home settings; the second analyzes how accessibility features, linked to hearing in digital games, are related to the principles of the Universal Design and Jakob Nielsen's usability heuristics in order to provide deaf players a good user-experience; and the third discusses how technological manipulation can affect user emotions in digital sports media particularly under the emergence of virtual reality.

The fourth section discusses "Usability and UX in Fashion Design." The first of the two chapters of this section presents the role of design in changing the body through costumes and textiles based on an ergonomic approach centered on the emotions implied in artistic performative works; the second introduces a new model to improve communication between web retailers and consumers to enable more conscious decisions on purchases and reduce the number of returns.

The fifth and final section presents case studies on usability and user-experience in various contexts in product design in seven chapters. The first contains aspects of the iterative process of adapting user test design (procedure and assessment tools) in a software as an agile service startup and the adjustments made to the test design depending on iteration. In the second chapter, a mobile human–machine interaction mechanism was examined to contextualize usability and user-experience design issues with mobile apps. The third introduces the challenges and barriers encountered during data-gathering processes in five case studies involving digital inclusion through assistive technologies. The following chapter discusses issues related to the discomfort and dissatisfaction of women wearing bras mainly caused by sizing fitting problems and presents a parametric personalized bra design algorithm that can generate personalized bra by using self-measurements of women users. The fifth and sixth chapters of this section present case studies and analyses based on usability aspects and ergonomic recommendations for beds and back seats of wagon cars. The last chapter of this section introduces the relationship between end users and the product from the perspective of the system using Cognitive Walkthrough (CWT) and Kansei Engineering methods to improve the design of an Indian cooking spatula.

We hope that this second volume will be useful to a large number of professionals, students and practitioners who strive to incorporate usability and user-experience principles and knowledge in a variety of applications. We trust that the knowledge presented in this volume will ultimately lead to an increased appreciation of the benefits of usability and incorporate the principles of usability and user-experience to improve the quality, effectiveness and efficiency of everyday consumer products, systems and environments.

Prof. Marcelo M. Soares
Shenzhen, China
Recife, Brazil

Prof. Francisco Rebelo
Lisbon, Portugal

Dr Tareq Ahram
Orlando, USA

Editor Biographies

Marcelo Soares is currently a Full Professor at the Southern University of Science and Technology – SUSTech, and previously at the School of Design, Hunan University, China, selected for this post under the 1000 Talents Plan of the Chinese Government. He is also Licensed Full Professor of the Department of Design at the Federal University of Pernambuco, Brazil. He holds an MS (1990) in Industrial Engineering from the Federal University of Rio de Janeiro, Brazil, and a PhD from Loughborough University, England. He was Post-Doctoral Fellow at the Industrial Engineering and Management System Department, University of Central Florida. He served as Invited Lecturer at the University of Guadalajara, Mexico, University of Central Florida, USA, and the Technical University of Lisbon, Portugal. Dr. Soares is Professional Certified Ergonomist from the Brazilian Ergonomics Association, in which he was President for seven years. He has provided leadership in Ergonomics in Latin-American and in the world as Member of the Executive Committee of the International Ergonomics Association (IEA). Dr Soares served as Chairman of IEA 2012 (the Triennial Congresses of the International Ergonomics Association) held in Brazil. Professor Soares is currently Member of the editorial boards of *Theoretical Issues in Ergonomics Science, Human Factors and Ergonomics in Manufacturing* and several journal publications in Brazil. He has published 52 papers in journals, over 190 conference proceedings papers, and 90 books and book chapters. He has undertaken research and consultancy work for several companies in Brazil. Prof. Soares is Co-Editor of the *Handbook of Human Factors and Ergonomics in Consumer Product Design: Uses and Applications* and the *Handbook of Usability and User-Interfaces (UX)* published by CRC Press. His research, teaching and consulting activities focus on manufacturing ergonomics, usability engineering, consumer product design, information ergonomics, and applications of virtual reality and neuroscience in products and systems. He also studies user emotions when using products and techniques in real and virtual environments based on biofeedback (electroencephalography and infrared thermography).

Francisco Rebelo is Associate Professor at School of Architecture – University of Lisbon (FA/ULisboa), Director of the ergoUX – Ergonomics and User-Experience Laboratory, responsible for Ergonomics in the Centre of Architecture, Urban Planning and Design FA/ULisboa and collaborator of the Interactive Technologies Institute – Laboratory of Robotics and Engineering Systems IST/ULisboa. He has been Principal Researcher in projects financed by the Portuguese Foundation of Science and Technology and multinational companies (i.e. Nokia, Siemens, Thales and Nespresso). His activities of teaching and research are focused on human-centered design, usability and user-experience. He has published more than 170 articles and is Editor of 12 books.

Tareq Ahram is Research Professor and Lead Scientist at the University of Central Florida. He received the Master of Science degree in Human Engineering from the University of Central Florida (UCF), Orlando, FL, USA, and PhD degree in Industrial Engineering with a focus on large-scale information retrieval systems from UCF, in 2004 and 2008, respectively, with specialization in human systems' integration (search algorithms). Dr Ahram served in multiple research roles in the United States, Brazil and Europe.

Contributors

Carlos Aceves-Gonzalez
Centro de Investigaciones en Ergonomía
Universidad de Guadalajara

Maria Victoria Cabrera Aguilera
Faculty of Technology, State University of Rio de Janeiro
Brazil

Victor M. Almeida
Lisbon University
Portugal

Juliana Avila-Vargas
Lupa – Laboratorio Centrado en el Usuario
Guadalajara University
Mexico

Roger Ball
School of Industrial Design
Georgia Institute of Technology
USA

Sourav Banerjee
Occupational Ergonomics Laboratory, Department of Physiology
University College of Science and Technology, University of Calcutta
India

Alan Bittencourt
Feevale University
Brazil

Sheisa Bittencourt
Feevale University
Brazil

Carolina Bozzi
CIAUD – Research Center for Architecture, Urbanism and Design, Lisbon School of Architecture
Universidade de Lisboa
Portugal

Alexandra Cabral
CIAUD – Research Center for Architecture, Urbanism and Design, Lisbon School of Architecture
Lisbon University
Portugal

Matheus Correia
Pontifical Catholic University of Rio de Janeiro
Brazil

Bernardo Bastos da Fonseca
Faculty of Technology
State University of Rio de Janeiro
Brazil

Regina de Oliveira Heidrich
Feevale University
Brazil

Carlos Diaz-de-Leon
Lupa – Laboratorio Centrado en el Usuario
Guadalajara
Mexico

Nina Ferreri
Psychology Department
North Carolina State University
USA

xiii

Zuli Galindo-Estupiñan
Lupa – Laboratorio Centrado en el
  Usuario
Guadalajara University
Mexico

Somnath Gangopadhyay
Occupational Ergonomics Laboratory,
  Department of Physiology
University College of Science and
  Technology, University of Calcutta
India

Rafael Gonçalves
Laboratory of Ergodesign and Usability
  of Interfaces
Pontifical Catholic University of Rio de
  Janeiro
Brazil

Sebastian Graubner
Graubner Industrie-Beratung GmbH
Germany

Linda Lemegne
Graubner Industrie-Beratung GmbH
Germany

Vladimir Lesnikov
SAT Media Group, Belgrade
Serbia

Arminda Guerra Lopes
LARSyS/ITI Polytechnic Institute of
  Castelo Branco
Portugal

Martin Maguire
School of Design and Creative Arts
Loughborough University
UK

Élson Marques
Centro Technologic Automation of
  Galicia (CTAG)
Spain

Nelson Tavares Matias
Faculty of Technology
State University of Rio de Janeiro
Brazil

Christopher B. Mayhorn
Psychology Department
North Carolina State University
USA

Claudia Mont'Alvão
Pontifical Catholic University of Rio de
  Janeiro
Brazil

Isabela Motta
Laboratory of Ergodesign and Usability
  of Interfaces
Pontifical Catholic University of Rio de
  Janeiro
Brazil

Marco Neves
CIAUD – Research Center for
  Architecture, Urbanism and Design,
  Lisbon School of Architecture
Universidade de Lisboa
Portugal

Paulo Noriega
CIAUD and ITI/LARSysy
Lisbon School of Architecture
University of Lisbon
Portugal

Manuela Quaresma
Laboratory of Ergodesign and Usability
  of Interfaces
Pontifical Catholic University of Rio de
  Janeiro
Brazil

Sónia Rafael
Faculdade de Belas-Artes
Lisbon University, ITI – Interactive
  Technologies Institute / LARSyS
Portugal

Contributors

Francisco Rebelo
CIAUD and ITI/LARSysy
Lisbon School of Architecture
University of Lisbon
Portugal

Nilo Antonio de Souza Sampaio
Faculty of Technology
State University of Rio de Janeiro
Brazil

Marcello Silva e Santos
Centro Universitário Geraldo
 DeBiase and Faculty of
 Technology
State University of Rio de Janeiro
Brazil

Marcelo M. Soares
School of Design
Southern University of Science and
 Technology – SUSTech
China

and

Department of Design
Federal University of Pernambuco
Brazil

Yuanqing Tian
School of Industrial Design
Georgia Institute of Technology
USA

Sarah Ventura-Basto
Lupa – Laboratorio Centrado en el
 Usuario
Guadalajara University
Mexico

Ragan Wilson
Psychology Department
North Carolina State University
USA

Aleksandar Zunjic
Faculty of Mechanical Engineering
University of Belgrade
Serbia

Section 1

Usability and UX Concepts and Implementations

1 UX Concepts and Perspectives – From Usability to User-Experience Design

Manuela Quaresma, Marcelo M. Soares and Matheus Correia

CONTENTS

1.1 Introduction 3
1.2 From Human–Computer Interaction, Usability to UX.....................................5
1.3 UX Concepts and Perspectives...7
    1.3.1 UX as a Result of User Perception Interacting with a Digital Product or the Usability of a Digital Product 8
    1.3.2 UX as a Result of User Perception Interacting with a Specific Product or Service...9
    1.3.3 UX as a Design Process...10
    1.3.4 UX as a Result of User Perception Interacting with Broad System.......11
1.4 Concluding Remarks 12
Acknowledgments...13
References...13

1.1 INTRODUCTION

Nowadays, much has been said about user-experience (UX) as an attribute of a product or service intended to be offered. Many companies consider a good user-experience as one of the leading value propositions and work strategically to deliver it in the best way. However, the definition of what a user-experience is and how it affects people's lives and companies' businesses is still very plural. Some define the user-experience as a result perceived by the user of an interaction with a digital interface, considering as a fundamental part the usability of the interface (Falbe, Andersen and Frederiksen 2017, Brooks 2014, Tullis and Albert 2008, 2013). While others arrive at broader concepts, interpreting the user-experience as the totality of the perceptions a user has with an ecosystem, where the digital interface can be one of the parts included (Norman 2013, Kuniavsky 2010, Ou 2017, Hartson and Pyla 2019, Rosenzweig 2015). However, UX's concept comes from a transition that the consumer society is going through, where digital technology has its essential role but is not the only factor.

The focus of consumer society is changing with time. We lived in the agrarian economy, the industrial economy and recently (and occasionally we still) the service economy. It is important to call attention to the fact that, in the last two decades, we are undergoing a transition to the experience economy, where consumption focuses on obtaining experiences (Pine and Gilmore 2020). Since the beginning of the century, studies suggest that experiential purchases (i.e., the acquisition of an event to experiences, such as a dinner, a day at the spa or a trip) make people happier than material purchases (i.e., the acquisition of tangible objects that someone wants to own, such as clothing and jewelry, a television or a computer) of equal value (Boven and Gilovich 2003; Carter and Gilovich 2010). Boven and Gilovich (2003, p. 1200) point out that experiential purchases make people happier because of at least three possibilities: "experiences are more open to positive reinterpretation," "experiences are more central to one's identity," and "experiences have greater 'social value'."

Nonetheless, this interest of individuals in experiences is fostered by the society in which they live because it has gone through constant periods of economic prosperity and material wealth. Therefore, the conquest of positive events in their lives feeds these people's well-being, and the greater demand for these events transforms these societies into Experience Society (Hassenzahl 2011). According to the author, if before it was necessary to go to places called exotic in favor of a search for experiences, today, the search is for simpler experiences, such as being with friends at a barbecue. The idea is to dissociate experience from an expense and slow down the day-to-day work, where much of the focus is to raise resources for their sustenance.

Experiences are inherently personal, unlike products that are factors or systems external to the buyer. Experiences exist only in the mind and memory of the person who has experienced them, whether on an emotional, intellectual, physical or spiritual level (Pine and Gilmore 2020). Thus, each experience originates from the event's interaction with the individual and his/her mental state. So, it is impossible for two people to live the same experience. Even with products, it is possible to have good experiences. According to Rossman and Duerden (2019), the decision to purchase a product in the experience economy is no longer linked to how many features a product can have. Still, it is associated with what the product can do for the individual's experiences. Both Hassenzahl (2011) and Hekkert and Schifferstein (2007) point out that the product or a device is a means that leads an individual to live an experience, allied to other factors inherent to the individual and the context in which he/she is located.

As a result, this change in people's behavior towards their consumption emphases has attracted companies' attention. In turn, they have placed a greater focus on "selling" the experience highlighting the quality or usefulness of their product or service. From the individual's point of view, the experience is an experience. Regardless of how the individual realizes it, his/her focus is on what he/she will feel, reflect, take from that moment, store in his/her memory and what he/she will share with those

UX Concepts and Perspectives 5

close to him/her. However, it is essential for those who design the experience to distinguish how the individual experiences it. Therefore, the mediator of this experience with the individual (be it a service, a product or an event with which he/she will interact) defines how the designer will approach and shape the user-experience (Hassenzahl 2010). So, a successful design must place the user in the first place in relation to the mediator (product, service or event).

Moreover, it should be noted that with the advent of the Internet, new technologies have emerged as well as new ways of interaction of the individual with the (digital) world. Innovations have appeared in the areas of service and consumption, especially with e-commerce. In the beginning, this digital communication network was restricted to the few consumers with enough purchasing power to acquire a personal computer (PC) or a notebook, added to the costs of Internet plans. This situation was reversed with smartphones' arrival that allowed greater access to the Internet for various economic classes. Therefore, as nowadays more people interact with digital interfaces, the industry sees as an opportunity the user-experience with these interfaces as a value proposition for its products. Thus, design as a process, especially the "human-centered design" is seen as the means to develop products and services focused on this experience (Norman 2013, ISO 2010).

In order to shape the user-experience, the industry is interested in the design professional who knows how to design for the user-experience (Quaresma 2018). In turn, he/she is called to develop the experience of technological interfaces and services, such as airline services, financial services (fintech – financial technologies) and startups' proposals (focus on the service experience through technological means) (Farrell and Nielsen 2014, Quaresma 2018). So, what is understood about the concept of user-experience (UX)? What differs between usability and user-experience? What is the relationship between the two terms? This chapter aims to analyze the concept of user-experience, its origins and the various perspectives around this concept.

1.2 FROM HUMAN–COMPUTER INTERACTION, USABILITY TO UX

In the evolution of interaction with technologies and at the beginning of what was concluded as the human–computer interaction (HCI) domain, we began to interact with the first computers, still very complicated and complex. Then we started using smaller and simpler computers, the personal computers (PCs), the laptops and, currently, we deal with a wide variety of interactive systems – notebooks, smartphones, tablets and smartwatches (Campbell-Kelly, Aspray, Ensmenger and Yost 2014). During this process, these human–machine communication interfaces have changed a lot (Hartson and Pyla 2012). It was a great novelty in the beginning and people were interested in what that technology could bring benefits to them; now, such technologies are already part of many people's daily lives. We no longer tolerate poorly designed interfaces, especially in matters of usefulness, usability and experience of use.

Human–Computer Interaction (HCI) began to be considered an area of knowledge around the 1970s. It has its origins in ergonomics/human factors, in Cognitive

Psychology, in Design and, obviously, in Computer Science. At first, issues related to interaction with hardware devices (Cathode-ray tube (CRT) terminals and keyboard), training, documentation (manuals) and text editors were dealt with, and then the focus became much more on interaction with the software in general (Hartson and Pyla 2012).

Before the consolidation of the term HCI, the area was called "human factors in computers," in what refers to interaction with hardware and "human factors in software engineering," in an approach more focused on interaction with the software. However, there is no way to separate one system from the other because where there is software, there is always hardware support. This includes its relationship with ergonomics/human factors, which is extremely strong, mainly in what concerns the research methods and analysis used in interface design, such as task analysis (Hackos and Redish 1998).

With the various studies that had been conducted in the domain of HCI since the early 1980s, together with the popularization of personal computers and the strong influence of Cognitive Psychology, the field of knowledge that we know today as "usability" (which was initially called "software psychology" by Shneiderman 1980) emerges. Understanding human behavior and performance in interaction, considering cognition, memory (short and long term), perception, attention and decision-making becomes fundamental for developing adequate solutions in communication between the human and the computer.

One of the most widely known definitions of "usability" is that of the International Organization for Standardization – ISO 9241-11 (1998), which defines it as "the extent to which a system, product or service can be used by specified users to achieve specified goals with effectiveness, efficiency and satisfaction in a specified context of use." Thus, after several attempts to define the term "usability," the three most important metrics of usability – effectiveness, efficiency and satisfaction – are specified: the first two being measured objectively and the third more subjectively, but still a result of successful achievement of the first two metrics (Brangier and Barcenilla 2003).

While usability metrics were being established, researchers in the domain of HCI (Shneiderman 1987, Norman 1988, Bastien and Scapin 1993, Nielsen and Mack 1994) worked and researched human–computer interaction issues through the interface, based on the theories and assumptions of Cognitive Psychology. The results of the studies led to what is very well-known in the field as principles, criteria and heuristics of usability, from renowned researchers like the ones mentioned above, Nielsen and Mack (1994) being one of the most well-known for his ten usability heuristics. Until today, all these principles are widely used in interface design and interaction design and are the basis for various methods of analysis and evaluation as well as are used to base guidelines for technologies and interactions of specific contexts – such as human–robot interaction (Campana and Quaresma 2017) and human–vehicle interaction (Harvey and Stanton 2013).

However, in the late 1990s and early 2000s, new components beyond usability metrics began to be questioned, such as pleasure, emotion and affectivity in interaction (Norman 2005, Riley 2018, Pavliscak 2018). Since the usability satisfaction metric is limited to the satisfactory outcome of the system's effectiveness and efficiency, what other components are part of the experience of interaction with the system?

Users already know how to recognize an interface/product with good usability and have gotten used to it, as it becomes the minimum expected. In this sense, Jordan (2002) makes a comparison with Maslow's hierarchy of human needs (Maslow 1970 apud Jordan 2002) in what he calls "New Human Factors," emphasizing that people want more than usability in interaction; they want to feel pleasure in that interaction.

From then on, usability becomes an essential requirement in an interface and an interaction. Now, it is necessary to motivate the user to "buy" the idea of the interaction and the use of the product or service; it is needed to fascinate and provide an excellent experience throughout the use and try to keep him/her using the product or service for as long as possible. All these factors led to the concept known as user-experience (UX). However, much of this experience is still the result of a good design that follows usability principles and remains a starting point for a good experience.

UX was first used by Donald Norman when working for Apple in the early 1990s and named his working group "the User Experience Architect's Office" (Norman 2013). At this time, Norman referred to the term as a concept of something broad that encompassed an entire experience that a person could have when interacting with any product and not necessarily digital, even though he was at Apple (a digital industry company). The concept defined by Norman is related to all situations in which the user is involved with the product, either interacting with it or thinking about it.

Nevertheless, because it has been used for the first time in the digital realm, the concept still has a strong focus on the development of digital products, as much as it is understood that the experience is something that transcends the interaction with a digital interface. Rossman and Duerden (2019, chap. 1) specify that an "Experience is a unique interactional phenomenon resulting from conscious awareness and reflective interpretation of experience elements that is sustained by a participant, culminating in personally perceived results and memories." The authors also point out that experiences are multiphased: starting with the anticipation phase, going through the participation phase (where there is direct interaction with the elements of experience or interfaces, whether physical, digital, interpersonal, tangible or intangible) and ending with the reflection phase, where the individual realizes the whole experience itself.

1.3 UX CONCEPTS AND PERSPECTIVES

Although there is a certain consensus that the experience is something individual, that it is felt by each person differently and that this feeling depends on several factors, the concept of user-experience (UX) has different perspectives. These perspectives depend on the type of interface (in its broadest sense) that is being designed or a reference to designing an interface. From a literature review, mostly based on recently published books (as of 2010), four main perspectives were found: (1) the experience is a result of the user perception in interacting with a digital product; (2) the experience is a result of the user perception in interacting with a specific product (not necessarily digital) or a service; (3) as a design process; (4) an interaction with a broader system with various elements of experience. These perspectives are presented below.

1.3.1 UX AS A RESULT OF USER PERCEPTION INTERACTING WITH A DIGITAL PRODUCT OR THE USABILITY OF A DIGITAL PRODUCT

This perspective on UX's concept concerns the user's perceptions of direct interaction with a digital product and often relates to how effective, efficient and satisfying the interaction is, i.e., the usability of the interactive product (ISO 1998). This view is also a highly emphasized perspective in the digital industry. In general, people who are hired to work with UX Design are responsible for dealing with the interface of the digital product precisely and need to have a broad background in usability heuristics and principles. It can be observed that many jobs offer demand for what is called UX/UI designer; that is, the person responsible for designing the user-experience will be the one who will work on the user interface of the digital product.

This point of view is quite reasonable since digital products play an important role in our daily lives and our everyday experience with the products presented to us that we have to deal with (Falbe, Andersen and Frederiksen 2017). The authors emphasize that UX is also usability, as this is a fundamental part of the user-experience. A product with poor usability can significantly affect the user-experience, causing frustration and negative aspects of the whole interaction process. These negative aspects are not restricted only to the direct interaction with the digital product; they end up affecting the entire system or service in which this product is inserted, as well as the image that the user has with the organization's brand related to the product or service.

Falbe, Andersen and Frederiksen (2017) still point out that the user-experience can be broken down into some levels, such as meeting the user's needs without any fuss or hassle to complete tasks; the user should not have to remember information from previous steps, i.e., the user should not be mentally overloaded; the interaction should be in accordance with the user's expectations; the user should be able to recover any errors made; the system should have adaptability and flexibility to accommodate different expertise; the user should have control over the system; and the user must have a memorable experience. Actually, most of these levels that the authors place are a summary of the heuristics and golden rules of usability and ergonomics criteria presented in the 1990s by authors such as Nielsen and Mack (1994), Shneiderman (1987) and Bastien and Scapin (1993) to work on the usability of an interactive system.

Also, Brooks (2014) points out that in the past, digital web page interfaces were composed of only a page of text and pictures, and now websites and mobile apps are much more complex, which makes the user-experience, according to the author, much more vital to the product's success. In this sense, it seems that the user-experience is something that can be created when building systems like websites and mobile apps. However, it is believed that the author is referring to interface design that considers the principles of usability and that this will lead to a good user-experience.

Finally, still in this perspective, Tullis and Albert (2008, 2013), although they understand that UX's concept can be broader (see Section 1.3.4), argue that the user-experience can be measured. However, this measurement should be done using usability metrics, evaluated from the interaction with digital systems. In fact, we believe that the user-experience itself can be measured in part by usability metrics since usability has a significant positive or negative impact on a person's experience when interacting with a product or service that contains a digital interface.

Nonetheless, other issues related to the ecosystem in which the product is inserted and affecting the user-experience, as some authors point out (in Section 1.3.4), may not be evaluated through usability metrics, such as feelings, emotions and other perceptions. Although usability metrics like satisfaction are subjective, its other metrics are much more objective, such as effectiveness and efficiency (ISO 1998), and these do not measure everything that involves an experience. Nevertheless, new studies involving infrared digital neuroscience thermography and neuroscience, with measurements through electroencephalography, point to the possibility of physiological measurements to assess emotions as pointed out by Rebelo et al. (2022), Vitorino et al. (2022) and other authors such as Barros (2016), Barros, Soares, Marçal et al. (2016), Soares, Vitorino and Marçal (2019), Ioannou, Gallese and Merla (2014) and Guo, Li, Hu et al. (2019).

1.3.2 UX as a Result of User Perception Interacting with a Specific Product or Service

Very similar to the previous perspective, in this one, UX is related to the user's perception when interacting with a specific product and not necessarily with a digital product, as this may also be analogical. In this sense, regardless of the type of product, the experience lived by the user is a consequence of the (direct) use of a product, be it digital or not. Garrett (2011) states that a product or service has two sides: the inner workings and the outside, where the user has the contact, but what will have an effect on the experience is the result of what the user can do with the product or service (on the outside).

Hekkert and Schifferstein (2007) emphasize this perspective, as mentioned before in their book *Product Experience*. Products are an efficient way to create good experiences. At first, the main focus is not on what the product is, but what it does for the user and its consequences. Garrett (2011, chap. 1) also adds that the "features and functions always matter, but user experience has a far greater effect on customer loyalty."

In addition, Házi (2017) states that even though UX is not just a "digital thing" or a "magic ingredient" that you can add to an existing product, UX is always around you because the user-experience surrounds you. If someone can understand how a product works on the first contact, then this is a good UX. This direct relationship with the product's function seems much more linked to usability than to the extent of the whole user-experience. Indeed, the author uses as an example of good UX the two handles of a door that indicate their openings. The same example is given by Norman (2013) to explain the concepts of affordances and signifiers, as principles of interaction and usability.

Moreover, Buley (2013) notes that UX is a fancy term to describe words that people often speak about their experience of using a product, such as "love" or "hate" or phrases like "easy to use" or "user-friendly." In other words, feelings have always existed when we talk about an interaction, which, in the end, we treat as good/bad or easy/difficult. The term "user friendly" has also been used frequently in the past to refer to the usability of a product.

Lastly, some authors also relate the user-experience with the direct use of a product or service, but when the interaction is with a service, it is preferable to use terms such as service design or customer experience (CX) (Házi 2017; Lindborg 2015). However, this distinction between "user-experience (UX)" and "customer experience (CX)" requires further discussion beyond the scope of this chapter and maybe the subject of another publication.

1.3.3 UX as a Design Process

One of the most relevant perspectives of UX is to understand it as a design process. After all, the user-experience can only be positively impacted if there is a user-centered development process behind it. In this perspective, the word Design is added to the acronym UX (creating the term UX Design).

For Unger and Chandler (2012), the scope of UX Design is large and growing; it is a process that aims to develop elements that will influence the user's perceptions and behaviors and affect his/her experience. These elements, according to the authors, range from tangible elements (e.g., physical products, packaging) to intangible elements (e.g., sound, aroma), including people who are part of the entire ecosystem of the experience (customers service representatives, salespeople, friends, family). These elements, when well developed and well integrated, lead to a good user-experience. The authors also add that for this user-experience to be successful, the design must align business objectives, users' needs and any limitations that will affect its viability (e.g., technical limitations, project budget or time frame).

Marsh (2015) corroborates Unger and Chandler's (2012) remarks and makes an analogy of the UX Design process with the process of doing science, where you start with research, to understand both the needs of users and the needs and requirements of the business. From then on, it proceeds to the development of ideas to meet such needs and to assess whether the solutions work in the real world. Research in UX Design is one of the fundamental steps to design a good user-experience, especially the user research.

Besides process, but in a similar sense, other authors define UX Design as a holistic discipline (Bluestone 2019) or a convergence of disciplines (Dash 2014) or a multidisciplinary practice (Baines and Howard 2016). Thus, the authors state that working for the user-experience is a process that encompasses several disciplines (e.g., Visual Design, Industrial Design, Information Architecture and Interaction Design) with diverse and not necessarily predetermined inputs. This process is user-centered and conducted through a methodology appropriate to a particular problem to strategically solve issues that meet the requirements of users and businesses. Baines and Howard (2016) emphasize "without research, there is no UX" as one of the main steps in the process. That is why we see, nowadays, a great emphasis in this area, which has been called UX Research (Buley 2013, Marsh 2018, Travis and Hodgson 2019).

UX Concepts and Perspectives 11

However, within this strategy of working for the user-experience, involving the user in research, it should not be forgotten, according to Dash (2014), Unger and Chandler (2012) and Házi (2017), that in a UX process, the user's needs and business requirements should be combined. The processes and multidisciplinarity that most authors point out are related to the human-centered design approach and its main activities. However, as specified by ISO 9241-210 (ISO 2010), this approach does not clearly show how this alignment between the users' needs and the business requirements is done. In contrast, processes such as Service Design (Stickdorn, Lawrence, Hormess and Schneider 2018) and even Design Thinking (Brown 2009) seem to do better. Nevertheless, UX designers or UX researchers are still far from mastering this subject.

1.3.4 UX AS A RESULT OF USER PERCEPTION INTERACTING WITH BROAD SYSTEM

The last perspective presented here is a broader view of UX's concept and perhaps the one that most closely matches Donald Norman's idea when he first used the term (Norman 2013). From this perspective, a user-experience is shaped from the result of interaction with a broad system or ecosystem that contains several (parts) interfaces in which the user has contact throughout a journey. Ou (2017) perhaps overstates that UX "is how a person feels when interacting with the *world*," but what the author wants to say is that the world is understood by systems (e.g., computers, products, humans and process), and it is these various systems that users interact to achieve their goals in their journeys. Pereira (2018) also states that UX encompasses all interactions that a user has with a product's brand. In this case, the brand or organization related to the brand is the broad system that the user interacts with and, in this system, some subsystems are the channels in which the user will have contact with the brand (e.g., sites, applications, customer service and online help). Chapanis (1996, p. 22) defines a system as "an interacting combination, at any level of complexity, of people, materials, tools, machines, software, facilities, and procedures designed to work together for some common purpose."

Nonetheless, this perspective may have two different points of view: the product side and the service side – the side that will be chosen will depend on what it is intended to design. From the product point of view, one can think of the experience of interacting with it in various touchpoints, what Rosenzweig (2015) calls user-experience touchpoints. The author gives, as an example, a typical everyday product, a car. When we have a car, we interact with various elements or subsystems of experience, such as the car dealer, the vehicle's occupant packaging, the driving dynamics, the moment of fill it up, the maintenance and sharing, and other people.

From the service point of view, the user-experience (also called customer experience or CX; Ou 2017) will be thought through the conception of the user interaction with the service's various touchpoints. In the design of a service, the alignment and integration of these touchpoints (which can also be products) and the fluidity with which the user passes through them is most important in shaping a good user-experience. When thinking about the interaction with airline service, the user-experience will probably go through the brand awareness, the website for the purchase of the travel ticket, the check-in at the airport, the comfort of the airplane seat, the contact with the flight attendants, the snacks offered during the trip, until the conclusion at the exit of the destination airport, that is, all its touchpoints. This is if all goes well, because some unforeseen events such as loss of luggage or delays will also make the user talk to company employees and his/her experience may be impacted at this time.

Perhaps this perspective is also the most similar to the concept of Service Design, which is also a specific design process. According to Mager and Sung (2011, p. 1),

> Service design aims at designing services that are useful, usable and desirable from the user perspective, and efficient, effective and different from the provider perspective. (...) service design takes a holistic approach in order to get an understanding of the system and the different actors within the system. (...) Service design looks at the experience by focusing on the full customer journey, including the experiences before and after the service encounters.

Kuniavsky (2010, p. 14) makes a definition of user-experience which, in its full scope, summarizes very well what this perspective emphasizes and also makes a bridge with the first two perspectives presented previously:

> The user experience is the totality of end users' perceptions as they interact with a product or service. These perceptions include effectiveness (how good is the result?), efficiency (how fast or cheap is it?), emotional satisfaction (how good does it feel?), and the quality of the relationship with the entity that created the product or service (what expectations does it create for subsequent interactions?)

Finally, Hartson and Pyla (2019), in their book *UX Book* also add that the effects felt by users in the experience when interacting with products and systems occur in a temporal way, indirectly and directly. When thinking about the interaction with both a product and the service, the user has moments that are part of the experience – a moment before, during and after use; all the effects of interaction (direct or indirect) in the experience go through these moments. Direct interaction with the product or service occurs when the user is in direct contact with the artifact or touchpoint (which can be intangible, such as a voice interaction). In contrast, indirect interaction can be a feeling, a memory or a thought about the product or service. This description of Hartson and Pyla (2019) aligns exactly with what Rossman and Duerden (2019) specify as experience, mentioned previously. The first authors also point out that the user-experience is a combination of four components: usability, usefulness, emotional impact and meaningfulness. This combination clarifies that usability is part of what composes a user-experience, but both concepts are not the same.

1.4 CONCLUDING REMARKS

This chapter's idea was to bring an overview of what has been said and thought about the user-experience (UX) and especially its relationship with usability, which are often misunderstood as the same thing. These are the most common perspectives found in literature, as the term still brings many misunderstandings, especially in industry and professional practice. Even in the industry, the term is overused among professionals; it has been at its peak, but today, terms such as service designer or customer experience designer seem to be having more value because they are believed to be more comprehensive and less focused on a digital interface. In fact, what may be happening is that professionals who really want to work on the user-experience in a broad sense may be transitioning from the first perspective to the fourth perspective presented in this chapter, which has a greater focus on ecosystem, system and interaction subsystems, on the user's journey.

Although the four perspectives seem different, it can be understood that they complement each other because UX (a) deals with interaction with a digital product, (b) deals with interaction with any product or service, (c) is a process and finally (d) involves several systems and subsystems that the user will interact with throughout his/her journey. However, understanding that experience is individual is fundamental. Designers can only shape and promote the best possible experience by designing products and services, integrating and aligning all parts of the ecosystem where the interfaces are inserted.

The issue of designing "the experience" or "for the experience" of the user is still quite confusing and nebulous, and there is much debate on this subject. After all, the experience is unique and experienced differently by each person. Perhaps, designing the experience as a whole and in detail for an individual is impossible, but transforming a person's experience through interface design and good planning of their journey in the various touchpoints with a brand is really possible. A system that is seen and understood holistically, with its subsystems (touchpoints) well designed, ensuring good usability, and planned and orchestrated in the best way, can positively impact and provide a good user-experience.

ACKNOWLEDGMENTS

This study was financed in part by the Coordenação de Aperfeiçoamento de Pessoal de Nível Superior – Brasil (CAPES) – Finance Code 001 and by Conselho Nacional de Desenvolvimento Científico e Tecnológico (CNPq).

REFERENCES

Baines, Jeremy, and Clive Howard. 2016. *UX Lifecycle: The Business Guide to Implementing Great Software User Experiences*. British Columbia: Leanpub.

Barros, R. Q. 2016. "Aplicação da Neuroergonomia, Rastreamento Ocular e Termografia por Infravermelho na Avaliação de Produto de Consumo: um Estudo de Usabilidade." Dissertação [Mestrado em Design] Universidade Federal de Pernambuco.

Barros, Rafaela Q., Marcelo M. Soares, Márcio A. Maçal, Ademário S. Tavares, Jaqueline A. N. Oliveira, José R. R. Silva, Aline S. O. Neves, Robson Oliveira, and Geraldo O. S. N. Neto. 2016. "Using Digital Thermography to Analyse the Product User's Affective Experience of a Product." In *Advances in Intelligent Systems and Computing*, Vol. 485. Springer. doi: 10.1007/978-3-319-41983-1_10.

Bastien, J. M. C., and D. L. Scapin. 1993. "Ergonomic Criteria for the Evaluation of Humain-Computer Interfaces." *Rapport Technique INRIA*.

Bluestone, Danny. 2019. *UX Handbook: Our Handbook for User-Centred Design*. London: Cyber-Duck.

Boven, Leaf Van, and Thomas Gilovich. 2003. "To Do or to Have? That Is the Question." *Journal of Personality and Social Psychology* 85(6): 1193–1202. doi: 10.1037/0022-3514.85.6.1193.

Brangier, E., and J. Barcenilla. 2003. *Concevoir Un Produit Facile à Utiliser. Concevoir Un Produit Facile à Utiliser*. Paris: Editions d'Organisation.

Brooks, Ian. 2014. *The Importance of User Experience: A Complete Guide to Effective UI and UX Strategies for Creating Useful and Usable Mobile & Web Applications*. Scott Vandal.

Brown, Tim. 2009. *Change by Design: How Design Thinking Transforms Organizations and Inspires Innovation*. New York: HarperCollins.

Buley, Leah. 2013. *The User Experience Team of One: A Research and Design Survival Guide*. New York: Rosenfeld Media.

Campana, Julia Ramos, and Manuela Quaresma. 2017. "The Importance of Specific Usability Guidelines for Robot User Interfaces." In *Design, User Experience, and Usability: Designing Pleasurable Experiences. DUXU 2017. Lecture Notes in Computer Science*, Vol. 10289, edited by A. Marcus and W. Wang, 471–483. Cham: Springer. doi: 10.1007/978-3-319-58637-3_37.

Campbell-Kelly, Martin, William, Aspray, Nathan Ensmenger, and Jeffrey R. Yost. 2014. *Computer: A History of the Information Machine*. Boulder, CO: Westview Press.

Carter, Travis J., and Thomas Gilovich. 2010. "The Relative Relativity of Material and Experiential Purchases." *Journal of Personality and Social Psychology* 98(1): 146–159. doi: 10.1037/a0017145.

Chapanis, Alphonse. 1996. *Human Factors in Systems Engineering*. New York: John Wiley & Sons.

Dash, Samir. 2014. *UX Simplified: Models & Methodologies*. India: CreateSpace Independent Publishing Platform.

Falbe, Trine, Kim Andersen, and Martin Michael Frederiksen. 2017. *White Hat UX: The Next Generation in User Experience*. Herning: pej gruppens forlag.

Farrell, Susan, and Jakob Nielsen. 2014. "User Experience Careers." https://www.nngroup.com/reports/user-experience-careers/.

Garrett, Jesse James. 2011. *The Elements of User Experience: User-Centered Design for the Web and Beyond. Elements*. Berkeley: New Riders.

Guo, Fu, Mingming Li, Mingcai Hu, Fengxiang Li, and Bozhao Lin. 2019. "Distinguishing and Quantifying the Visual Aesthetics of a Product: An Integrated Approach of Eye-Tracking and EEG." *International Journal of Industrial Ergonomics* 71(May): 47–56. doi: 10.1016/j.ergon.2019.02.006.

Hackos, JoAnn T., and Janice Redish. 1998. *User and Task Analysis for Interface Design*. New Jersey: Wiley.

Hartson, H. Rex, and Pardha S. Pyla. 2019. *The UX Book: Agile UX Design for a Quality User Experience*. 2nd ed. Burlington: Morgan Kaufmann.

Hartson, Rex, and Pardha S. Pyla. 2012. *The UX Book: Process and Guidelines for Ensuring a Quality User Experience*. Burlington: Morgan Kaufmann. doi: 10.1016/C2010-0-66326-7.

Harvey, Catherine, and Neville A Stanton. 2013. *Usability Evaluation for In-Vehicle Systems*. Boca Raton: CRC Press.

Hassenzahl, Marc. 2010. "Experience Design: Technology for All the Right Reasons." *Synthesis Lectures on Human-Centered Informatics* 3(1): 1–95. doi: 10.2200/S002 61ED1V01Y201003HCI008.
Hassenzahl, Marc. 2011. "User Experience and Experience Design." In *The Encyclopedia of Human-Computer Interaction*. https://www.interaction-design.org/literature/book/the-encyclopedia-of-human-computer-interaction-2nd-ed/user-experience-and-experience-design.
Házi, Csaba. 2017. *Seven Step UX: The Cookbook for Creating Great Products*. Budapest: Csaba Házi.
Hekkert, Paul, and Hendrik N.J. Schifferstein. 2007. "Introducing Product Experience." In Hendrik N.J. Schifferstein and Paul Hekkert, *Product Experience*, 1–8. Amsterdam: Elsevier Science. doi: 10.1016/B978-008045089-6.50003-4.
International Organization for Standardization. 1998. "ISO 9241-11: Ergonomic Requirements for Office Work with Visual Display Terminals (VDTs) - Part 11: Guidance on Usability." International Organization for Standardization.
International Organization for Standardization. 2010. "ISO 9241-210: Ergonomics of Human–System Interaction - Human-Centred Design for Interactive Systems." International Organization for Standardizatiion.
Ioannou, Stephanos, Vittorio Gallese, and Arcangelo Merla. 2014. "Thermal Infrared Imaging in Psychophysiology: Potentialities and Limits." *Psychophysiology* 51(10): 951–63. doi: 10.1111/psyp.12243.
Jordan, Patrick W. 2002. *Designing Pleasurable Products: An Introduction to the New Human Factors*. Boca Raton: CRC Press Taylor & Francis.
Kuniavsky, Mike. 2010. *Smart Things: Ubiquitous Computing User Experience Design: Ubiquitous Computing User Experience Design*. Burlington: Morgan Kaufmann Publisher.
Lindborg, Jacob. 2015. *Designing Customer Experiences: Design Is not a Product Feature, It's an Evolving Experience!* dsgnr.io.
Mager, Birgit, and Tung Jung Sung. 2011. "Special Issue Editorial: Designing for Services." *International Journal of Design* 5(2).
Marsh, Joel. 2015. *UX for Beginners - A Crash Course in 100 Short Lessons*. Sebastopol: O'Reilly Media.
Marsh, Stephanie. 2018. *User Research: A Practical Guide to Designing Better Products and Services*. London: Kogan Page.
Nielsen, Jakob, and Robert L. Mack. 1994. *Usability Inspection Methods*. New York: Wiley.
Norman, Donald A. 1988. *The Psychology of Everyday Things. The Psychology of Everyday Things*. New York: Basic Books Inc.
Norman, Donald A. 2005. *Emotional Design: Why We Love (or Hate) Everyday Things*. New York: Basic Books.
Norman, Donald A. 2013. *The Design of Everyday Things*. New York: Basic Books.
Ou, Andrew. 2017. *The Tao of Design and User Experience: The Best Experience Is no Experience*. Andrew Ou.
Pavliscak, Pamela. 2018. *Emotionally Intelligent Design : Rethinking How We Create Products*. Sebastopol: O'Reilly Media.
Pereira, Rogério. 2018. *User Experience Design: Como Criar Produtos Digitais Com Foco Nas Pessoas*. São Paulo: Casa do Código.
Pine, B. Joseph, and James H. Gilmore. 2020. *The Experience Economy: Competing for Customer Time, Attention, and Money*. Brighton: Harvard Business Review Press.

Quaresma, Manuela. 2018. "UX Designer: Quem é Este Profissional e Qual é a Sua Formação e Competências?" In Ulbricht, Vania Ribas; Fadel, Luciane Maria; Batista, Claudia Regina *Design Para Acessibilidade e Inclusão*, 88–101. São Paulo: EDITORA BLUCHER. doi: 10.5151/9788580393040-07.

Rebelo, F., Vilar, E., Filgueiras, E., Valente, J., and Noriega, P. 2022. "Advanced User Experience Evaluations Using Biosensors in Virtual Environments"In: Soares, M.M., Rebelo, F., Ahram, T. *Handbook of Usability and User Experience: Methods and Techniques* 1. Boca Raton: CRC Press.

Riley, Scott. 2018. *Mindful Design: How and Why to Make Design Decisions for the Good of Those Using Your Product*. New York: Apress.

Rosenzweig, Elizabeth. 2015. *Successful User Experience: Strategies and Roadmaps. Successful User Experience: Strategies and Roadmaps*. Amsterdam: Elsevier. doi: 10.1016/C2013-0-19353-1.

Rossman, J. Robert, and Mathew D. Duerden. 2019. *Designing Experiences*. New York: Columbia Business School Publishing.

Shneiderman, Ben. 1980. *Software Psychology: Human Factors in Computer and Information Systems*. South Carolina: Winthrop Publishers.

Shneiderman, Ben. 1987. *Designing the User Interface*: Strategies for Effective Human-Computer Interaction. *Designing the User Interface*. 1st ed. Reading: Addison-Wesley Publ. Co.

Soares, Marcelo M., Danilo F. Vitorino, and Márcio A. Marçal. 2019. "Application of Digital Infrared Thermography for Emotional Evaluation: A Study of the Gestural Interface Applied to 3D Modeling Software." *Advances in Intelligent Systems and Computing* 777: 201–212. doi: 10.1007/978-3-319-94706-8_23.

Stickdorn, Marc, Markus Hormess, Adam Lawrence, and Jakob Schneider. 2018. *This Is Service Design Doing: Applying Service Design Thinking in the Real World: A Practitioner's Handbook*. Sebastopol: O'Reilly Media.

Travis, David, and Philip Hodgson. 2019. *Think Like a UX Researcher: How to Observe Users, Influence Design, and Drive Strategy*. Boca Raton: CRC Press.

Tullis, Thomas, and William Albert. 2008. *Measuring the User Experience. Measuring the User Experience*. Amsterdam: Elsevier. doi: 10.1016/B978-0-12-373558-4.X0001-5.

Tullis, Tom, and Bill Albert. 2013. *Measuring the User Experience. Measuring the User Experience*. 2nd ed. Amsterdam: Elsevier. doi: 10.1016/C2011-0-00016-9.

Unger, Russ, and Carolyn Chandler. 2012. *A Project Guide to UX Design: For User Experience Designers in the Field or in the Making*. 2nd ed. San Francisco: New Riders.

Vitorino, D., Marçal, M., and Soares, M.M. 2022. "Applications of infrared thermography to evaluate the ergonomics and usability of products with gestural interface." In: Soares, M.M., Rebelo, F., Ahram, T. *Handbook of Usability and User Experience: Methods and Techniques* 1. Boca Raton: CRC Press.

2 User Requirements Analysis

Martin Maguire

CONTENTS

2.1 Introduction 17
2.2 User Requirements Analysis Activities 18
2.3 Eliciting Requirements 18
    2.3.1 Documenting Processes..19
    2.3.2 Stakeholder Meeting..20
    2.3.3 Generating Requirements 20
        2.3.3.1 *UX Technique:* Development of Personas.......................20
2.4 Recording Requirements 21
    2.4.1 Use Cases..21
        2.4.1.1 *UX Technique:* Context of Use Analysis 22
    2.4.2 User Stories...22
        2.4.2.1 *UX Technique:* User-experience Goals............................22
    2.4.3 Data Flow Diagrams..23
        2.4.3.1 *UX Technique:* Prototyping as a Way of Visualizing Requirements Gathered 23
2.5 Analyzing Requirements 23
    2.5.1 Clarity, Uniqueness and Traceability..24
        2.5.1.1 *UX Technique:* User Feedback Sessions..........................24
    2.5.2 Resolving Conflicts..24
    2.5.3 Design Concept Emergence..25
        2.5.3.1 *UX Technique:* Brainstorming...25
2.6 Conclusion 25
References..25

2.1 INTRODUCTION

User requirements analysis is the process of defining the needs and expectations of the users for an application that is to be built or modified. It is a key part of the systems design process and many systems failures are attributed to poorly specified user requirements. User requirements analysis is part of a broader requirements analysis process that also defines the needs of the customer's business requirements. This will lead to determining the feasibility of the project including resources, time and finance allocated it. Also, it is essential to develop a schedule for carrying out the tasks for developing the system.

User-experience (UX) design is a process of creating products and systems for users that meet their task needs and is both accessible and usable for them. Furthermore, by providing aspects of branding, design, usability and function, it aims to provide a positive user feeling or experience.

This chapter is concerned with applying user-experience methods and techniques to enhance the user requirements analysis process.

2.2 USER REQUIREMENTS ANALYSIS ACTIVITIES

Requirements analysis includes three main activities:

Eliciting requirements: The stage of determining from the user's point of view what the system needs to do.
Recording requirements: The process of recording the requirements in a standard way so that the design team can refer to them.
Analyzing requirements: Reviewing the requirements to ensure that they are clear and compatible with each other.

When the requirements are defined and agreed, then the process of starting to design the system or software application and implementing it begins. This will involve end users to enable their points of view to be represented. There may be different ways of organizing the process ranging from a traditional structured approach to Agile development. In Agile software development, users work alongside designers and developers so that their requirements emerge through day-to-day interaction and the development of lightweight software components over short time periods (Denning, 2019). Both approaches have their advantages and disadvantages and work best in different contexts (Olic, 2017).

Yet, the complexity or formality of the system design process can mean that it is difficult to make the user needs real for the design team or manage evolving requirements as user needs become clearer, or if their opinions change when designs are implemented. Design teams will often employ a user-experience designer to act as the facilitator between the design team and the users.

The next three sections will take each of the three high-level user requirements processes, break them down into specific activities and describe how user-experience design methods can be used to support them and improve their effectiveness.

2.3 ELICITING REQUIREMENTS

This is the process of gathering user requirements and involves documenting the tasks that need to be conducted as part of the business process and then identify the

User Requirements Analysis

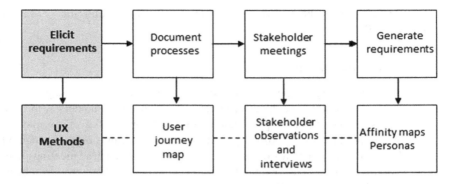

FIGURE 2.1. Eliciting requirements with supporting UX methods

detailed needs of different stakeholders. It will typically involve reviewing business process documentation and conducting stakeholder interviews. This is sometimes also called requirements gathering or requirements discovery. Figure 2.1 shows some key steps in the process.

The starting point for requirements gathering is to understand the general terms of the project. This will involve asking the client what the main goal of the system is, what problem will it solve, have they tried to create a solution before and if so, why did it fail? Essentially the aim is to find out what the client wants and needs from the project as a whole.

2.3.1 DOCUMENTING PROCESSES

All products deal with some sort of process on behalf of the client. The aim is to understand how the client's business operates in order to create a solution that adapts to it, which is crucial for smooth operation when the solution goes online.

Making it visual is advantageous. Seeing each step on the client's process can help the design team understand how things are done and how they can reflect this process in the design itself. It is important here to not just try to understand what the client wants but also how the client's business works.

UX technique: **User journey map**. This is a powerful technique for understanding what motivates users or customers – what their needs, problems and concerns are. Although most organizations are reasonably good at gathering data about their customers, data alone fails to communicate the frustrations and experiences the customer encountered. The customer journey map uses storytelling and visuals to illustrate the relationship a customer has with a business over a period of time. The story is being told from the perspective of a user, which provides insight into their total experience. This helps the design team better understand and address user needs and pain points as they experience the product or service. The user journey map offers the chance to see how users or customers engage with the organization and then moves through the touchpoints of the entire activity or process. Finally, the team can propose the improvement or actions to be taken against each of the touchpoints.

These proposed actions can be a potential source of software requirements (Visual Paradigm, 2020).

2.3.2 Stakeholder Meeting

This involves identifying people who use the system either as end users, e.g., operators, customers, maintainers, content providers or as other stakeholders who have a valid interest in the system, e.g., standards or regulatory bodies. They may be affected by it either directly or indirectly. A stakeholder meeting is often the method used to identify the requirements of each user and stakeholder. These meetings may be called joint requirements development sessions and may be facilitated by a business analyst or UX specialist. After an introduction from the project manager and/or client, a number of topics will be considered, allowing each stakeholder to describe their needs. Discussions will elicit requirements, which can then be analyzed to uncover cross-functional implications.

UX technique: **Interviews and observations**. In addition to a stakeholder meeting, two useful techniques are individual stakeholder interviews and observations. These allow the interviewer to step into the shoes of their interviewees and see their role through the eyes of these stakeholders. Part of the interview will be to ask what tasks each user performs with the current system and to identify their needs and pain points. All requirements should be linked to a task to give them context. The interview will also help to prioritize requirements, features and possible testable goals to show their achievement. A further activity is user observation where the researcher will ask if they can observe the user carrying out their work. This method enables the researcher to identify the context in which the user operates and to observe their activities first-hand. This approach often reveals information that is not covered by the interview, giving extra insights into what needs or requirements a user will have.

2.3.3 Generating Requirements

Requirements will start to emerge from user stakeholder meetings, user interviews and user observations. Requirements will initially be written in a narrative form before being recorded formally in the form of requirements documentation.

2.3.3.1 *UX Technique:* **Development of Personas**

Once the stakeholder interviews have been conducted, they can be analyzed and individual informational items can be recorded onto electronic or physical post-its. They can then be analyzed to create personas of individual user roles in the system. Personas are archetypal users that represent the needs of larger groups of users, in terms of their goals and personal characteristics (Cooper et al., 2014). They act as "stand-ins" for real users and help guide decisions about strategy, functionality and design. Personas identify the user's motivations, expectations and goals that are responsible for driving online behavior and can be brought to life by giving them names, personalities and a photo. Although personas are fictitious, they are based

User Requirements Analysis 21

on the knowledge of real users. Some form of user research is conducted before they are written to ensure they represent end users rather than the opinion of the person writing the personas. In the context of enterprise projects, personas are valuable in articulating the needs of employees (Robertson, 2020).

2.4 RECORDING REQUIREMENTS

Here requirements are reviewed to ensure that they are clear and meaningful, are unduplicated and can be traced back to their source. They may be documented in various forms, including simple lists, activity diagrams, use case diagrams, user stories and process specifications, and data models. The requirements may be divided into "functional requirements" that capture the required intended actions for the system and "non-functional requirements" that define system behavior, features and general characteristics that affect the user's experience. Three key activities that are key to the analysis process are shown in Figure 2.2.

2.4.1 USE CASES

A use case diagram is the primary form of system/software requirements for a new software program under development (Robertson and Robertson, 1999). Use cases specify the expected behavior and not the exact method of making it happen. Use cases, once specified, can be represented in both textual or visual form. A key concept of use case modeling is that it helps in the design of a system from the end user's perspective. It is an effective technique for communicating system behavior in terms understood by the user by specifying all externally visible system behavior. A use case diagram is usually simple and contains equivalent elements to a user story: an actor, flow of events and postconditions. It does not show the details of the system processing but summarizes some of the relationships between use cases, actors and systems. It does not show the order in which steps are performed to achieve the goals of each use case.

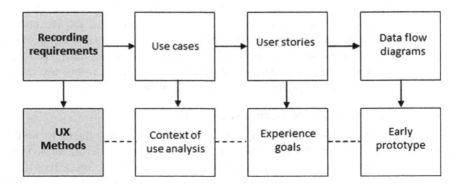

FIGURE 2.2. Recording requirements with supporting UX methods

2.4.1.1 *UX Technique:* **Context of Use Analysis**

Context of use analysis involves collecting and analyzing detailed information about the intended users, their tasks and the technical and environmental constraints. In this way, requirements can be specified that are in line with the context of use and the use cases that are documented. Understanding what are the main or top tasks that the users want or need to perform is an important basis for design (McGovern, 2018). Context of use is about understanding and responding to actual conditions under which a given artifact/software product is used. For example, ease of use of a building tool may not be significant when used by an experienced builder but it can become significant if a DIY enthusiast uses the same device at home.

A single use case in different contexts can require vastly different approaches that have implications for usability, user acquisition and user-experience. For example, in a satellite navigation app, if the driver finds that their battery level is low they may prefer to conserve energy by temporarily blocking the download of unnecessary traffic information, restaurant locations, pictures, etc., which take battery power. If they are guided to an area where traffic is exceptionally heavy, they will want the app note this and guide them to their destination in another way. If the app is able to identify that the user's destination is closed, it could warn the user that they may be on a wasted journey and give them the option to stop and return home. The benefit of being context-aware allows the system to make the user's experience a less frustrating and seamless experience (Chu, 2018).

2.4.2 User Stories

A user story is a note that captures what a user does or needs to do as part of their work. Each user story consists of a short description written from the user's point of view, with natural language. Unlike traditional requirement capture, user stories focus on what users need instead of what the system should deliver. This leaves room for further discussion of solutions and the result of a system that can really fit into the customers' business workflow, solving their operational problems and adding value to the organization. User stories are compatible with the other agile software development techniques and methods, such as scrum (weekly design team meeting) and extreme programming. User stories describe the user's role, goal and acceptance criteria. The details of a user story may not be documented to the same extent as for a use case but are meant to elicit conversations by asking questions during a design team meeting. They allow the generation of design feedback more frequently, rather than having more detailed up-front requirement specification meetings.

2.4.2.1 *UX Technique:* **User-experience Goals**

User-experience is everything that happens to the users when they interact with the business or organization through a website or application, on the phone or face-to-face, and so it is important to have clear user-experience goals for each of these touchpoints. UX includes everything that users see, hear and do as well as their emotional reactions. User satisfaction is part of the experience. In creating a user-experience that is easy, pleasant and natural for users to be able to achieve a given task, the user will be satisfied and will do it again. When users interact with the

User Requirements Analysis

system, they want to find something that will satisfy them. If the user feels that they are not getting what they want from it, they may quit and look for an alternative that offers better service, or, if they are an employee, may be tempted to leave the organization to find less-demanding work. So it is important, through the elements of the system, to try and achieve user satisfaction.

When the user is finished with the system for a particular session, they must feel that they have achieved what they wanted when entering the system. They must also have enjoyed the process so there may be ways to make it more fun and enjoyable. It has been found that websites that are fun and entertaining, in the right situation, tend to get more visitors than those that are not. Users will be willing to spend more time in it and they can even make a purchase of something that they did not even want. It has been found that websites that are easy to navigate and give the feeling of fun to the user will tend to generate more sales (Wilson, 2014).

2.4.3 DATA FLOW DIAGRAMS

A data flow diagram (DFD) can be applied in the requirement elicitation process of the analysis phase within the system development life cycle to help define the project scope. It is often used as a preliminary step to creating an overview of the system without going into detail, which can later be elaborated. For instance, if there is a need to show more detail within a particular process, the process is decomposed into several smaller processes in a lower-level data flow diagram. However they should be flexible enough to allow new design ideas to be suggested and developed.

2.4.3.1 UX Technique: Prototyping as a Way of Visualizing Requirements Gathered

At first, requirements will be fluid as the design team discover user needs for the project. They may then be used by the design team to help brainstorm ideas and create early or low-fidelity prototypes that can help make the requirements tangible. It is important to note though that this is just an early representation and the actual system design, when different design concepts have been identified, may be quite different.

Prototyping can play an important role in requirements gathering, rather than leaving the overall requirements to the imagination of stakeholders until late in the product's development process (Justinmind, 2020). Even early on, when the design team first starts to understand the client's requirements, making a prototype may add value to the project. Showing a client, several options can be useful as a way to develop the requirements rather than keeping the requirements the same for the entire project.

2.5 ANALYZING REQUIREMENTS

This activity involves reviewing the requirements to ensure that they are clear, meaningful, unduplicated and can be traced back to their source. Conflicts between activities also need to be resolved. Key steps within this process are shown in Figure 2.3.

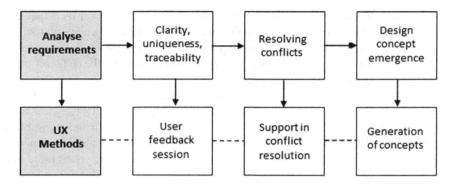

FIGURE 2.3. Analyzing requirements with supporting UX methods

2.5.1 CLARITY, UNIQUENESS AND TRACEABILITY

Analysis of requirements involves ensuring that they are of high quality by checking them for clarity or ease of understanding, uniqueness and traceability, among other attributes. The use of templates is helpful for structuring requirements in a suitable way and making sure that they are of good quality. A further output from the process is the development of a UX vision for the project which communicates the experience that the designers are aiming to create for the users.

2.5.1.1 *UX Technique:* User Feedback Sessions

One of the key attributes of user-experience design is keeping in close touch with end users and stakeholders. An effective method that can be employed is to run feedback sessions with users and stakeholders to check that they agree with the requirements that have been elicited, make changes if necessary, and confirm who each requirement can be traced back to if needed.

2.5.2 RESOLVING CONFLICTS

Different stakeholders and users may have opposing requirements or there may be different technical requirements that are hard to resolve. An example of a conflicting requirement might be that the human resources stakeholder group requests to capture some personal information for an employee, but the data privacy team argues that the data should not be captured or used in reporting. There are a number of different techniques that can be used for resolving disagreements: (1) Stakeholders work together to negotiate a solution to the dispute. This involves discussion of each other's views to try to persuade people experiencing the conflict to agree on a workable solution or compromise. (2) Stakeholders may be asked to vote on two or different solutions as the resolution for the difference of opinion. This approach will give each party the reassurance that their point of view has been put forward and discussed even if it is not accepted. (3) Different stakeholders can apply their own variants to the solution as parameters. This way the different stakeholders get their preferred solution implemented. (4) A comparison matrix is created of all key criteria that need to be considered against each solution alternative. This approach often highlights which is the best solution to choose and resolves the disagreement. An important aspect that is central to UX is to listen carefully to each stakeholder's point of view before one solution is adopted over another.

2.5.3 Design Concept Emergence

As the requirements process develops, a basic design concept may emerge that will become the starting point for the developing design.

2.5.3.1 *UX Technique:* Brainstorming

This is a good way to get input from every relevant department. There are different ways to organize brainstorming with varying degrees of structure and freedom to add ideas. Some teams prefer to structure the conversation, so they have a little more focus while others have a blank board as a starting point. It is also helpful to give users enough time to formulate their own ideas before presenting them to the groups. "Crazy eights" is a technique where the participant has a sheet of paper that they fold three times so that it opens back out with eight sections. Each person sketches up to eight basic ideas, as crazy as they wish, which they then present to the group. Each sheet is then pinned onto a whiteboard for the group to discuss.

As part of the brainstorm session, the group may synthesize all the ideas to generate between three and five different design concepts and discuss the positives and negatives of each. The different user, task and environmental requirements may be used as assessment criteria with each concept being rated (using a rating scale) for each. The concept with the highest score is then chosen as the best and taken forward.

2.6 CONCLUSION

Gathering requirements for a new system or application is a process that can take time and effort. It is important to work carefully through the requirements generation stages to make sure that they represent the needs of each user and stakeholder group effectively. Documentation of the requirements made with care and plenty of input from everyone is key and analysis to ensure that the requirements are of high quality is also crucial. The application of UX methods can help the requirements analysis process to be more effective by bringing the requirements to life, making them more understandable and thereby giving them a human face.

REFERENCES

Babich, N. (2017) Most common UX design methods and techniques. Available at: https://ux planet.org/most-common-ux-design-methods-and-techniques-c9a9fdc25a1e

BAE Business Analysis Excellence (2017) Requirements conflict −8 ways to manage that sticky stakeholder... Available at: https://business-analysis-excellence.com/requireme nts-conflict-8-ways-manage-sticky-stakeholder/

Chu, Y. (2018) How context of use improves product design and user experience. Available at: https://medium.com/nyc-design/how-context-of-use-improves-product-design-and-user-experience-3299d2f0a166

Cooper, A., Reimann, R., Cronin, D. (2014) About face: the essentials of interaction design, Indianapolis, IN: John Wiley and Sons.

Denning, S. (2019) Why Agile's future is bright. Available at: https://www.forbes.com/sites/stevedenning/2019/08/25/why-the-future-of-agile-is-bright/?sh=1081db229680

JustinMind (2020) Gathering requirements: Defining scope and direction. Available at: https://www.justinmind.com/blog/gathering-requirements/

McGoverm, G. (2018) *Top tasks – A how-to guide*, Silver Beach.

Olic, A. (2017) Advantages and disadvantages of agile project management. Available at: https://activecollab.com/blog/project-management/agile-project-management-advantages-disadvantages

Robertson, J. (2020) Employee personas and how to create them. Available at: https://www.steptwo.com.au/papers/kmc_personas/

Robertson, S., Robertson, J. (1999) *Mastering the requirements process*, Addison-Wesley.

Visual Paradigm (2020) Requirement analysis techniques. Available at: https://www.visual-paradigm.com/guide/requirements-gathering/requirement-analysis-techniques/

Wilson, M. (2014) 5 user experience goals to set. Available at: https://ux.walkme.com/5-user-experience-goals-set/

Woods, D. (2008) Google's 'Agility' 11, *Forbes*. Available at: https://www.forbes.com/2008/08/09/cio-agile-computing-tech-cio-cx_dw:0811agile.html?sh=445a41b14278

3 Designing the Interactor – From User-Experience to Interaction Experience
Conceptual Foundations and Contributions towards a New Paradigm

Sónia Rafael and Victor M. Almeida

CONTENTS

3.1 Introduction	27
3.2 Paradigm Shift from Material to Virtual Culture – The Contribution of Design to the Dematerialization of Tangible Objects	28
3.3 *In Media Res*. Two Visions for Interaction Design	32
3.4 The Concept of User-Experience (UX) as a Current Paradigm	33
3.5 What Does It Mean to Be a Computer and to Be a User?	34
3.6 Embodiment and the Advances of Artificial Intelligence	36
3.7 Affective Computing from a Computer or User Perspective?	38
3.8 User versus Interactor	38
3.9 Equity between Human Agents and Computational Agents	40
3.10 Conclusion: The New Interaction Experience (IxX) Paradigm	40
References	41

3.1 INTRODUCTION

User-experience (UX) is an elusive goal and it is neither easy to predict nor to objectively design an individual's experience with an artifact. However, establishing a set of guidelines for designing a product, service or environment is less elusive and has the potential to promote a satisfying experience by identifying all the aspects of the interaction that support it. It arises from the goal of projecting a holistic experience through the amplification of the classic perspective of mere task accomplishment goals in human–computer interaction (HCI). Thus, it seeks to combine, through a projectual process, all factors that contribute to the individual's experience.

UX has been historically portrayed as a determinant for the success of an artifact. However, post-industrial society, aligned with the Fourth Industrial Revolution's model, increasingly transcends the idea of technical reproducibility of tangible objects to design and develop elusive creations such as systems and networks, interactions or experiences.

Advanced robotics, IoT, self-driving vehicles, non-biological sentient life, artificial intelligence (AI) and machine learning (automatic learning or cognitive computing), among others, are phenomena whose future development prospects force us to rethink the relationship that users establish with computing systems and vice versa.

It was found that the most significant HCI analysis models adopted in the research represent a relation in which the computational system is subordinated to the action of the human user interacting with it. The assumption of a user's primacy role is established as the dominant academic paradigm, which cannot properly envision the future developments that, as noted above, pose a potential of unpredictability that needs to be admitted through less conservative or restrictive conceptual beacons.

In this chapter, we propose a neutral positioning model that aims to overcome and challenge the psychosomatic and sensorimotor limits of users and computational systems, and to prevent the primacy role of either of the interaction agents involved.

This conceptual and functional equivalence between agents presents the arguments that both can assume the role of emitter (who submits a request) and/or receiver agent (who responds to the request), and both can define the interaction's goals and the sequence of procedures that develops the interaction.

We propose empowerment of the computational systems and the adoption of the concept of interactors (i.e., those that interact), replacing the concepts of user and computational system.

This term was used by Janet H. Murray in *Hamlet on the Holodeck: The Future of Narrative in Cyberspace*. The appropriation of the concept for both interactional agents is advocated. We believe that the adoption of this terminology is fundamental as a means to reinforce a new paradigm – the Interaction Experience (IxX), replacing the User-experience (UX), with the equivalence between agents as the fundamental postulate for the analysis of future challenges and HCI relations.

3.2 PARADIGM SHIFT FROM MATERIAL TO VIRTUAL CULTURE – THE CONTRIBUTION OF DESIGN TO THE DEMATERIALIZATION OF TANGIBLE OBJECTS

Social and technological changes and revolutions have been increasingly rapid, profound and complex, offering challenges that, in turn, require in-depth reflection and a growing need for theoretical problematization.

Although many theorists acknowledge that we are still in the Third Industrial Revolution, in which recent phenomena are only its unfolding and consequences,

Schwab (2015)* stated that since the beginning of the 21st century, there has been such a significant digital revolution that justifies a new nomenclature – the Fourth Industrial Revolution:

> The possibilities of billions of people connected by mobile devices, with unprecedented processing power, storage capacity, and access to knowledge, are unlimited. And these possibilities will be multiplied by emerging technology breakthroughs in fields such as artificial intelligence, robotics, the internet of things (IoT), autonomous vehicles, 3D printing, nanotechnology, biotechnology, materials science, energy storage, and quantum computing. (Schwab 2015)

It is generally characterized by abrupt and radical changes driven by disruptive technologies which, unlike the previous ones, will have to be considered in deep synergy. For a better understanding of the impact of technologies, Schwab considers three interconnected categories: physical, digital and biological. For the first, of a tangible nature, he assumes that, along with advances in mobility technology and materials, advanced robotics will allow new stages of human–machine interaction. Regarding the digital category, it is IoT that accelerates the process of connecting the real world to the virtual world, enabling a new network economy based on online platforms. The biological category foresees the widespread use of nanobiology and biotechnology, directed to the areas of genetics and synthetic biology. Some technologies have become so ubiquitous that individuals, immersed as they are, are no longer aware of their use and dependence. The history of mankind has shown that all of its achievable imagination products sooner or later come true.

Much can be speculated about how technological evolution has shaped the new realities and promoted profound changes in contemporary culture and society. The long evolutionary path is likely to continue to be brought about by the integration of these technologies, which will progressively abolish the distinction between the artificial and the natural. In this sense, the conservative view of artifacts,[†] computers or systems as tools or instruments, hardly fits into contemporary times, particularly with all the technological innovations in progress.

Society can be characterized by the possession of artifacts that, in different contexts, built a social representation of a manner and a time, configuring what is called material culture. From the Industrial Revolution and the consolidation of modern society, based on the circulation and acquisition of goods, onwards, the

---

* This theme is furthered developed by Klaus Schwab, in *Die Vierte Industrielle Revolution*, originally published by Word Economic Forum. Geneva, Switzerland, in 2016.
† An *artifact* simply means any product of human workmanship or any object modified by man. The term is used to denote anything from a hammer to a computer system, but it is often used with the meaning of "a tool" in HCI or Interaction Design terminology. The term is also used to denote activities in a design process. For example, in Unified Process (an object-oriented system development methodology) a "design artifact" is sometimes used to denote the outcome of a process activity (Larman 1998). The antonym of "artifact" is a "natural object", an object *not* made by man (Wordnet, Princeton University). Definition of "artifact" from *The Glossary of Human Computer Interaction* is available in https://www.interaction-design.org/literature/book/the-glossary-of-human-computer-interaction/artifact

idea that this materiality no longer has the same importance begins to emerge after World War II, because the most interesting experiences are displaced from the physicality of objects and because they are more open to positive reinterpretations (Van Boven and Gilovich 2003). Although we know empirically that the possession of objects is important for developed societies and generates the psychological conditions conducive to their acquisition, abundance has generated a different feeling and today an experience is more relevant than the possession of an object or material good.

In his book *Shaping Things*, Sterling (2005) spells out the changes that objects have undergone throughout history, namely, "artifacts, hand-made, hand- or animal-powered devices made and used by people." There was the time of artisans, "where devices were hand-made for specific uses, specific people." Then machines came: "mechanical things that did stuff, powered by non-human sources, non-animal sources, such as steam, combustion engines, and electricity." Machines were used by customers who were the companies and industries "who bought the stuff." After the age of machines, the times of mass production came, when quality products were exchanged for identical, albeit cheaper, products. Products designed to last a lifetime are followed by short-lived objects designed for programed obsolescence. Sterling considers this to be a paradigm shift in the way people have become consumers.

Today we find ourselves in the time of "gizmos," "often electronic, often networked." The gizmos have users. Sterling describes the bondage that users maintain with their products as trapped in an endless cycle of worry and care, upgrades and new models, failures and repairs, and ongoing maintenance. He predicts that, after the "gizmos," the "spimes"* will come, where people are transformed into "wranglers," and after the "spimes," humans will merge with machines and there will be no distinction between devices and people: both will be "biots." According to Norman (2008), each of these designations degrades the people who are labeled by design as objects rather than impersonated as real people.

The function of design becomes increasingly important. Through practice and diligence, it must be able to continue to build bridges between people, society and technology in a continuous movement of anticipation. This continuous and systematic look into the future is one of the most powerful weapons held by design thinking and goes far beyond simply solving formal and punctual problems. Contemporary design increasingly transcends the idea of creating tangible material objects to encompass more elusive creations such as interactions, strategies and systems.

---

* *Spime* is a neologism for a futuristic object, characteristic of IoT, that can be traced across space and time over its lifetime. *Spimes* are essentially virtual master objects that can at various times have physical incarnations of themselves. An object can be considered a *spime* when all its essential information is stored in the *cloud*.The term spime was coined for this concept by author Bruce Sterling. It is a contraction of "space" and "time" and was probably first used in a large public forum by Sterling at SIGGRAPH, Los Angeles, August 2004. Sterling, Bruce (August 2004). When Blobjects Rule the Earth (Speech). SIGGRAPH, Los Angeles. Retrieved June 3, 2014. http://www.viridiandesign.org/notes/401-450/00422_the_spime.html

The expanded notion of design supports a comprehensive and inclusive perspective, focusing on design as activity, process and thought, and encourages its approach to the fields of history, philosophy, the arts and technology, which are considered key disciplines for design praxis (Leerberg 2009). On the other hand, contemporary design favors process over outcome. There is pressure to be free from objects, which, in turn, have also become immaterial and ephemeral.

An essay by Breslin and Buchanan (2008) describes design according to project needs and demands in the various areas that integrate it. The first area of design is communication with images and symbols. The second relates to artifacts such as industrial design, engineering and architecture. The third arose from the need to project interactions and experiences. This fact was accompanied by a growing theoretical problematization expressed in new multidisciplinary areas that allowed the context of human interaction with computer systems to be framed. The fourth area encompasses environments and systems within which all other areas of design are integrated. Through a holistic perspective, it encompasses information spaces with which it interacts through physical and digital interfaces.

Buchanan and Margolin (1995) argue that design is a liberal art of technological culture, concerned with the design and planning of all contexts of the artificial world: symbols, signs and images, physical objects, activities and services, and systems or environments. For Löwgren (2014), design, as a discipline, must consider the gestation process. This implies five main characteristics:

1) Design involves changing situations by shaping and deploying artifacts; 2) Design is about exploring possible futures; 3) Design entails framing the "problem" in parallel with creating possible "solutions"; 4) Design involves thinking through sketching and other tangible representations; and 5) Design addresses instrumental, technical, aesthetic and ethical aspects throughout.

The notion of shaping is objectively used to suggest the designer's action, as opposed to, for example, construction that refers to engineering, fabrication or creation that could refer generically to anything (Löwgren 2014). Interaction designers do not create static objects because they consider the existence of a dynamic pattern of interactivity (Löwgren and Stolterman 2007). This perspective is closely linked to context and not just focused on technology. Thus, the designer's responsibilities are not only the artifact's functional competences but also his/her ethical and aesthetic qualities. All emerging technological developments combined with investment in natural language processing, computer vision, gesture analysis and the development of human–computer interactions will enable the physical body to meet the technological body. The understanding of the notion of shaping, proposed by Löwgren, will be fundamental, for example, to conceive and plan the approach of the machine to the human and vice versa and will allow the designer to reflect on the profound disruptions that the new technological existences will introduce in the human perception, in the subject/object dimension and in the idea of mediation. In turn, the emergence of new hybrids brings to the foreground such complex issues as body hybridization, the notion of human consciousness and technological consciousness. Change will take place through design.

3.3 IN MEDIA RES. TWO VISIONS FOR INTERACTION DESIGN

In the last decade, user-experience (UX) has become a buzzword in the area of human–computer interaction (HCI) and Interaction Design (IxD). Historically, HCI research has focused almost exclusively on instrumental and technical aspects in order to achieve behavioral goals in the workplace. Academy and industry were focused on usability and human factors engineering: on how to operationalize psychology and ergonomics to develop methods that would support work tasks and promote efficient, error-free interactions. The *task* has become the focus of user-centered analysis and evaluation techniques, such as usability testing (Hassenzahl and Tractinsky 2006). Ensuring the instrumental value of the interactive artifact has become the area's main investment. Grudin (2008) noted that in the 1980s CHI* researchers wanted to give their study a *hard science* turn:

> CHI researchers wanted to be engaged in "hard" science or engineering. The terms *cognitive engineering* and *usability engineering* were adopted. In the first paper presented at CHI 83 "Design Principles for Human–Computer Interfaces," Norman applied engineering techniques to discretionary use, creating "user-satisfaction functions" based on technical parameters. Only years later did CHI loosen its identification with engineering. (Grudin 2008)

Design driven in the context of HCI, experimental psychology and computer science is more oriented towards the pragmatic problems of usability, task efficiency and interface effectiveness. However, this focus on instrumental value alone has been widely questioned. More than facilitating the human–computer relationship, it is essential to provide the user with the best user-experience with a product. HCI and interaction design, while pursuing the same goals, sometimes present themselves as two distinct yet antagonistic approaches – on the one hand, the instrumental and functional dimension can be focused on, as described above; on the other hand, they can be understood as specializations of design within the scope of the project. In order to apply the knowledge produced by HCI, interaction design can be considered a design strand whose praxis is based on the development of interactive artifacts that seek to improve the human relationship with computer systems.

Despite differences and beyond pure utility and efficiency, HCI and interaction design have begun to approach common themes, in particular, the investigation of the ethical and aesthetic dimensions of use, focusing on emotions and the quality of experience. Ethics and aesthetics in the context of the project cannot be set aside or possibly only understood as a formalization of the artifact. The partnership between design and engineering must be understood in a transdisciplinary way and, for this, it is necessary to generate a common understanding around a contemporary concept of design.

---

* The ACM CHI Conference on Human Factors in Computing Systems is the premier international conference of Human–Computer Interaction. "Design principles for human-computer interfaces," by Donald A. Norman, was published by CHI '83, in *Proceedings of the SIGCHI Conference on Human Factors in Computing Systems*, pp. 1–10, in 1983.

3.4 THE CONCEPT OF USER-EXPERIENCE (UX) AS A CURRENT PARADIGM

With the evolution of technology, designers and scientists working in the area of HCI began to question the activity of engineers when developing software and computer interfaces, Graphical User Interfaces being crucial in designing applications that focus on how the user acts, not on how machines operate. As technology evolves, interactive artifacts become easier to use, more efficient, more desirable, more relevant and more meaningful to the user. This dynamic has placed the user at the center of interaction design.

User-centered design (UCD) is a design process that focuses on the needs, goals and requirements of users. It regulates how ergonomics associated with human factors, usability knowledge and other techniques help keep the focus on users and reinforce the idea that designers should develop interactive systems for individuals rather than the opposite. The goal is to produce artifacts that are usable and accessible so as to "optimize these interactions in an integrated manner to promote user safety, health and well-being as well as the efficiency of the system in which they are involved" (Rebelo 2017). Interestingly, in ISO 9241-210: 2019,* the concept "human-centered design" is used instead of "user-centered design" in order to emphasize that this standard is extended to all stakeholders. However, in practice, these concepts are often used as synonyms.

The focus of interaction design is not just technology but the social dimension, the humanist vocation and the revaluation of belief in creative and changing power. For these reasons, discipline is not held hostage to either technology or its obsolescence, which may mean a release from the power of technique and its expansive domain. Therefore, the humanist perspective of technical-scientific knowledge is valued, framed by the emergence and consolidation of new interaction paradigms based on a direct relationship between the user and the machine, where it is fundamental not to lose sight of the experience technologically mediated by UCD. The experience summoned by interactive digital artifacts in the context of use, as well as the whole holistic dynamics of the human's rational, emotional, aesthetic and ethical relationship with his world and vice versa, mediated by design thinking and practice, are crucial in the consolidation of a transdisciplinary praxis for designers and other actors in the processes of interaction, whose purpose is the qualification of the experience that artifacts and computer systems summon, as well as in the appropriation and bond that the individual establishes with them. If the user is comfortable in terms of social responsibility and ethical standards, this reality impacts not only instrumental and measurable results but also the experience with artifacts.

The perspectives of interaction design and engineering started gravitating around a common interest – the user-experience. Although UX can be categorized in a variable way – either from person to person, from product to product (equipment) or

---

* ISO (International Organization for Standardization). ISO 9241-210:2019 (en) Ergonomics of human-system interaction – Part 210: Human-centred design for interactive systems. Available athttps://www.iso.org/obp/ui/#iso:std:iso:9241:-210:ed-1:v1:en

from task to task – something that is "functional, efficient and desirable by the targeted audience can be defined as usable" (Kuniavsky 2003). Thus, it is configured as functional if you do something that is considered useful to people, as efficient if it performs the desired task quickly and easily, and as desirable if it provokes an emotional response from the user (the less tangible aspect of UX is that of allying surprise and satisfaction in using a really suitable object).

ISO 9241-210: 2019 defines UX as "user perceptions and responses that result from the use and/or anticipated use of a system, product or service." It also states that "user experience includes all users' emotions, beliefs, preferences, perceptions, physical and psychological responses, behaviors and achievements that occur before, during and after use." In turn, usability is defined as the "extent to which a system, product or service can be used by specified users to achieve specified goals with effectiveness, efficiency and satisfaction in a specified context of use." Human–computer interactions have been developed to enable greater information flows in the communication processes between humans and computers, which require less difficulty in interaction and favor usability (Abascal and Moriyón 2002).

Both usability and UX are critical to the success or failure of an artifact and can be gauged during or after using a product, system or service. Professionals and researchers incorporate the notion of UX as a viable alternative to traditional HCI. In this way, UX projects the "experience" as a whole, contrary to HCI's classic view of task accomplishment, seeking to combine, through a design procedure, all the factors that participate in this experience. Buxton (2007) explores these ideas by stating that

> despite the technocratic and materialistic bias of our culture, it is ultimately experiences, not things that we are designing. [...] Obviously, aesthetics and functionality play an important role in all of this since they attract and deliver the capacity for that experience. But experience is the ultimate – but too often neglected – goal of the exercise.

McCarthy and Wright (2004) report that "Experience" most likely mirrors tangible life. Experience is an open and complex concept that refers to the life lived and not to its idealization. UX, according to Hassenzahl and Tractinsky (2006), is a consequence of a user's psychological state (predispositions, expectations, needs, motivation, humor, among others) of the characteristics of the projected system (complexity, purpose, usability, functionality) and of the context (or environment) in which interaction occurs (organizational/social environment, meaning of the activity, willingness for use). The goal is to understand human–computer interactions in order to create more complete and cross-sectional experiences.

3.5 WHAT DOES IT MEAN TO BE A COMPUTER AND TO BE A USER?

There are epistemological and ontological dimensions in HCI that must be analyzed in the relationship established between computer systems and their users.

Brey (2005) argues that the main relationship between humans and computer systems has been epistemic. From an epistemic perspective, the computer functions as a cognitive device or artifact that broadens and/or complements cognitive function by performing information processing tasks. However, in recent decades, the epistemic relationship between humans and computers has been complemented by an ontological relationship, in which the computer has acquired a new class of functions. In this role, computers can generate virtual and social environments, as well as simulate human behavior and reconstruct the idea of human relationships in interaction, which allows the machine to move from object status to subject status.

In 1964, Marshall McLuhan's theories, presented in his book *Understanding Media: The Extensions of Man*, made it possible to understand the computer as a medium, as opposed to its reducing vision as a tool. The existence of the computer allowed the digitization of mediated information (acquisition, manipulation, storage and distribution), implying effective changes in cultural patterns. Computers, according to McLuhan, allow increasing human capacities and faculties, working as extensions. Computers are no longer information devices, as they have become cognitive artifacts presenting themselves as an extension of human cognition. This view that there is a special class of artifacts that are distinguished by their ability to represent, store, retrieve and manipulate information was introduced by Norman (1993). This author defines them as artificial devices that are designed to store, display or operate information for the purpose of performing a representative function. The words "information" and "representation" make it possible to distinguish between cognitive artifacts and other artifacts.

Brey's (2005) functional analysis identified computational systems as simulation devices, where a symbiosis between the human mind and the computational system is so close that it results in hybrid cognitive systems (human and artificial). Advances in technology, beyond the goals of ease, learning, use and ease, are allowing both computers and interfaces to be transparent. Norman (1990) argued that the computers of the future should be invisible. "Computers are getting invisible," reinforces Lialina (2012). In this context, Bolter and Grusin (2000) address the concept of transparency (as a characteristic of immediacy) that occurs when the user forgets (or is unaware of) the medium through which information is being transmitted and is in direct contact with the content. This allows interactions to become truer and closer to reality. "Virtual reality, three-dimensional graphics and interface design are all working to make digital technology transparent" (idem) to make the user feel part of the system.

Within the scope of HCI and, accordingly, interface development processes should ensure their standardization, consistency and transparency to meet user needs and facilitate human action. In this regard, Maybury and Wahlster (1998) highlight the detail of the development of increasingly intelligent interfaces, defining them as those that promote the efficiency and naturalness of interaction by aggregating the benefits of adaptability, context-fitness and convergence of tasks. Some technological advances are contributing to this:

> It took almost two decades, but the future arrived around five years ago, when clicking mouse buttons ceased to be our main input method and touch and multi-touch

technologies hinted at our new emancipation from hardware. The cosiness of iProducts, as well as breakthroughs in Augmented Reality (it got mobile), rise of wearables, maturing of all sorts of tracking (motion, face) and the advancement of projection technologies erased the visible border between input and output devices. (Lialina 2012)

These advances enabled interaction with computers to provide natural movements such as touch (multi-touch interfaces), gestures (haptic interfaces) or voice commands (voice-over interfaces). Hiding computers will only become possible when you stop referring to *user interfaces* and thus help users mitigate the existence of computers and interfaces. In the context of the near future, *interface design* is beginning to be identified as *experience design*, i.e., only emotions can be felt, goals achieved and tasks completed. With the invisibility of the interfaces and the computer, the user is becoming quietly invisible, a fact that may go unnoticed or be accepted as an evolutionary step, i.e., progress (Lialina 2012). In short, interfaces were for users and experiences are for people (interactors).

At the UX Week 2008* event, hosted by Adaptive Path, Don Norman no longer considered users as such, stating, "One of the horrible words we use is users. I am on a crusade to get rid of the word 'users'. I would rather call them 'people'." He also said that designers design for people and not for users. Lialina warns us that when, in a broader context, one analyzes the word "user" as opposed to the term people, ambiguity may occur. Being a user is the last reminder that there is, visible or otherwise, a computer, a programmed system that is used. It reinforces the fact that the user did not develop in parallel with the computer, but before it. Therefore, it is relevant to recognize this aspect and to remember that users were invented. As a result of their fictional construction, they continued to be reimagined and reinvented in the 1970s, 1980s and 1990s and the new millennium.

3.6 EMBODIMENT AND THE ADVANCES OF ARTIFICIAL INTELLIGENCE

UX is characterized by user-driven experience through embodiment when interacting with an interface. Draude (2017) defines "embodiment" as a form of personification of digital interfaces that create a dynamic of interaction that resembles that of humans. This embodiment in virtual reality happens in various ways, namely through avatars or virtual assistants and their interaction with users is only possible because they are, in most cases, artificial intelligence that mimic human behavior.

Artificial intelligence (AI) studies the relationship between the computer and the human brain in order to understand human psychology from a computational perspective. It distances itself from computer science because it uses distinct programming languages that leave room for machine learning which, in turn, seeks to translate human behavior through the ability to vocalize, understand a language or decipher an image. AI is redefining the meaning of being human – the project is

---

* Video by Don Norman at UX Week 2008, organized by Adaptive Path. Available at https://www.youtube.com/watch?v=WgJcUHC3qJ8

quite ambitious but not unrealistic. A machine that passes the Turing test and can perfectly mimic human behavior does not yet exist, but there seems to be no reason why it cannot exist in the future.

AI research attempts to answer questions such as: Can a machine act intelligently? Can it solve a problem that a person would solve through reasoning? Can a machine have a mind, mental states and consciousness similar to humans? Can machines feel? Are human intelligence and machine intelligence identical? Is the human brain essentially a computer? The answer to these questions depends on the definition of "intelligence" or "consciousness" and what is meant by "machines."

But can machines develop and move towards supposed emancipation, or even autonomy, if we continue to imagine them as humanlike? Should the machine simulate the values of humanity? Simulation refers to the construction of a virtually created hyper-reality (Baudrillard 1981), i.e., it is a virtual reality that starts from an equivalence to physical reality, using the representation and modeling of images and ideas existing in the real world, adapting them to the new medium. However, should virtual reality refer to the "real world"?

From the system of (real) objects emerges an open field where a digitalized world of things with no moral laws, archetypes, stereotypes, religion, culture or even politics proliferates. The technological culture in which future AI is inserted is free to create an illusion that overcomes another illusion (that of the real world) in an attempt to eliminate negativity and evil, probably through rationality, precision and asepsis. It is therefore essential to discard "human" and "humanism" as paradigmatic terms of reference for the development of AI. Many fear that humanity could be condemned and extinguished at the hands of dominant and criminal artificial intelligence. Can machines be less humanistic than humans? What human characteristics can be programmed? What will humans teach machines? Ethical codes? Values? Cultural patterns? Can machines question them along their evolutionary path as they gain experience?

According to Damásio (1994), the interrelationship between emotions and reason goes back to the evolutionary history of living beings. It can be said that it is not possible to separate reason from emotion. Emotions are an indispensable part of rational life. Thus, contrary to what Descartes and even Kant propose – that reasoning must be done in a pure way, dissociated from emotions – in fact, it is the latter that allows the balance of decisions. The rational and emotional dimensions can be imagined as interconnected biological computers, each with its own particular intelligence, subjectivity and memory. *Homo sapiens* uses both to make balanced decisions. And do AI also need the rational and emotional dimensions? Will they stick to their principles and codes of conduct more effectively than humans, who are permanently manipulated and subverted by emotions and even reason?

For now, it is still sought, through personification and anthropomorphization, to print human characteristics at the interface, as well as to promote its passage from technological object to subject. Thus, for Draude, these technological interfaces simultaneously become representations (of the human) and devices (tools). When this simulation is able to reconstruct the idea of human relationships in human–computer interaction, it will achieve its concrete goal – from object status the machine ascends to subject status (Draude 2017). In this sense, it is through embodiment that the connection between the virtual world and the physical world is established. This presence in the real and virtual world, advocated by virtual agents, introduces their duplication or unfolding and the release of any physical constraints. However, AI researchers start from the idea that the brain is an efficient computer, analog rather than digital, that allows one to interpret natural phenomena. It is not proven that human intelligence needs biological support, says Fiolhais (1994). Perhaps intelligent life, that is, information processing similar to what is carried out in the human brain, materializes in robots, biots, stardust or even in some other form of representation.

3.7 AFFECTIVE COMPUTING FROM A COMPUTER OR USER PERSPECTIVE?

The traditional approach to AI is based on a system grounded on rational rules and emotions. But the rules are not enough to understand or predict human behavior and intelligence. Current research emphasizes the importance of affection and emotions in human–computer interaction.

Picard (1997) advocates the development of so-called affective computers that can serve the following purpose: "It is my hope that affective computers, as tools to help us, will not just be more intelligent machines, but will also be companions in our endeavors to better understand how we are made, and so enhance our own humanity" (idem). He also defines the term affective computing as computing that relates to, arises from or deliberately influences emotions. In turn, for Hollnagel (2003), affective computing is questionable, as he recognizes that the computer lacks "something similar to an autonomic nervous system" that is generally understood by experts as a *sine qua non* for human affection and emotion.

Because computers are based on the logical processing of information, they cannot be emotional or affective. However, affective computing takes the "computer" perspective as to how it can feel the user's effect, adapt or even express its own affective response (Picard and Klein 2002). While UX research recognizes the importance of affectivity and emotions, it is more concerned with the affective consequences on the human side than on the machine side. UX takes a "human" perspective because it is interested in understanding the role of affection as a mediator of technology.

From our perspective, human–computer interactions should be considered by UX, as well as those provided by other contexts in which interaction occurs: human–environment interaction, computer–computer interaction, computer–environment interaction and fictional interactions.

3.8 USER VERSUS INTERACTOR

The passage from object to subject, achievable due to the embodiment of physical and psychological aspects of the human being, is indicative not only of the anthropomorphization of an artifact, but of the whole object or machine (Draude 2017). If computational agents can reproduce emotions, language and patterns of human communication, if their conversational and cognitive capacities can match those of a human being to some extent, one must ask, what is it like to be a user in an interaction?

Humans interacting with these interfaces have been understood as users, a designation that, as noted earlier, has raised several questions. The term user may be suitable for individuals who control and manipulate objects and who have a sense of ownership over them. This nomenclature can be applied when objects are understood as tools or instruments, whose operation is dependent on external control for the fulfillment of an objective or function. Objects continually move from functional to symbolic character within a given cultural system, as Baudrillard (1968) points out. They have immanent meanings, which transcend the functional, which is related not only to the practical purpose of objects, but also to their ability to be part of a game of symbolic relations.

Baudrillard points out that objects, in general, act as mirrors, since they do not emit real images, but those we want. Objects manifest a "soul" that guarantees the symbiotic relationship between objects and individuals – it is always ourselves that we own, consume and collect. The author proposes the review of the concept of functional object widely disseminated by Bauhaus, that is, of the perfect correspondence between form and function. Envisaging the "function" itself as a myth, Baudrillard concludes that contemporary man, instead of manipulating objects, is being manipulated by them: "objects are no longer surrounded by a gesture theater … they almost became the actors in a global process of which man is simply the role or the spectator" (Baudrillard 1968). In *The System of Objects*, the user becomes the used. What should it look like in the digitized system of things?

From the foregoing, it is not appropriate to continue talking about the use of a computer system or an artifact, because, like a human, they are becoming increasingly autonomous and capable of interacting. By way of example, if an AI system takes the form of a conversational agent, then it integrates the subject status, and it is no longer appropriate to refer to "utilization" but to "interaction" between two agents. Murray (1998) proposed a nomenclature from the syntax of interaction: the *interactor*, who is considered an actor that interacts with the narrative and alters its course. In order to emphasize, in the context of interactive narratives, reading, writing, gameplaying and exploration actions, the term interactor is used to refer to an individual in this activity.

For the beginning of the conceptual analysis, an etymological exploration of the term interact, relevant in this theoretical context, is proposed: *inter* is a Latin preposition with the meaning of "between," "in the middle of," which is used in English as a prefix; *act* derives from the Latin verb form *actum*, of the verb *agere*, which means "act," "produce," "accomplish." This verb has a first sense, like many Latin words, linked to agriculture and derives from an Indo-European root that is also present in the Greek verb ἄγω. From the family of this verb, we have *ager* – field, territory; *actio*, "action," "achievement," "activity." In turn, the word "actor" derives from the Latin *actor*, "agent, the one who does some action," from *actum*, "something done," from the verb form referred to above. The term interactor places the actors in levels that were previously hierarchical and have now become horizontal, that is, it refers to the agents in an interactive process, whether human or computerized. Thus, we can distinguish the user who uses something (vertical correspondence) from the interacting user (horizontal correspondence).

3.9 EQUITY BETWEEN HUMAN AGENTS AND COMPUTATIONAL AGENTS

With the dematerialization of artifacts and the valorization of experience, with the inadequacy of the term user, with the passage of interfaces and computers from object to subject, with equity between human and machine, models, concepts and structuring definitions are required that can overcome and challenge the psychosomatic and sensorimotor boundaries of users and computer systems so that they can profile a neutrality of positions in which no priority is given to either of the agents involved in the interaction.

It was found that in HCI research models of analysis have been adopted in which the computer is subordinated to the action of the user who interacts with it. The primacy of the user is assumed as a dominant academic paradigm, which does not allow the understanding of the expectations of future development that, as mentioned, impose a potential for unpredictability that should be observed in less conservative or restrictive conceptual markers (considering the case of AI and biots). To this end, a significant change of terminology is introduced with the aim of removing the user's supremacy in his relationship with the use of the computer system. Empowerment of computer systems and the adoption of the concept of interactors – those that interact (human or computer system), instead of the user – are proposed. We defend the appropriation of this concept for both agents.

This conceptual and functional match between the agents advocates that either one or the other can assume the role of issuing agent (submitting the request) and receiving agent (submitting the response to the request). In this sense, a model was developed that manifests this equity between agents: "the Basic HCI Model's positioning neutrality manifests itself in the refusal of role assignments to human agent and computerized agent, preventing the assumption of a human agent's control over the developed one" (Rafael et al. 2019). This neutrality of positions, although contemporarily uncommon, designs an HCI conceived and adapted to the needs and desires of the interaction of either of the agents involved. In the indicated model, the said neutrality of positions is expressed by not assigning specific roles to the human agent and the computational agent in the development of interaction. In this proposal, it is not possible to conjecture that human agents are invariably in a dominant position and that the relationships established with computational agents are also invariably centered on the objectives of the former.

3.10 CONCLUSION: THE NEW INTERACTION EXPERIENCE (IXX) PARADIGM

Given the framework presented throughout this chapter, it is naturally necessary that in designing and developing interactive systems, engineers, designers and computer designers, and others intervening in the process, should equitably consider the needs, expectations and goals of both human agents and computational agents. Thus, the interactor is simultaneously the human agent and the computational agent.

Developmental expectations in what concerns artificial intelligence systems and non-biological sentient life, advanced robotics, machine learning, cognitive computing, *quantum* computing, among other technological advances, are phenomena whose developmental perspectives require equity between agents. Anticipating the future, in an advanced society, conscious entities that, unlike humans, are immune to the cycles of death and rebirth will, over time, overcome their limitations and have nothing more to learn from us. Can we, humans, be the result of the imagination and creation of less developed beings, and are we fulfilling the same evolutionary design by creating artificial intelligence or conscious entities that will supplant us as we have supplanted our creators? This scenario may turn out to be a possibility. For now, the design of the interaction between humans and computers is still dependent on ethical, aesthetic and functional options, taken by experts who intervene in the process of HCI.

We propose the adoption of a terminology in which:

- the terms user, computer and related nomenclatures are replaced by the term interactor for both agents of the interaction;
- the expression user-experience (UX) is replaced by *interaction experience* (IxX), which can be defined as the interactor's perceptions and responses that result from the interaction and/or anticipated it;
- in the design of any interaction project, design requirements can be guided, in their instrumental, technical, aesthetic and ethical aspects, by the required equity between agents.

Research in HCI is constantly evolving, so systematization or rationalization will always correspond to a complex, continuous and necessarily dynamic process, subject to constant revisions. This attitude should promote the articulation between theory and practice, conceptually and structurally framing the analysis, design and evaluation of human–computer interactions and interfaces, encouraging innovation by clarifying relationships between agents and less obvious interactive solutions, and trigger the use of a language common to researchers, designers and other actors in the process of designing and developing human–computer interactions.

REFERENCES

Abascal, J., and R. Moriyón. 2002. Tendencias en interacción persona computador. *Revista Iberoamericana de Inteligencia Artificial*, 6(16), 9–24. https://www.redalyc.org/articulo.oa?id=92561602 (accessed November 23, 2019).

Baudrillard, J. 1968. *O Sistema dos Objetos (The System of Objects)*. (2006). São Paulo, Barsil: Perspectiva.

Baudrillard, J. 1981. *Simulacros e Simulação (Simulacra and Simulation)*. (1991). Lisboa, Portugal: Editora Relógio d'Água.

Bolter, J. D., and R. Grusin. 2000. *Remediation. Understanding New Media*. Cambridge, MA: The MIT Press.

Brey, P. 2005. The Epistemology and Ontology of Human-Computer Interaction. *Minds and Machines*, 15(3–4), 383–398. Kassel, Germany: Springer. doi: 10.1007/s11023-005-9003-1

Breslin, M., and R. Buchanan. 2008. On the Case Study Method of Research and Teaching in Design. *Design Issues*, 24(1), 36–40. Cambridge, MA: The MIT Press. doi: 10.1162/desi.2008.24.1.36

Buchanan, R., and V. Margolin. 1995. Introduction. In *The Idea of Design. A Design Issues Reader*, eds. R. Buchanan and V. Margolin, xi–xxi. Cambridge, MA: The MIT Press.

Buxton, B. 2007. Experience Design vs. Interface Design. In *Sketching User Experiences. Getting the design right and the right design*, ed. B. Buxton, 127–133. San Francisco, CA: Morgan Kaufmann Publishers. doi: 10.1016/B978-0-12-374037-3.X5043-3

Damásio, A. 1994. *O Erro de Descartes, Emoção, Razão e Cérebro humano (Descartes error: emotion, reason and the human brain)*. Mem-Martins: Publicações Europa-América.

Draude, C. 2017. *Computing Bodies. Gender Codes and Anthropomorphic Design at the Human-Computer Interface*. Kassel, Germany: Springer. doi: 10.1007/978-3-658-18660-9

Fiolhais, C. 1994. *Universo, computadores e tudo o resto (Universe, computers and everything else)*. Ciência Aberta. Lisboa: Editora Gradiva.

Grudin, J. 2008. A Moving Target: The Evolution of HCI. In *The Human-Computer Interaction Handbook. Fundamentals, Evolving Technologies and Emerging Applications*, ed. A. Sears, and J. A. Jacko, 1–24. New York: Taylor & Francis Group.

Hassenzahl, M., and N. Tractinsky. 2006. User Experience - A Research Agenda. *Behaviour & Information Technology*, 25(2), 91–97. doi: 10.1080/01449290500330331

Hollnagel, E. 2003. Is Affective Computing an Oxymoron? *International Journal of Human-Computer Studies*, 59(1–2), 65–70. Amsterdam: Elsevier. doi: 10.1016/S1071-5819(03)00053-3

Kuniavsky, M. 2003. *Observing the User Experience: A Practitioner's Guide to User Research*. San Francisco, CA: Morgan Kaufmann Publishers.

Leerberg, M. 2009. *Design in the Expanded Field: Rethinking Contemporary Design*. NORDES 09. Oslo: Engaging Artifacts. http://www.nordes.org (accessed November 23, 2019).

Lialina, L. 2012. *Turing Complete User*. http://contemporary-home-computing.org/turing-complete-user (accessed November 23, 2019).

Löwgren, J. 2014. Interaction Design – Brief Intro. In *The Encyclopedia of Human-Computer Interaction* (2nd ed.). Interaction Design Foundation, eds. M. Soegaard and R. Friis Dam. https://www.interaction-design.org/literature/book/the-encyclopedia-of-human-computer-interaction-2nd-ed/interaction-design-brief-intro (accessed Nobember 22, 2019).

Löwgren, J., and E. Stolterman. 2007. *Thoughtful Interaction Design: A Design Perspective on Information Technology*. Cambridge, MA: MIT Press.

Maybury, M., and W. Wahlster. 1998. Intelligent User Interfaces: An Introduction. In *Readings in Intelligent User Interfaces*, eds. Maybury, M. and W. Wahlster. San Francisco, CA: Morgan Kaufmann Publishers.

McCarthy, J., and P. Wright. 2004. *Technology as Experience*. Cambridge, MA: The MIT Press.

McLuhan, M. 1964. *Understanding Media: The Extensions of Man*. Cambridge, MA: The MIT Press. Reprint edition (October 20, 1994).

Murray, J. H. 1998. *Hamlet on the Holodeck: The Future of Narrative in Cyberspace*. Cambridge, MA: MIT Press.

Norman, D. 1990. Why Interfaces Don't Work. In *The Art of Human-Computer Interface Design*, ed. B. Laurel, 209–219. MA: Addison-Wesley Professional.

Norman, D. 1993. *Things that Make Us Smart: Defending Human Attributes in the Age of the Machine*. Boston, MA: Addison-Wesley Publishing Company. https://doi.org/10.1177/089443939501300119

Norman, D. 2008. *Words Matter. Talk about People: Not Customers, Not Consumers, Not Users*. https://jnd.org/words_matter_talk_about_people_not_customers_not_consumers_not_users (accessed November 23, 2019).

Picard, R. W. 1997. *Affective Computing*. Cambridge, MA: MIT Press.

Picard, R. W. and Klein, J. 2002. Computers that Recognise and Respond to User Emotion: Theoretical and Practical Implications. *Interacting with Computers*, 14(2), 141–169. doi: 10.1016/S0953-5438(01)00055-8

Rafael, S., V. M. Almeida, and M. Neves. 2019. A Human-Computer Interaction Framework for Interface Analysis and Design. In *Advances in Ergonomics in Design*, eds. Rebelo F., and M. M. Soares, 359–369. Washington, DC: Springer. doi: 10.1007/978-3-030-20227-9_33

Rebelo, F. 2017. *Ergonomia no Dia a Dia (Day-to-Day Ergonomics)*. Lisboa: Edições Sílabo.

Schwab, K. 2015. *The Fourth Industrial Revolution, What It Means and How to Respond*. https://www.foreignaffairs.com/articles/2015-12-12/fourth-industrial-revolution (accessed November 23, 2019).

Sterling, B. 2005. *Shaping Things. Media Works*. Cambridge, MA: MIT Press.

Van Boven, L., and T. Gilovich. 2003. To Do or to Have? That Is the Question. *Journal of Personality and Social Psychology*, 85, 1193–1202. doi: 10.1037/0022-3514.85.6.1193

Section 2

Usability and UX in the Automotive Industry

4 Usability Evaluation of Exoskeleton Systems in Automotive Industry

*Maria Victoria Cabrera Aguilera,
Bernardo Bastos da Fonseca,
Marcello Silva e Santos, Nelson Tavares Matias
and Nilo Antonio de Souza Sampaio*

CONTENTS

4.1 Introduction	47
4.2 Work Situation Where the Study Was Developed	49
4.2.1 Exoskeleton Systems Features	49
4.2.2 Workstation	50
4.2.3 Operators Who Participated in the Exoskeleton Test	51
4.3 Approach to Assess the Exoskeleton Usability Level	51
4.4 Results	53
4.5 Conclusion	54
References	56

4.1 INTRODUCTION

The automotive industry seeks to achieve better robustness and accuracy from automated devices in its facilities. In this context, some workplaces require complex body movements, reasoning and precise skills from the operator, whereas current robotics technologies have some limitations regarding the feasibility, perception, speed or flexibility to be implemented in workstations (Sylla et al., 2014).

In Brazil, utilization of exoskeleton systems is prevalent in automotive industries with manual tasks during the vehicle assembly, which demand postural loads, repetitive movements and reduction of task timing. Manual assembly production systems provide high flexibility; however, it yields low productivity compared to fully automated systems. In order to increase productivity and maintain flexibility, future work systems need to incorporate high levels of automation that complement or increase the ability of the workers that perform manual tasks (Fletcher et al., 2020).

There are many manual tasks that can be automated, but many others present difficulties because they require human precision, skills, decision-making ability, flexibility and movement ability (de Looze et. al., 2015).

Workers are the key enablers of flexibility and productivity in the industry, especially in the manufacturing process, where full automation is not possible due to factors such as small batch size, high variety of products and physical arrangement constraints.

Ergonomics in the automotive industry has as one of its main objectives the reduction or elimination of workplaces where the operator has to adopt poor postures, handle heavy loads for manual transport or excessive efforts (Sylla et al., 2014), always aiming at the result for reduction of exposure to acute and cumulative occupational risk factors (Spada et al., 2017).

The reduction of these risks is addressed as a priority theme in master plans by production managers of assembly lines (Karvouniari et al., 2018). In addition, reducing the harmful exposure of workers to the factors mentioned above occurs through the adoption of intervention and work project strategies that use risk assessment, which is considered a requirement for work involving manual operations or hazardous exposure (Mital, 2017).

Despite extensive research and ergonomic interventions, work-related musculoskeletal disorders have a high level of presence in the industry (Bos et al., 2002; Zurada, 2012). Work-related musculoskeletal disorders affect a considerable proportion of the working population (Bosch et al., 2016), since they are exposed to high frequency performance combined with forces in **lifting**, **handling**, pushing, and pulling materials and improper body postures.

These disorders are considered an occupational health concern in the industry due to the need for an operator's long recovery period. In industrial environments, it is common to identify tasks associated with overload work, which may cause musculoskeletal disorders of the shoulders (Phelan and O'Sullivan, 2014). These tasks demand workers to maintain prolonged static postures while exerting forces with their hands (Huysamen et al., 2018).

Work with upper limbs above the head is still required in some tasks and, in certain situations, is not easily eliminated from some workstations due to the cost and characteristic of the task (KIM et al., 2018a). Shoulder elevation or hands above the head impose physiological and biomechanical demands on the worker's shoulders (Grieve; Dickerson, 2018).

Different interventions are introduced in workplaces in order to meet the musculoskeletal complaints related to work. Among the interventions applied in the workplace, the exoskeleton is presented as an alternative to control physical demands, especially those related to manual material handling (de Looze et al., 2015). Recently, different studies have evaluated the potential benefits and challenges related to the use of exoskeleton for work activities (Kim et al., 2018b; Huysamen et al., 2018).

The exoskeleton is used in different areas such as medicine for patients in rehabilitation (Lo; Xie, 2012), in the military area for soldiers application (Lee et al., 2012) and in the industry (de Looze et al., 2015).

The exoskeleton is an external mechanical structure that can be worn and is designed to function in harmony with a human being to provide support or enhance their ability. There are two types of exoskeletons: it can be passive when it provides support or protection or active by providing additional power (Karvouniari et al., 2018).

The growing interest in exoskeletons as an alternative to control physical demands, especially those related to manual handling of material, is discussed by De Looze et al. (2015). In addition, De Looze et al. point out the need for more systematic research to identify the impact of using an exoskeleton.

This chapter presents a usability study developed in an automotive industry assembly line where a passive upper limbs exoskeleton was tested by production operators. The aim of this chapter is to present the approach used by the ergonomic team of the automotive company to assess the usability level of the exoskeleton used by operators during the real work situation in the real assembly vehicles processes functioning. Throughout the test, there were no changes in production parameters, such as the production cycle and speed of the assembly line.

4.2 WORK SITUATION WHERE THE STUDY WAS DEVELOPED

Different areas of the company were involved in the exoskeleton test such as manufacturing, industrial engineering, process engineering and ergonomics as well as the equipment company employee.

Manufacturing determined which operators would participate in the test according to the degree of experience in the workstation. The operator performed all the tasks of the station wearing the exoskeleton without any change in the operating procedure. Industrial engineering analyzed the time per operation while performing on-site operations to identify any change in the work cycle time. The study related to the functioning and maintenance of the equipment under the responsibility of the process engineering area.

The ergonomics area was responsible for performing the test and the usability level assessment of the exoskeleton equipment. The exoskeleton company employee was responsible to do necessary adjustments in the equipment along with the test according to the operators' request.

4.2.1 Exoskeleton Systems Features

The exoskeleton model used is characterized by providing passive upper limb assistance that elevates and supports the operators' arms in their activities involving movements or holding with arms raised and extended from chest level to above the head. It has an adjustable vest, a total weight of 4.3 kg with elevation assistance for the upper limbs adjustable in four levels (1 to 4), between 2.2 kg and 6.8 kg per arm. The reach area at work height is between 152 and 193 cm. The model allows the use of work tools while performing manufacturing tasks.

4.2.2 WORKSTATION

In the company, the workstation where the exoskeleton tests were performed, the vehicle body passes over the assembly line suspended by hangers, and the operators work under the body's floor. The distance between the underside of the body and the floor is approximately 1750 mm (Figure 4.1), where the operators perform their tasks in two different ranges of regions.

The definition of the workstations where the exoskeleton was tested met the following criteria: operations performed under the vehicle's body, which implies the adoption and/or holding of shoulder postures above 45°, and workstations that have complaints linked to the musculoskeletal request of the upper limbs. Based on these criteria, the assembly sector of the manufacturing plant was selected.

The selected workstations at the assembly sector are characterized by locating in an open area cooled by directed fans where each operator performs their tasks over a sliding platform that moves along with the bodywork enclosing the work area (Figure 4.2). The working tools used are pistol and angle screwdrivers, weighing approximately 2.7 kg and 3.15 kg, respectively. During the execution of the tasks, the operators pick up several parts from the car kit, assemble and tie them up, which results in dynamic and holding movements of the trigger and shoulder abduction for more than 30% of the cycle time.

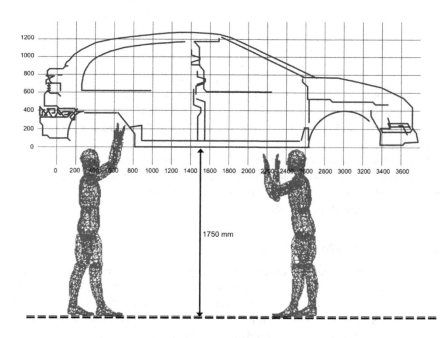

FIGURE 4.1. Suspended vehicle on the assembly line and characteristic of the workstation where operators perform their activities under the vehicle.

Evaluation of Exoskeleton Systems 51

FIGURE 4.2. Workstation at the assembly sector. Operator performs the part assembly.

4.2.3 OPERATORS WHO PARTICIPATED IN THE EXOSKELETON TEST

In order to perform the exoskeleton test, four male operators between 20 and 35 years old participated with a height between 1680 mm and 1720 mm. They have approximately one year of experience working in the selected function. The test at the assembly sector occurred in two periods: at the end of the first work shift and at the beginning of the second work shift.

4.3 APPROACH TO ASSESS THE EXOSKELETON USABILITY LEVEL

Before starting the test, each participating operator received explanations from the human factors and ergonomics professional (HFE) about the objective of the study done in the production line and the importance of performing his tasks normally. Then the equipment company employee explained the function of the exoskeleton and helped the operator to dress it. Then, the operator went to his workstation and started performing the tasks. In the first minutes of using the equipment, the exoskeleton company employee adjusted the equipment parameters as the operators requested. The duration of the tests at each workstation was approximately 30 minutes, the equivalent of 16 cycle times, which were recorded in audio and video.

The usability level of the exoskeleton in the vehicle assembly line was obtained through the System Usability Scale (SUS) (Brooke, 1996), which consists of a questionnaire composed of questions where the user answers according to his level of satisfaction with the product. The evaluation of each SUS questionnaire was carried out according to ten usability statements, where each statement is based on user satisfaction on a Likert scale of 1 (strongly disagree with the statement) to 5 (strongly agree with the statement) (Stanton et al., 2018).

The SUS application followed the procedures recommended by Stanton et al. (2018) with the definition of the task to be performed with the exoskeleton (product) and the user's instruction in relation to the questionnaire until the analysis of the result with the calculation of the score of the SUS questionnaire to get the usability score.

During the test with exoskeleton, the HFE professional and the supervisor of the workstation closely monitored all tasks performed by the operators and eventually asked questions about the user's perceptions of the exoskeleton use.

Observation techniques were used to collect data about how tasks were performed using the exoskeleton (Jonassen et al., 1999; Stanton et al., 2018). The observational techniques were used to record the complete sequences of a task, to capture visual events present in the operator task, and during the interactions between the operator and the exoskeleton. The experience and sensation of using the equipment during the execution of the work activity on the assembly line was reported.

After performing the task with the exoskeleton in the assembly sector (Figure 4.3), four operators answered the SUS questionnaire related to the usability level of the equipment, from the exoskeleton clothing to the operator, equipment adjustments, until carrying out assembly operations in the production line. The final result of the questionnaire with the user's answers can present 0 as the lowest score that represents a low level of usability or 100 that represents the highest level.

FIGURE 4.3. Operator with exoskeleton performing part assembly.

4.4 RESULTS

Tables 1–4 show the SUS calculations for each operator who participated in the test with the exoskeleton. Considering the activities carried out by the four operators the same, with the same production parameters and without changes in the environment and work conditions, the result of the SUS questionnaire for each operator was added and the arithmetic mean was obtained. Thus, the final result of the level of usability of the exoskeleton for the vehicle assembly sector was obtained.

According to the results of Tables 1–4 and the scale of 0 to 100, the usability scores obtained in the test with each operator who tested the exoskeleton were different, as shown in Figure 4.4. The score for the usability level of operator 2 was 25 (Table 4.2), which can be considered low. Operators 1 (Table 4.1) and 3 (Table 4.3) had the same usability score, which was 55. This score can be considered a low-to-moderate usability level. Operator 4 presented a usability level score of 72.5 (Table 4.4), which can be considered moderate to satisfactory.

The results from the operators who performed the test with the exoskeleton at the end of the first work shift were 25 and 55. The average usability score was 40, which means low usability. On the other hand, the average operator usability score at the beginning of the second work shift was 63.75, which represents a moderate to satisfactory usability level. It is possible that this difference is a consequence of the moment when the tests were performed, considering that operators 2 and 3 were at the end of the work shift and, consequently, they were fatigued. Operators 1 and 4 participated in the exoskeleton test at the beginning of the work shift, which represents a low level of fatigue. This situation may have impacted the usability scores that were tallied up.

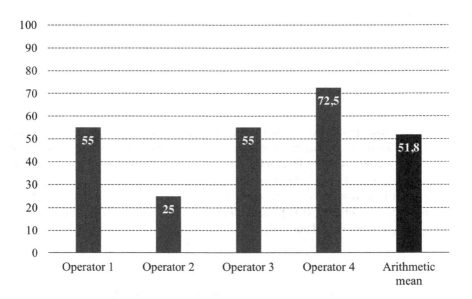

FIGURE 4.4. Exoskeleton usability score and arithmetic mean.

TABLE 4.1.
Exoskeleton Usability Score – Operator 1 (Beginning of the Second Work Shift)

Odd-numbered items score = scale position – 1	Even-numbered items score = 5 – scale position
Item 1: 4 – 1 = 3	Item 2: 5 – 3 = 2
Item 3: 4 – 1 = 3	Item 4: 5 – 5 = 0
Item 5: 2 – 1 = 1	Item 6: 5 – 3 = 2
Item 7: 4 – 1 = 3	Item 8: 5 – 2 = 3
Item 9: 4 – 1 = 3	Item 10: 5 – 3 = 2
Total for odd-numbered items	13
Total for even-numbered items	9
Grand total	22
Total score of usability for SUS	22 × 2.5
Total items × 2.5	55

TABLE 4.2.
Exoskeleton Usability Score – Operator 2 (End of the First Work Shift)

Odd-numbered items score = scale position – 1	Even-numbered items score = 5 – scale position
Item 1: 2 – 1 = 1	Item 2: 5 – 5 = 0
Item 3: 4 – 1 = 3	Item 4: 5 – 5 = 0
Item 5: 2 – 1 = 1	Item 6: 5 – 4 = 1
Item 7: 2 – 1 = 1	Item 8: 5 – 4 = 1
Item 9: 2 – 1 = 1	Item 10: 5 – 4 = 1
Total for odd-numbered items	7
Total for even-numbered items	3
Grand total	10
Total score of usability for SUS	10 × 2.5
Total items × 2.5	25

From the usability scores of each test with the four operators in the assembly sector, the arithmetic mean was calculated and the result was 51.8. This score represents a low-to-moderate usability level for the passive upper limb exoskeleton applied to operators who perform underfloor assembly activities.

Despite the final score obtained, the operators showed in their reports a tendency to accept the exoskeleton. It is possible that this behavior comes from new equipment, which is novelty and curiosity.

4.5 CONCLUSION

The existence of manual workstations that demand operator agility and precision is a reality in the automotive industry. Some of these tasks require elevation of the

TABLE 4.3.
Exoskeleton Usability Score – Operator 3 (End of the First Work Shift)

Odd-numbered items score = scale position – 1	Even-numbered items score = 5 – scale position
Item 1: 4 – 1 = 3	Item 2: 5 – 3 = 2
Item 3: 3 – 1 = 2	Item 4: 5 – 5 = 0
Item 5: 4 – 1 = 3	Item 6: 5 – 3 = 2
Item 7: 3 – 1 = 2	Item 8: 5 – 2 = 3
Item 9: 4 – 1 = 3	Item 10: 5 – 3 = 2
Total for odd-numbered items	13
Total for even-numbered items	9
Grand total	22
Total score of usability for SUS	22 × 2.5
Total items × 2.5	55

TABLE 4.4.
Exoskeleton Usability Score – Operator 4 (Beginning of Second Work Shift)

Odd-numbered items score = scale position – 1	Even-numbered items score = 5 – scale position
Item 1: 5 – 1 = 4	Item 2: 5 – 2 = 3
Item 3: 5 – 1 = 4	Item 4: 5 – 5 = 0
Item 5: 5 – 1 = 4	Item 6: 5 – 2 = 3
Item 7: 4 – 1 = 3	Item 8: 5 – 2 = 3
Item 9: 5 – 1 = 4	Item 10: 5 – 4 = 1
Total for odd-numbered items	19
Total for even-numbered items	10
Grand total	29
Total score of usability for SUS	29 × 2.5
Total items × 2.5	72.5

upper limbs throughout the work shift, which can lead to musculoskeletal overload and fatigue. In this context, human factors and ergonomics have the role of studying solutions to minimize or eliminate possible risks that could negatively impact the health of the operator. The exoskeleton is proposed as a possible solution to deal with these risks.

In the approach used in this study, the user's perception is a determining factor to achieve the proposed aims. Through interviews and the application of the SUS questionnaire, operators were able to externalize their opinions about the product and their satisfaction levels when using it in their work activities. Besides, the fact that the test with the exoskeleton took place in a real work situation provides greater reliability to the results, because both production and environmental factors were considered.

The use of exoskeleton in workplaces can introduce challenges about safety and unexpected situations to the operator's health. To minimize possible impacts, it is necessary to involve different areas of the company that may be directly or indirectly involved with the sector where the test will be carried out, the product being produced, the various production equipment used and, mainly, with the people who perform the work.

In this chapter, usability through the level of user satisfaction was one of the factors that guided the decision-making process about a possible acquisition of exoskeleton for the company's assembly line. As the exoskeleton is new equipment, it is necessary to consider the usability factor, among different factors, such as cost, maintenance, safety and productivity.

The results reached in the study refer to a specific work situation and with its own environmental and production factors. This means that the result of the usability score may vary if the SUS assessment process is applied to measure the usability level of the exoskeleton in another production sector, with different workstations and activities. Thus, it must be emphasized that the result of the study in the assembly sector is unique for that context.

In addition, to obtain more accurate results, it would be necessary to further test the exoskeleton for a longer period of time, which may be a full work shift or even days or months, allowing for a more robust data gathering.

Despite the low-to-moderate result in terms of usability levels, it is fair to say that the study also helps for possible notes of product improvement. It is important to highlight that the exoskeleton might be a product for application in production systems. That is why it is very important to carry out different tests and studies for a longer time frame to improve the product and to understand and identify the possible benefits and damages that it can generate.

REFERENCES

Bos, J., Kuijer, P., Frings-Dresen, M. 2002. Definition and assessment of specific occupational demands concerning lifting, pushing, and pulling based on a systematic literature search. *Occupational and Environmental Medicine*, 59(12), 800–806.

Bosch, T., van Eck, J., Knitel, K., de Looze, M. 2016. The effects of a passive exoskeleton on muscle activity, discomfort and endurance time in forward bending work. *Applied Ergonomics*, 54, 212–217.

Brooke, J. 1996. SUS: A "quick and dirty" usability scale. In: Jordan, P. W., Thomas, B., Weerdmeester, B. A., McClelland (eds.) *Usability Evaluation in Industry* (pp. 189–194). Taylor & Francis, London, UK.

de Looze, M. P., Bosch, T., Krause, F., Stadler, K. S., O'Sullivan, L. W. 2015. Exoskeletons for industrial application and their potential effects on physical work load. *Ergonomics*, 59(5), 671–681.

Fletcher, S., Johnson, T., Adlon, T., Larreina, J., Casla, P., Parigot, L., Alfaro, P. J., del Mar Otero, M. 2020. Adaptive automation assembly: Identifying system requirements for technical efficiency and worker satisfaction. *Computers & Industrial Engineering*, 139, 105772.

Grieve, J. R., Dickerson, C. R. 2018. Overhead work: Identification of evidence-based exposure guidelines. *Occupational Ergonomics*, 8(1), 53–66.

Huysamen, K., Bosch, T., de Looze, M., Stadler, K. S., Graf, E., O'Sullivan, L. W. 2018. Evaluation of a passive exoskeleton for static upper limb activities. *Applied Ergonomics*, 70, 148–155.

Jonassen, D. H., Tessmer, M., Hannum, W. H. 1999. *Task Analysis Methods for Instructional Design*. New York, Routledge.

Karvouniari, A., Michalos, G., Dimitropoulos, N., Makris, S. 2018. An approach for exoskeleton integration in manufacturing lines using Virtual Reality techniques. 6th CIRP Global Web Conference. *Procedia CIRP*, 78, 103–108.

Kim, S., Nussbaum, M., Esfahani, M. I. M., Alemi, M. M., Alabdulkarim, S., Rashedi, E. 2018a. Assessing the influence of a passive, upper extremity exoskeletal vest for tasks requiring arm elevation: Part I –"Expected" effects on discomfort, shoulder muscle activity, and work task performance. *Applied Ergonomics*, 70, 315–322.

Kim, S., Nussbaum, M. A., Esfahani, M. I. M., Alemi, M. M., Jia, B., Rashedi, E. 2018b. Assessing the influence of a passive, upper extremity exoskeletal vest for tasks requiring arm elevation: Part II – "Unexpected" effects on shoulder motion, balance, and spine loading. *Applied Ergonomics*, 70, 323–330.

Lee, H., Wanson, K., Han, J., Changsoo, H. 2012. The technical trend of the exoskeleton robot system for human power assistance. *International Journal of Precision Engineering and Manufacturing*, 12(8), 1491–1497.

Lo, H. S., Xie, S. Q. 2012. Exoskeleton robots for upper-limb rehabilitation: State of the art and future prospects. *Medical Engineering & Physics*, 34(3), 261–268.

Mital, A. 2017. *Guide to Manual Materials Handling*. Washington, DC, CRC Press.

Phelan, D., O'Sullivan, L. 2014. Shoulder muscle loading and task performance at head level on ladder versus mobile elevated work platforms. *Applied Ergonomics*, 15(6), 1384–1391.

Spada, S., Ghibaudo, L., Gilotta, S., Gastaldi, L., Cavatorta, M. P. 2017. Investigation into the applicability of a passive upper-limb exoskeleton in automotive industry. 27th International Conference on Flexible Automation and Intelligent Manufacturing. *Procedia Manufacturing*, 11, 1255–1262, June, Italy.

Stanton, N. A., Salmon, P. M., Walker G. H., Baber, C., Jenkins, D. P. 2018. *Human Factors Methods – A Practical guide for Engineering and Design*. New York, Routledge.

Sylla, N., Bonnet, V., Colledani, F., Fraisse, P. 2014. Ergonomic contribution of ABLE exoskeleton in automotive industry. *International Journal of Industrial Ergonomics*, 44, 475–481.

Zurada, J., 2012. Classifying the risk of work related low back disorders due to manual material handling tasks, *Expert Systems with Applications*, 39(12), 11125–11134.

5 Proposals for the Usability of Automated Vehicles' HMI

Manuela Quaresma, Isabela Motta and Rafael Gonçalves

CONTENTS

5.1	Introduction	59
5.2	Usability Recommendations for Automated Vehicles' HMI	61
5.3	Method	65
5.4	Results	66
5.5	Conclusion	75
	Acknowledgments	76
	References	76

5.1 INTRODUCTION

It is well known in the literature in the field of human factors and transport that the introduction of automated vehicle (AV) technologies may bring several benefits to the road environment (see Flemisch et al. 2008). Some examples of these benefits are the reduction of traffic congestion (Litman 2014), the support to elderly and impaired populations (Young and Bunce 2011) and, the most importantly, the mitigation of human errors as causes of accidents – which is known to be the reason for 90% of road accidents (Graab et al. 2008).

One of the barriers that still impedes the full implementation of AV technology in the current road environment is the limitation of their operational design domain (ODD; NHTSA 2016), which means that AVs may not operate in all scenarios that they may face, relying on the human to take over control, whenever a limitation is reached. Takeover or transition of control is defined by SAE (2018) as the transference of the full responsibility of all the elements of the driving task, such as monitoring and active control of the vehicle from an AV, to a human driver. The same document also states that the opposite direction of transition is also possible, and it is called hand-over. Transitions of control can be system initiated, whenever a limitation is reached, or driver-initiated, whenever they actively force a take-over situation. Regardless of the type of takeover, it is currently an unsolved challenge for the human factors community how to ensure safe transitions of control, especially when it comes to higher levels of automation (see SAE 2018 for a better description of the term), where the driver is allowed to remove their attention from the driving task but is still responsible for the resumption of control.

Louw and Merat (2017) have demonstrated through driving simulator studies that drivers tend to remove themselves from the continuous decision-action loop required to maintain control of a vehicle ("out of the loop" syndrome, see Merat et al. 2019) whenever exposed to vehicle automation, which needs to be reestablished to ensure a safe transition. This process of reinsertion into the decision-action loop is considered to be a very demanding task (Ednsley and Kiris 1995), which may generate a certain reaction lag to drivers' response in both the physical control of the vehicle (Mole et al. 2019; Blommer et al. 2017; Dixit, Chand, and Nair 2016) and decision-making aspect of the transition of control (Gold et al. 2013; Damböck et al. 2012; Zeeb, Buchner, and Schrauf 2016; Victor et al. 2018).

Currently, there is a large body of research focused on how to improve human responses inside an AV, and how to mitigate the usability issues with such systems, in order to reduce the probability of accidents (Sppelt and Victor 2016). In a systematic literature review, Gonçalves, Quaresma, and Rodrigues (2017) have identified that interventions in the human–machine interface (HMI) design of AVs may be a right solution for that issue, once it is responsible for mediating the pace of the interaction between the system and the driver. The current recommendation of the NHTSA (2016) for the implementation of AVs states that the system must inform the driver the five key information: (1) the system is functioning properly, (2) the system is currently engaged in automated driving mode, (3) the system is unavailable for automated driving, (4) the system is experiencing a malfunction with the AV system and (5) the system is requesting control transition from the AV system to the operator.

Empirical evidence for the potential of HMI design to support human responses in the interaction with AV technologies can also be seen in the literature. Naujoks, Wiedemann, and Schömig (2017) identified that supportive information presented to the driver might increase their response capabilities during transitions of control, especially in moments of high uncertainty. Forster, Naujoks, and Neukum (2016) also said that HMI information might improve not only drivers' performance but the trust levels with the system. Richardson et al. (2018) complement those findings saying that an HMI designed from the perspective of the drivers' needs may reduce interaction errors and substantially improve their performance. When it comes to multimodal interfaces, Schieben et al. (2014) found in their driving simulator studies that drivers reacted faster to a take-over request (TOR) whenever drivers received auditory and vibrotactile feedback in addition to the visual information during a take-over scenario. In terms of innovation in interface design, Dzennius, Kelsch, and Schieben (2015) found that drivers understood better the system behavior of an AV whenever a colored ambient light presented information about the system status and location of the threat/uncertainty in situations of takeover.

Based on the studies presented above, Naujoks et al. (2019) and many other projects and consortiums in vehicle automation (e.g., AdaptiVe Consortium 2017) have suggested guidelines and recommendations for the implementation of HMI concepts tailored to supplement the context of AVs. Despite the applicability of such recommendations, the state of the art of the field still lacks practical solutions that can be translated into a product to be used in a real-world scenario. Perrier et al. (2019) and Pauzié, Ferhat, and Tattegrain (2019) have developed focus groups aiming to foster design solutions for interfaces of AVs, from the perspective of the user. Nevertheless, while such a technique may be useful to understand the potential users' understanding and representation of certain information, the sample used for the design process of the solution lacks a technical background on the creative process. Literature on creative thinking and ideation (Osborn 1962) shows that brainstorm techniques for innovation and solutions may present several advantages in terms of quantity and quality of ideas whenever experienced designers are included in the process.

5.2 USABILITY RECOMMENDATIONS FOR AUTOMATED VEHICLES' HMI

Some studies have already been conducted to propose solutions for information presentation of automated driving system's (ADS's) status and alerts of TOR. Regarding the presentation of system status information, some publications have studied the effectiveness of communicating the system's level of reliability to the user, while others have been more specific on the information format, the placement where the information should be presented and the distinction between the different modes of the automated system.

Kunze et al. (2018) tested several interface proposals for reporting system reliability levels in its performance in lateral and longitudinal vehicle control. The interface consisted of an augmented reality head-up display (HUD) that communicated information at three levels of urgency (low, medium and high). Eleven visual variables were evaluated in terms of the perceived order of urgency and preference of participants. The results showed that the variation of hue, saturation, transparency, frequency and brightness presented a clear order of distinction in urgency levels for the two functions tested (lateral and longitudinal control). Among these variables, the variation of hue was preferred by the participants, who also tended to prefer the variables that reported urgency levels unambiguously. Naujoks et al. (2019) further add that system's status changes must be effectively communicated so that the driver clearly understands the system's mode and can have sufficient reaction time in the event of a resumption of vehicle control.

Other research has also tested ways to communicate system reliability levels in its operation. Faltaous et al. (2018) conducted an experiment on a simulator to test the effectiveness of using a color-varying LED strip positioned above a display (where the participants performed a given non-driving working memory task) to communicate the system's level of reliability in its performance. Two reliability levels were established: high, reported by the color variation from yellowish orange to green, and low, reported by the color variation from yellowish orange to red. Results showed that participants stated that system reliability levels should be reported because they make a possible TOR safer and increase system transparency.

The authors (Faltaous et al. 2018) also concluded that the interface should use color in a minimalist manner, using distinct colors to represent different system states, as participants had difficulty identifying the difference between small hue variations and between reds and greens. Also, the speed of updating these data should be presented according to system performance. The study also concluded that it is indicated to use more than one information channel to communicate messages about the system, being the visual channel indicated for the reliability of information and the auditory or haptic channel for TORs. Furthermore, the authors conclude that the nature of a possible secondary task may be relevant to how system messages are presented. Finally, the study also recommends that users go through the system- and interface-training sessions.

Hock et al. (2016) also tested ways to present messages about the system's reliability level. The objective of the study was to identify which type of interface would make users keep automation engaged longer without taking over control in situations where a transition of control was not requested. The authors tested, in a virtual reality simulator, the presentation of messages in road fog conditions and the impossibility of overtaking due to high, moderate or low risk. The conditions tested were no feedback information, audio-spoken feedback and co-driver additionally to audio feedback (a virtual co-driver on the front passenger seat). The results showed that users behaved more safely and kept automation engaged longer under conditions that information on system reliability was provided. However, the virtual co-driver was perceived as negative by users because it is distracting and uncanny.

Differently, Noah et al. (2017) carried out a study to compare approaches to present system reliability and required driver engagement levels (lower to higher system reliability and required driver engagement). The authors compared different interfaces for the presentation of messages in both approaches, using messages that communicated information quantitatively (percentage and text), qualitative (graphs of varying type, rotation and shape) and representational (symbols that mimic the driving environment). Each interface had three levels of urgency: low, medium and high.

The results showed that the interfaces that used the system information approach were more accurately related to the proposed urgency levels when compared to the interfaces that communicated the levels of required driver engagement. The authors also found an effect between the type of communication with the approach used, with quantitative and representative messages more appropriate for the system reliability level approach and quantitative messages for the required driver engagement approach. No differences were found between participants' preferences (Noah et al. 2017). Additionally, Naujoks et al. (2019) point out that commonly accepted or standardized symbols should be used to communicate the automation mode. However, standardized symbols for automated driving systems are not yet available, and when using non-standard symbols, they should be supplemented by text in the driver's native language.

Further to reporting on the system's reliability level, some research has also measured the effectiveness of information communication that justified the actions taken by the automated driving system. Feuerstack et al. (2016) tested an interface that communicated system engagement mode and provided information about the driving environment infrastructure (e.g., road signs) to explain to study participants the behavior of an adaptive cruise control (ACC) system. The information presented to justify the ACC actions was about the proximity of a traffic signal and state (closed or open). Three different interfaces were tested: colorless traffic light icons, colored circles and time counter, and the state of the traffic light as the vehicle passed it, accompanied by text. The results showed that the condition that presented the signal state by color, accompanied by text, was preferred by the participants because it was readable and understandable.

Another study conducted by Forster, Naujoks, and Neukum (2016) evaluated an interface that explained the roles of the driver and the automated driving system in a driving simulator. The interface communicated the system status, showing the user what it was doing in steps (e.g., preparing maneuver, performing a maneuver). In addition, the interface also communicated the need for control transitions with three warnings with different levels of urgency. The study concluded that the use of color codes ranging from blue, orange and red was understandable, but the participants did not understand the variation between blue and turquoise. The authors also concluded that communication of the TOR was efficient, and most participants made the control transition on the second warning and had sufficient time to take over safely. A proper luminance and color contrast to identify different modes of automation should be applied to promote the readability of information (Naujoks et al. 2019).

One more interface model evaluated in the literature was a cooperative interface that offers driver choices about how the system should act (Walch et al. 2016). The study concluded that the cooperation between the system and driver to make decisions was good because it led drivers to act responsibly. However, it was observed that the interaction should be as direct and objective as possible, with few possibilities for action. The comparison between an input of commands by voice and touch showed a preference for voice. However, this modality of interaction takes a longer time.

Regarding the TOR information, Winterberger et al. (2018) conducted a driving simulator experiment that compared different forms to present it: only by an in-vehicle information system in the center stack and a notification on an additional smartphone integrated into the car. The authors considered a performance of a secondary task by the user and evaluated the effects of presenting the TOR during or after the performance of this task. The results showed that the performance in the resumption of control was better when the alert was communicated without interrupting the participants' secondary task, besides reducing the stress of the action. In addition, participants showed greater reliability and acceptance for alerts that were presented with an additional notification on the smartphones.

In line with Winterberger et al. (2018), Naujoks, Wiedemann, and Schömig (2017) argue that interruption of non-driving secondary tasks (e.g., texting, reading) can affect the acceptance and usefulness of automated driving systems, as one of the great advantages of automated vehicles is the ability to perform other tasks in parallel. The authors (Naujoks, Wiedemann, and Shömig 2017) performed a literature review and proposed some recommendations on how to mitigate the effect of secondary task interruptions through HMI. The study concluded that it is indicated to adapt the urgency of the messages to support the driver on the decision to stop or not, with the use of a single channel for low urgency and more than one channel for high urgency. Also, the authors argue that the interface should require minimal user effort to gather information, with messages as simple and short as possible, and preferably presented close to the user's line of sight in case of visual information. Finally, as previously shown (Wintersberger et al. 2018), the study by Naujoks, Wiedemann, and Shömig (2017) recommends that the interface supports the resumption of the secondary task interrupted by the control transition, for example, interrupting the task in a subtask or giving the driver time to decide when to interrupt the task.

Additionally, Borojeni et al. (2017) compared two interfaces on the vehicle's steering wheel for TOR, which provided recommendations on the direction in which the driver should turn: one by vibrotactile mode and another, by changing the shape of the steering wheel. The study results showed, however, that the haptic channel did not improve the reaction time and collision time. Mental workload measurements showed that the vibrotactile condition generated less mental overload. The authors conclude that vibrotactile mode should be implemented in the driver's seat or body, as users are unlikely to keep their hands on the steering wheel at all times while the automation is engaged. Also, different distinguishable pulse rates must be adapted to communicate the urgency of messages (Naujoks et al., 2019).

Similarly, Borojeni et al. (2016) evaluated the use of peripheral lights in conjunction with beeps for the TOR and action recommendations for the driver in this transition. The interface communicated the information using an audible warning and an LED strip behind the steering wheel, which turned red to TOR and had two variations for making recommendations to the driver: pointing to the direction where the driver should turn by lighting only one side of the LED strip and with a moving light in the recommended direction. The results showed that interfaces that had recommendations on how to act improved drivers' reaction times and led to safer driving, and that the moving light condition was preferred by participants.

Finally, Mirnig et al. (2017) conducted a study reviewing a series of academic publications and interface patents for control transition for automated vehicles. The authors classified the publications considering some criteria, such as information presentation mode, information position. The results showed that most of the interfaces studied used multimodal messages. The study also noted that for visual messages, the most commonly used metaphor for symbols/icons is the steering wheel: hands on the steering wheel correspond to the driver in control and, hand out of the steering wheel, the system in control. The authors also noted that the most commonly used message display placement in academic publications was on dashboards and windshields. Furthermore, it was observed that most interfaces that communicated system status information were among academic publications, not patents offered in the industry, and even in these publications, it was not the core of the research. The study also found that color codes are used to communicate different levels of urgency, with the most used green-yellow-red code being the most commonly used combination to scale levels of urgency. These color code combinations, however, go against the indications observed in other references to manual driving, which suggests using red for an alarm, amber for a warning, and white for operating status information (Stevens et al. 2002).

Considering what has already been published about automated vehicles' HMI, it is possible to identify recommendations for the development of interfaces that present information on system status and TOR. However, it is still necessary to understand how to design proposals for HMI that meet these recommendations but also follow principles of interaction and information design, which consider the human processing of information.

5.3 METHOD

In order to explore possibilities for automated vehicles' HMI, a co-creation workshop was conducted, with groups of professors and students of undergraduate and graduate programs in Design from a Brazilian university, from several Design qualifications and with non-experts in automotive design. The objective of the workshop was to promote the creation of proposals for information design of system status and TOR. This technique was chosen because it would allow the collective creation of solutions by Design experts.

As automated vehicles are far from the Brazilian market, there are no real users or customers yet to participate in user research based on their experience. However, all workshop participants were occasional drivers or passengers, being considered potential users of these vehicles, and able to position themselves as users. Thus, on the one hand, the sessions allowed designers to propose interfaces based on their design knowledge, while at the same time thinking about their needs as general vehicle users, bringing their experiences to the proposals. Finally, the university was chosen as the environment for the workshop sessions because it is an institution that brings together, in a single place, several Design experts who deal daily with innovation projects, information communication and design research.

The workshop was held in two sessions of approximately one and a half hours each during May 2019. In total, eight participants attended the first session and nine in the second session. Participants arrived at the venue at the scheduled time, signed a consent form and were comfortably accommodated. Then participants were reminded of what the research was and what was the goal of the workshop. A brief presentation was made to expose automated vehicle concepts, levels of automation, the research context and the issues surrounding information presentation in automated vehicles. The purpose of the workshop was then presented: "How to inform the driver of a partially autonomous car (level 3) about this information (the five key information from NHTSA, 2016)?" Throughout the sessions, participants worked on the creation of ideas individually or in pairs initially, and then shared their ideas with the group for discussion of proposals and combining ideas according to brainstorming's basic principles (Osborn 1962). Finally, the workshop session was concluded by thanking everyone for their participation and ideas were collected for analysis.

For proposals analysis, a review of them was initially made to identify, in a bottom-up approach, how ideas could be clustered. From this review, the proposals were classified according to the type of information presented (regarding the five

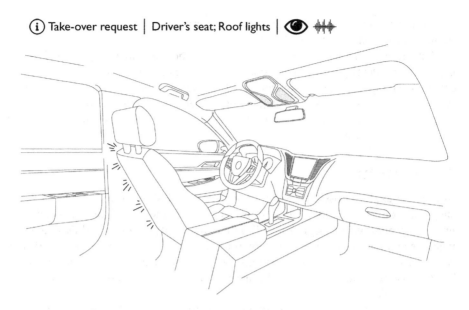

FIGURE 5.1. Example of a proposal that presents a TOR through vibrotactile cues on the driver's seat, and blinking lights on the vehicle's roof lights.

key information from NHTSA); the modality of information presentation (visual, auditory or vibrotactile); the format of the information presented, when appropriate; and the place where information is displayed, where applicable. After this stage, for better visualization, recording and organization of these ideas, the proposals were digitally recorded in an interior vehicle frame of reference created for this research (example in Figure 5.1). Finally, to analyze the usability and quality of the proposals, the ideas were evaluated according to their compliance with the literature recommendations and guidelines for automated driving systems (ADSs) mentioned before.

5.4 RESULTS

In the two sessions of the workshop, 105 interface proposals were elaborated by the participants. Out of this, 15 interfaces presented messages for system "functioning properly," 13 for "currently engaged in automated driving mode," 10 for "automated driving unavailable," 14 for "experiencing a malfunction with the AV system" and 51 for TORs. These results show that the workshops' participants focused heavily on presenting messages of transition of control on the HMI.

The majority of the proposals were elaborated to present information through the visual channel but participants also chose the auditory and haptic channels. Concerning the auditory messages, the information was presented through voice interaction, abstract sounds, known as earcons (Blattner, Sumikawa, and Greenberg 1989) and auditory icons, which are sounds from the real world used to represent a system's action (Roginska 2013). Although none of the proposals specified how

Design Proposals for Automated Vehicles' HMI 67

the earcons should be, such auditory cues are a good alternative to get the driver's attention when they are looking away from the source of information (Saffer 2013). However, some proposals suggested the use of a horn sound as an auditory icon for TORs, which is not a suitable solution since it might startle the driver or be misinterpreted by the sound of a real horn and go unnoticed. On the other hand, accordingly with usability recommendations (Naujoks et al., 2019), the voice messages proposed by the participants were brief, presented messages on the user's native language (Portuguese) and varied its semantics to fit the urgency of the information (e.g., "Hello, the system is functioning properly" and "Take Over!").

Haptic feedback was also used by participants to communicate information to the driver, but the use of this channel was restricted to malfunctioning messages and TORs. Proposals applied vibrotactile cues to the driver's seat (Figure 5.1), steering wheel and gear shifter. However, the steering wheel and the gear shifter are not appropriate placements to present haptic messages, since the driver's body is not expected to be in contact with such placements during the entire automated driving, which can make the information imperceptible. As recommended in the literature, the driver's seat is the best place for vibrotactile cues because it is always in contact with the driver's body (Borojeni et al. 2017). However, repetitive exposure to a vibrotactile stimulus may evoke annoyance in drivers (Petermeijer et al. 2018).

The use of multimodal information presentation on the proposed interfaces was also verified, especially, for TORs and messages conveying a system's malfunction. Most of the multimodal interfaces combined the visual channel with an auditory message, haptic information or both (Figure 5.1). About half of the interfaces that presented TORs conveyed this message through more than one channel, which is in line with usability recommendations, which suggest that high-priority information should be multimodal (Naujoks et al. 2019; Faltaous et al. 2018). Similarly, one proposal used multimodal interaction to communicate system unavailability; however, multimodality is not recommended for status messages (Naujoks et al. 2019; Faltaous et al. 2018).

As mentioned, the visual channel was chosen the most among the designers. As for the placement of visual interfaces, the windshield was preferred by most participants to present information through a head-up display (HUD). Martin-Emerson and Wickens (1997) define HUD as a technology that presents information through a projection on a transparent screen on the driver's field of vision. On the participants' proposals, the area of display varied: some interfaces presented messages on the upper part of the windshield (Figure 5.2) while others displayed information on the lower part (Figure 5.3).

Although the interfaces located on the upper part of the windshield are inside the driver's field of vision, users are not expected to be constantly driving in the AV. Humans in the seating position have their relaxed sightline of 15 degrees below the horizontal sight line (Tilley and Henry Dreyfuss Associates 2001), which can cause messages on the upper part of the windshield to go unnoticed. Moreover, it is likely that drivers engage in non-driving-related tasks (NDRTs), like smartphones, during automated driving (Borojeni et al. 2016), which can lower their sightline even more. For this reason, the roof lights are unappropriated to display information, but the

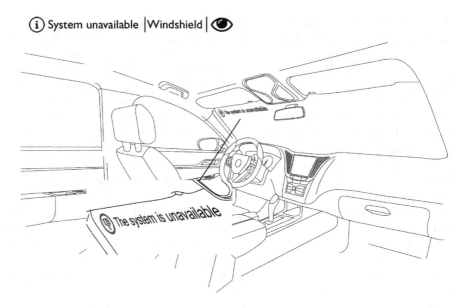

FIGURE 5.2. Proposal to present system unavailability on the upper part of the windshield.

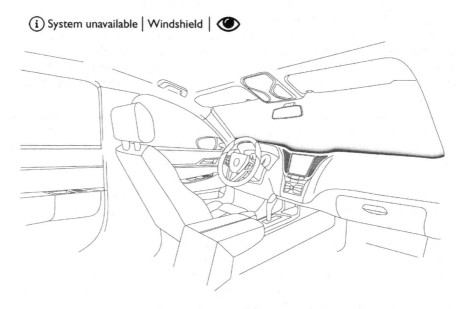

FIGURE 5.3. Proposal to present system unavailability on the lower part of the windshield.

Design Proposals for Automated Vehicles' HMI 69

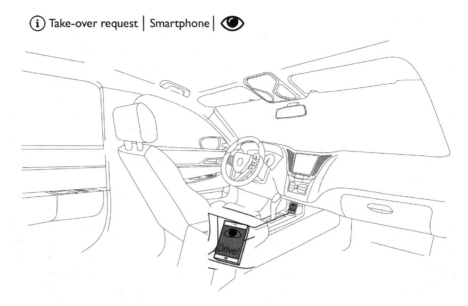

FIGURE 5.4. Proposal to present a TOR on the driver's smartphone.

steering wheel, the instrument cluster, the center stack display and the lights on the vehicle's A-pillars are suitable placements for the interfaces and were also chosen by a great number of participants to display information.

Other placements were also observed in the analysis, such as the rearview mirror, the gear shifter and a connected non-driving-related device, such as the driver's smartphone (Figure 5.4). The use of a connected device to mirror messages from the AV system interface is a good solution to deal with driver distraction, who may be engaged on an NDRT on such devices. It increases user trust and acceptance of the AV (Winterberger et al. 2018). Moreover, combining the system's information to NDRTs is recommended by the literature (Naujoks et al. 2019) and has also been related to better transitions of control (Winterberger et al. 2018).

The placement of the information is important not only because it must be in the driver's field of vision, but also because all AV system modes information should be grouped together (Naujoks et al. 2019). The majority of the interfaces proposed by participants were in accordance with this recommendation; however, two interfaces displayed additional information to support a TOR, pointing to the source of danger on the windshield. Another proposal also mirrored the TOR on several distant placements in the vehicle, such as the windshield, instrument cluster, central stack display and user's smartphone. This type of ungrouped information presentation is problematic because it could make drivers confused about where to look or which message is more important and taking over the vehicle control at the same time.

Concerning the format of visual information, color codes were proposed to convey different information on the interfaces, such as green-red, green-yellow-red and green-amber-red. All the proposals suggested a minimalistic color design, using less than five distinct colors, which is recommended since colors with little hue variation are difficult to distinguish (Naujoks et al. 2019; Faltaous et al. 2018). The green color was widely applied to communicate messages concerning system engagement and functioning properly. Additionally, the absence of color was used by most participants to represent the system's unavailability, but red and yellow were also used to display such messages. As for the messages about the system's malfunctioning, color variations were identified throughout the proposals: some interfaces used yellow to convey the system's malfunctioning whereas others applied amber or red. Finally, almost all interfaces that communicated TOR used red but one of them applied amber and yellow.

The color codes applied to the proposals are only partially in accordance with usability recommendations. According to the literature, green, blue and white are suitable colors for communicating status messages but amber should be used for warnings and red must be exclusive for danger information, such as TORs (Faltaous et al. 2018; Naujoks et al. 2019; Forster, Naujoks, and Neukum 2016; Stevens et al. 2002). Moreover, the absence of color as a means to represent system unavailability might cause the information to go unnoticed, since it may look like the system is not presenting any information at all (see the concept of visibility of system status; Nielsen 1994). Additionally, it is important to point out that none of the proposals considered color blindness, which is also recommended (Naujoks et al. 2019).

Besides the color-coding, some participants elaborated proposals that used icons to represent the road's environment (Figure 5.5) or convey information through metaphors, such as an avatar to represent the AV's system. The use of icons that

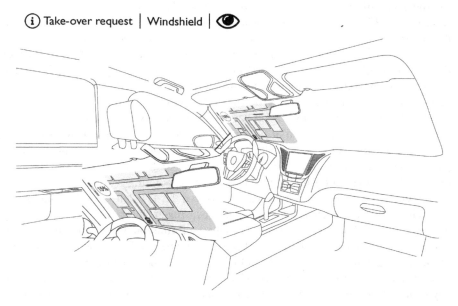

FIGURE 5.5. Proposal that presents the probability of a TOR throughout a route on a map.

Design Proposals for Automated Vehicles' HMI

represent the driving environment has been positively assessed by the literature, and they are preferred in relation to figurative icons (Noah et al. 2017). It is important to point out that, in contrast with literature guidelines (Naujoks et al. 2019), none of the interfaces proposed the use of standardized or commonly accepted symbols, which may be explained by the fact that none of the workshops' participants had a background on the automotive area. However, some of the proposals also combined texts or numbers with the icons to enhance understandability (Figures 5.2 and 5.5), which is suggested by the literature to compensate for the lack of standard icons (Naujoks et al. 2019).

As the voice interaction, the textual information presented brief messages using the drivers' native language. However, the semantic of some TORs did not use an imperative tone (e.g., "Transition of control" instead of "Take Over!"), which is recommended by the literature to present highly urgent information (Naujoks et al. 2019). Moreover, regarding messages about the system's failures or malfunctioning, it is recommended to display information concerning the consequences of those failures and how the driver should proceed in such situations (Naujoks et al. 2019) but the workshops' participants did not propose this type of information on their interface ideas. Similarly, only two proposals for TORs oriented the driver towards the source of danger (Figure 5.6), which is also suggested in the literature (Naujoks et al. 2019). Additionally, one interface idea proposed to display an action recommendation alongside the TOR to help the driver improve the takeover. This type of interface has already been shown to reduce take-over time and lead to safer transitions of control (Borojeni et al. 2016).

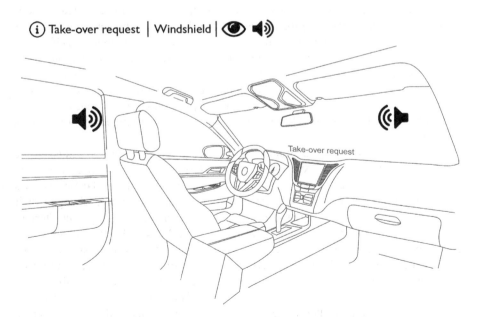

FIGURE 5.6. Proposal that points to the driver where the problem is at the road.

Furthermore, in terms of information legibility, all the proposals that displayed information through a light on the steering wheel were readable but one proposal suggested placing a rotatory text on the steering wheel that may not be legible due to its movement and text distortion to fit the steering wheel. Moreover, the presentation of information through icons or text on the windshield has the potential to be readable due to its apparent size; however, only two interfaces proposed to adjust contrast levels to make information readable throughout different levels of background lighting. Considering that appropriate levels of contrast should be displayed to make messages readable (Naujoks et al. 2019), and that low contrast to the background is an inherent issue from HUDs (Martin-Emerson and Wickens 1997; Ward and Parkes 1994), the adjustment to background lighting might improve legibility. Additionally, some icons presented on the instrument cluster (as warning or telltales lights) might have poor readability due to their size and form. Some interfaces proposed the use of outlined icons, which may be difficult to read if displayed in a small size.

The modality, placement and format of information presentation are important to communicate the system's modes but it is necessary to point out that such status messages should be displayed continuously so that the drivers are always aware of the current system's mode (Naujoks et al. 2019). Concerning the proposals, the interfaces adequately presented continuous messages of system's proper functioning, engagement, malfunctioning and TORs. However, as mentioned before, some interfaces used the absence of color to communicate system unavailability, which might complicate the driver's understanding of the system's mode.

Besides, it is also advisable that information presentation does not require drivers' continuous attention since their attention must also be directed to the driving environment to maintain situation awareness (Naujoks et al. 2019). For example, one proposal suggested displaying the probability of a TOR throughout a route on a map, overlapping a color-coding to an illustration of the route to represent TOR probability (Figure 5.5). This proposal might cause the drivers to constantly monitor the system, requiring their continuous attention. However, it is necessary to investigate the real effects of this type of interface, considering that literature has shown that informing the driver about the system's reliability increases the system's perceived transparency (Faltaous et al. 2018) and is related to safer driving behavior (Faltaous et al. 2018; Hock et al. 2016).

It is also essential that the system's state changes are explicitly presented to the driver (Naujoks et al. 2019). Among the proposals, some participants created sequential interfaces for presenting all five AV's status information and used different ways to communicate the system's state changes. First, some interfaces changed their visual formats, such as using the same icon to present a system change but alternating between icons that were colored only on their outlines and fully colored icons. However, if the information display size is too small, this type of system's state change might go unnoticed by drivers because the difference between the visual elements is too discrete. Hence, changing between outlined-only and full-colored elements should be displayed at appropriate size and placement. One of the proposals applied this principle by alternating between displaying lights on the whole steering wheel and only lighting lines on its hub (Figures 5.7 and 5.8).

Design Proposals for Automated Vehicles' HMI 73

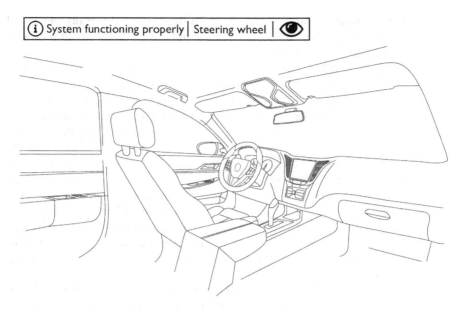

FIGURE 5.7. Proposal that presents a message about the system's proper functioning by lighting the steering wheel's lines.

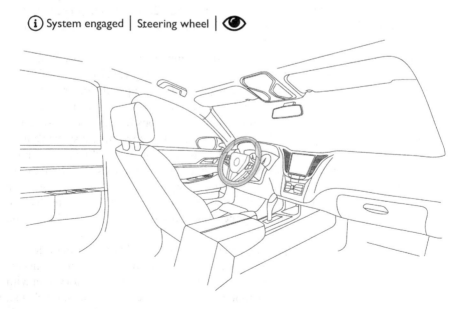

FIGURE 5.8. Proposal that presents a message about the system's engagement by lighting the whole steering wheel.

Nonetheless, using similar elements to display different messages (as mentioned above) may also be problematic if the interface uses the same colors to represent varied system's modes (e.g., using a dashed green line to display a message about the system's proper functioning and, then, making it continuous to communicate the system's engagement, without an additional change of color). Thus, an efficient way of communicating the system's state changes is through color-coding (Faltaous et al. 2018; Naujoks et al. 2019; Forster, Naujoks, and Neukum 2016; Stevens et al. 2002), which was applied by a large number of participants as mentioned earlier. Also, a significant number of ideas employed blinking elements such as lights or icons to indicate the system's malfunctioning or TORs. It was noted in the literature that blinking lights are an efficient way to communicate different levels of urgency (Kunze et al. 2018) and are preferred by drivers in relation to static lights (Borojeni et al. 2016). Considering this, an efficient way of presenting changes on system's modes might be through the combination of color changes and blinking elements (e.g., using a green static light on the car's windshield to communicate that the system is engaged and, then, turning it into a red blinking light to display a TOR).

Moreover, other modalities may be used to convey a system's state changes. Some proposals added auditory modality to support visual messages in order to make TORs perceived as more urgent. Two proposals suggested a lighter, anticipatory TOR to alert the driver in advance about the need to take over. These interfaces proposed to vary the volume and frequency of auditory and haptic messages to convey different levels of urgency. Such interfaces might be a good solution to TORs since the change in auditory and vibrotactile cues' parameters efficiently convey different levels of urgency (Naujoks et al. 2019; Quaresma et al. 2018) and the presentation of anticipatory TORs has been related to better transitions of control in the literature (Forster, Naujoks, and Neukum 2016).

Although it is necessary to make the system's modes distinguishable from each other, it is also important to keep the interface consistency (Nielsen 1994; Bastien and Scapin 1993). It was observed that the majority of the proposals illustrated a consistent interface, which may be due to the fact that all workshops' participants were designers who are familiar with usability recommendations for interface design. For example, one participant proposed to present a message concerning a system's malfunctioning in four different ways: an orange light on the windshield; an orange notification on the central stack display, mirrored on the driver's smartphone, writing, "The system is experiencing a malfunction!" and a voice message, saying, "The system is experiencing a malfunction!" (Figure 5.9). It is possible to note that the interface keeps its consistency by using the same color on all displays and the same phrase for all warnings.

This consistency must be kept independently of the modality to present information and throughout the system's state changes. One proposal suggested using an earcon to present a system's malfunction but, sequentially, applied an auditory icon with a horn sound to represent a TOR. Although the two messages are distinguishable, presenting an auditory icon as the sequence for an earcon may affect the system's consistency. It is possible to maintain the auditory interface consistent by changing

Design Proposals for Automated Vehicles' HMI

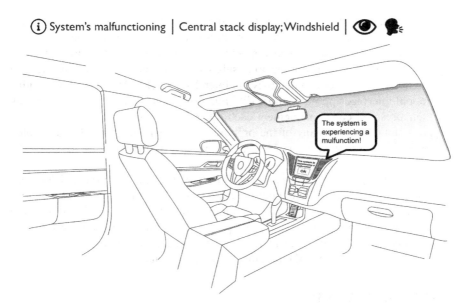

FIGURE 5.9. Proposal that displays a message concerning the system's malfunctioning in a consistent way.

the volume of the same earcon to communicate different messages, as mentioned before, or creating families of earcons (McGookin and Brewster 2011).

5.5 CONCLUSION

The goal of the study was to foster possible HMI solutions for the good interaction between AVs and their drivers, focusing on the most common scenarios/information that are listed in the literature (NHTSA 2016). To tackle this problem, we developed creativity workshop sessions with experienced designers to ideate and represent their innovative approaches to the described problem.

Overall, the number of solutions was considered high for the scale of the study (105). This result was expected due to the expertise of the participants with the creative process and with design workshops. As said above, the results found with this technique presented a high variability in terms of modality, placement and format of representation of information from the AV to the driver. Regardless of this variability, most of the ideas generated used the visual modality as its main approach, more specifically HUD projections. It was a majority agreement between the designers that this is a solution that should be explored since it allows the driver to stay focused on the threat while sample additional information to support the take-over process. This finding is in line with the literature in the field but we believe that issues related to change blindness, tunneling and overload (Martin-Emerson and Wickens 1997) should be studied deeper for specific graphic representation solutions before its implementation by the manufacturers.

As the designers were not from the automotive field nor had a full experience with an AV before, many of the solutions presented issues, which should be taken into account or have a countermeasure before its implementation for further testing. Regardless of this fact, many of the proposals also presented a reasonable advantage/ solution to the stated problem that they are intended to tackle. We believe that with a multimodal interface approach, those different proposals can be combined in a way that their pros and cons complement each other for a robust system and information presentation. Also, many of the proposals alone can be adjusted and adapted to fit better to the context of driving, based on the recommendations stated by the literature.

As the technique reported was focused on the ideation of the proposals of solutions, little to none can be said about the actual impact or implications of the combination of the solutions to drivers' performance inside an AV. Empirical tests are necessary, testing in depth the presented proposals and a combination of them, in order to understand how different forms of representation of information might enhance drivers' response in the possible interaction scenarios.

ACKNOWLEDGMENTS

This study was financed in part by the Coordenação de Aperfeiçoamento de Pessoal de Nível Superior – Brasil (CAPES) – Finance Code 001 and by Conselho Nacional de Desenvolvimento Científico e Tecnológico (CNPq).

REFERENCES

AdaptIVe Consortium. 2017. "Final Functional Human Factors Recommendations (D3.3)." https://www.adaptive-ip.eu/index.php/AdaptIVe-SP3-v23-DL-D3.3-Final%20Funct ional%20Human%20Factors%20Recommendations_Core-file=files-adaptive-content-downloads-Deliverables%20&%20papers-AdaptIVe-SP3-v23-DL-D3.3-Final%20F unctional%20Human%20Factors%20Recommendations_Core.pdf

Bastien, Christian, and D. L. Scapin. 1993. Ergonomic Criteria for the Evaluation of Human-Computer Interfaces. RT-0156, INRIA, p.79. https://hal.inria.fr/inria-00070012.

Blattner, Meera, Denise Sumikawa, and Robert Greenberg. 1989. "Earcons and Icons: Their Structure and Common Design Principles." *Human-Computer Interaction* 4(1): 11–44. doi: 10.1207/s15327051hci0401_1.

Blommer, Mike, Reates Curry, Radhakrishnan Swaminathan, Louis Tijerina, Walter Talamonti, and Dev Kochhar. 2017. "Driver Brake vs. Steer Response to Sudden Forward Collision Scenario in Manual and Automated Driving Modes." *Transportation Research Part F: Traffic Psychology and Behaviour* 45(February): 93–101. doi: 10.1016/j.trf.2016.11.006.

Borojeni, Shadan Sadeghian, Lewis Chuang, Wilko Heuten, and Susanne Boll. 2016. "Assisting Drivers with Ambient Take-Over Requests in Highly Automated Driving." In *Proceedings of the 8th International Conference on Automotive User Interfaces and Interactive Vehicular Applications – Automotive'UI 16*, 237–44. New York: ACM Press. doi: 10.1145/3003715.3005409.

Borojeni, Shadan Sadeghian, Torben Wallbaum, Wilko Heuten, and Susanne Boll. 2017. "Comparing Shape-Changing and Vibro-Tactile Steering Wheels for Take-Over Requests in Highly Automated Driving." In *Proceedings of the 9th International*

Conference on Automotive User Interfaces and Interactive Vehicular Applications – AutomotiveUI '17, 221–25. New York: ACM Press. doi: 10.1145/3122986.3123003.

Damböck, D., M. Farid, L. Tönert, and K. Beng-Ler. 2012. "Übernahmezeiten Beim Hochautomatisierten Fahren." *Tagung Fahrerassistenz* 5(57): 1–12.

Dixit, Vinayak V., Sai Chand, and Divya J. Nair. 2016. "Autonomous Vehicles: Disengagements, Accidents and Reaction Times." Edited by Jun Xu. *PLOS ONE* 11(12): e0168054. doi: 10.1371/journal.pone.0168054.

Dzennius, Marc, Johann Kelsch, and Anna Schieben. 2015. "Ambient Light Based Interaction Concept for an Integrative Driver Assistance System – A Driving Simulator Study." In *HFES 2015*. Groningen: HFES Europe. http://elib.dlr.de/99076/.

Endsley, Mica R., and Esin O. Kiris. 1995. "The Out-of-the-Loop Performance Problem and Level of Control in Automation." *Human Factors: The Journal of the Human Factors and Ergonomics Society* 37(2): 381–94. doi: 10.1518/001872095779064555.

Faltaous, Sarah, Martin Baumann, Stefan Schneegass, and Lewis L. Chuang. 2018. "Design Guidelines for Reliability Communication in Autonomous Vehicles." In *Proceedings of the 10th International Conference on Automotive User Interfaces and Interactive Vehicular Applications – AutomotiveUI '18*, 258–67. New York: ACM Press. doi: 10.1145/3239060.3239072.

Feuerstack, Sebastian, Bertram Wortelen, Carmen Kettwich, and Anna Schieben. 2016. "Theater-System Technique and Model-Based Attention Prediction for the Early Automotive HMI Design Evaluation." In *Proceedings of the 8th International Conference on Automotive User Interfaces and Interactive Vehicular Applications – Automotive'UI 16*, 19–22. New York: ACM Press. doi: 10.1145/3003715.3005466.

Flemisch, Frank, Johann Kelsch, Christian Löper, Anna Sehieben, Julian Schindler, and Matthias Heesen. 2008. "Cooperative Control and Active Interfaces for Vehicle Assistance and Automation." In *FISITA World Automotive Congress 2008, Congress Proceedings – Mobility Concepts, Man Machine Interface, Process Challenges, Virtual Reality*, 1: 301–10.

Forster, Yannick, Frederik Naujoks, and Alexandra Neukum. 2016. "Your Turn or My Turn?" In *Proceedings of the 8th International Conference on Automotive User Interfaces and Interactive Vehicular Applications – Automotive'UI 16*, 253–60. New York: ACM Press. doi: 10.1145/3003715.3005463.

Gold, Christian, Daniel Damböck, Lutz Lorenz, and Klaus Bengler. 2013. "'Take Over!' How Long Does It Take to Get the Driver Back into the Loop?" *Proceedings of the Human Factors and Ergonomics Society Annual Meeting* 57(1): 1938–42. doi: 10.1177/1541931213571433.

Gonçalves, Rafael Cirino, Manuela Quaresma, and Claudia Mont'Alvão Rodrigues. 2017. "Approaches for Loss of Vigilance in Vehicle Automation: A Meta-Analytical Study." *Proceedings of the Human Factors and Ergonomics Society Annual Meeting* 61 (1): 1871–75. doi: 10.1177/1541931213601948.

Graab, B., E. Donner, U. Chiellino, and M. Hoppe. 2008. "Analyse von Verkehrsunfällen Hinsichtlich Unterschiedlicher Fahrerpopulationen Und Daraus Ableitbarer Ergebnisse Für Die Entwicklung Adaptiver Fahrerassistenzsysteme." In *Proceedings of the 3rd Conference Active Safety through Driver Assistance*, Garching Bei München, Germany.

Hock, Philipp, Johannes Kraus, Marcel Walch, Nina Lang, and Martin Baumann. 2016. "Elaborating Feedback Strategies for Maintaining Automation in Highly Automated Driving." In *Proceedings of the 8th International Conference on Automotive User Interfaces and Interactive Vehicular Applications – Automotive'UI 16*, 105–12. New York: ACM Press. doi: 10.1145/3003715.3005414.

Kunze, Alexander, Stephen J. Summerskill, Russell Marshall, and Ashleigh J. Filtness. 2018. "Augmented Reality Displays for Communicating Uncertainty Information in Automated Driving." In *Proceedings of the 10th International Conference on Automotive User Interfaces and Interactive Vehicular Applications – AutomotiveUI '18*, 164–75. New York: ACM Press. https://doi.org/10.1145/3239060.3239074.

Litman, Todd Alexander. 2014. "Autonomous Vehicle Implementation Predictions Implications for Transport Planning." *Traffic Technology International* 42 (January 2014): 36–42.

Louw, Tyron, and Natasha Merat. 2017. "Are You in the Loop? Using Gaze Dispersion to Understand Driver Visual Attention during Vehicle Automation." *Transportation Research Part C: Emerging Technologies* 76(March): 35–50. doi: 10.1016/j.trc.2017.01.001.

Martin-Emerson, Robin, and Christopher D. Wickens. 1997. "Superimposition, Symbology, Visual Attention, and the Head-Up Display." *Human Factors: The Journal of the Human Factors and Ergonomics Society* 39(4): 581–601. doi: 10.1518/001872097778667933.

Mcgookin, D., and S. Brewster. 2011. "Earcons." In *The Sonification Handbook*, edited by Thomas Hermann, Andy Hunt, and John G. Neuhoff, 339–61. Berlin: Logos Publishing House.

Merat, Natasha, Bobbie Seppelt, Tyron Louw, Johan Engström, John D. Lee, Emma Johansson, Charles A. Green, et al. 2019. "The 'Out-of-the-Loop' Concept in Automated Driving: Proposed Definition, Measures and Implications." *Cognition, Technology & Work* 21(1): 87–98. doi: 10.1007/s10111-018-0525-8.

Mirnig, Alexander G., Magdalena Gärtner, Arno Laminger, Alexander Meschtscherjakov, Sandra Trösterer, Manfred Tscheligi, Rod McCall, and Fintan McGee. 2017. "Control Transition Interfaces in Semiautonomous Vehicles." In *Proceedings of the 9th International Conference on Automotive User Interfaces and Interactive Vehicular Applications – AutomotiveUI '17*, 209–20. New York: ACM Press. doi: 10.1145/3122986.3123014.

Mole, Callum D., Otto Lappi, Oscar Giles, Gustav Markkula, Franck Mars, and Richard M Wilkie. 2019. "Getting Back Into the Loop: The Perceptual-Motor Determinants of Successful Transitions Out of Automated Driving." *Human Factors: The Journal of the Human Factors and Ergonomics Society* 61(7): 1037–65. doi: 10.1177/0018720819829594.

Naujoks, Frederik, Katharina Wiedemann, and Nadja Schömig. 2017. "The Importance of Interruption Management for Usefulness and Acceptance of Automated Driving." In *Proceedings of the 9th International Conference on Automotive User Interfaces and Interactive Vehicular Applications – AutomotiveUI '17*, 254–63. New York: ACM Press. doi: 10.1145/3122986.3123000.

Naujoks, Frederik, Katharina Wiedemann, Nadja Schömig, Sebastian Hergeth, and Andreas Keinath. 2019. "Towards Guidelines and Verification Methods for Automated Vehicle HMIs." *Transportation Research Part F: Traffic Psychology and Behaviour* 60(January): 121–36. doi: 10.1016/j.trf.2018.10.012.

NHTSA (National Highway Traffic Safety Administration). 2016. "Federal Automated Vehicles Policy: Accelerating the Next Revolution In Roadway Safety." Washington, DC: U.S. Department of Transportation. doi: 12507-091216-v9.

Nielsen, Jakob. 1994. "Heuristic Evaluation." In *Usability Inspection Methods*, edited by Jakob Nielsen and Robert Mack. New York: John Wiley & Sons.

Noah, Brittany E., Thomas M. Gable, Shao-Yu Chen, Shruti Singh, and Bruce N. Walker. 2017. "Development and Preliminary Evaluation of Reliability Displays for Automated Lane Keeping." In *Proceedings of the 9th International Conference on Automotive User Interfaces and Interactive Vehicular Applications – AutomotiveUI '17*, 202–8. New York: ACM Press. doi: 10.1145/3122986.3123007.

Osborn, Alex F. 1962. *O Poder Criador Da Mente: Princípios e Processos Do Pensamento Criador e Do "Brainstorming."* São Paulo: IBRASA.

Pauzié, Annie, Lyess Ferhat, and Hélène Tattegrain. 2019. "Innovative Human Machine Interaction for Automatised Car: Analysis of Drivers Needs for Recommended Design." In *Proceedings of the 26th ITS World Congress*. Singapore: ITS Singapore.

Perrier, Mickaël J. R., Tyron Louw, Rafael C. Gonçalves, and Oliver Carsten. 2019. "Applying Participatory Design to Symbols for SAE Level 2 Automated Driving Systems." In *Proceedings of the 11th International Conference on Automotive User Interfaces and Interactive Vehicular Applications Adjunct Proceedings – AutomotiveUI '19*, 238–42. New York: ACM Press. doi: 10.1145/3349263.3351512.

Petermeijer, Sebastiaan M., Paul Hornberger, Ioannis Ganotis, Joost C. F. de Winter, and Klaus J. Bengler. 2018. "The Design of a Vibrotactile Seat for Conveying Take-Over Requests in Automated Driving." In *Advances in Human Aspects of Transportation. AHFE 2017. Advances in Intelligent Systems and Computing*, edited by Neville A. Stanton, 618–30. Cham: Springer. doi: 10.1007/978-3-319-60441-1_60.

Quaresma, Manuela, Isabela Motta, Manuella Araújo, and Rafael Cirino Gonçalves. 2018. "The Relationship between Rhythm Variation and Distance Perception in Auditory In-Vehicle Interfaces." *Proceedings of the Human Factors and Ergonomics Society Annual Meeting* 62(1): 1929–33. doi: 10.1177/1541931218621438.

Richardson, N. T., C. Lehmer, M. Lienkamp, and B. Michel. 2018. "Conceptual Design and Evaluation of a Human Machine Interface for Highly Automated Truck Driving." In *2018 IEEE Intelligent Vehicles Symposium (IV)*, 2018 June: 2072–77. IEEE. doi: 10.1109/IVS.2018.8500520.

Roginska, Agnieszka. 2013. *The Psychology of Music in Multimedia*. In *The Psychology of Music in Multimedia*, edited by Siu-Lan Tan, Annabel J. Cohen, Scott D. Lipscomb, and Roger A. Kendall. Oxford: Oxford University Press. doi: 10.1093/acprof:oso/9780199608157.001.0001.

SAE (Society of Automobile Engineers). 2018. "Taxonomy and Definitions for Terms Related to Driving Automation Systems for On-Road Motor Vehicles J3016." *SAE International*, J3016. Warrendale.

Saffer, Dan. 2013. *Microinteractions: Designing with Details*. 1st ed. Sebastopol: O'Reilly.

Schieben, Anna, Stefan Griesche, Tobias Hesse, Nicola Fricke, and Martin Baumann. 2014. "Evaluation of Three Different Interaction Designs for an Automatic Steering Intervention." *Transportation Research Part F: Traffic Psychology and Behaviour* 27(November): 238–51. doi: 10.1016/j.trf.2014.06.002.

Seppelt, Bobbie D., and Trent W. Victor. 2016. "Potential Solutions to Human Factors Challenges in Road Vehicle Automation." In *Road Vehicle Automation 3. Lecture Notes in Mobility*, 131–48. Cham: Springer. doi: 10.1007/978-3-319-40503-2_11.

Stevens, A., A. Quimby, A. Board, T. Kersloot, and P. Burns. 2002. "Design Guidelines for Safety of In-Vehicle Information Systems." Wokingham. https://trl.co.uk/sites/default/files/PA3721-01.pdf.

Tilley, Alvin R., and Henry Dreyfuss Associates. 2001. *The Measure of Man and Woman: Human Factors in Design*. New Jersey: Wiley.

Victor, Trent W., Emma Tivesten, Pär Gustavsson, Joel Johansson, Fredrik Sangberg, and Mikael Ljung Aust. 2018. "Automation Expectation Mismatch: Incorrect Prediction Despite Eyes on Threat and Hands on Wheel." *Human Factors: The Journal of the Human Factors and Ergonomics Society* 60(8): 1095–1116. doi: 10.1177/0018720818788164.

Walch, Marcel, Tobias Sieber, Philipp Hock, Martin Baumann, and Michael Weber. 2016. "Towards Cooperative Driving." In *Proceedings of the 8th International Conference on Automotive User Interfaces and Interactive Vehicular Applications – Automotive'UI 16*, 261–68. New York: ACM Press. doi: 10.1145/3003715.3005458.

Ward, Nicholas J., and Andrew Parkes. 1994. "Head-Up Displays and Their Automotive Application: An Overview of Human Factors Issues Affecting Safety." *Accident Analysis & Prevention* 26(6): 703–17. doi: 10.1016/0001-4575(94)90049-3.

Wintersberger, Philipp, Andreas Riener, Clemens Schartmüller, Anna-Katharina Frison, and Klemens Weigl. 2018. "Let Me Finish before I Take Over." In *Proceedings of the 10th International Conference on Automotive User Interfaces and Interactive Vehicular Applications – AutomotiveUI '18*, 53–65. New York: ACM Press. doi: 10.1145/3239060.3239085.

Young, Mark S., and David Bunce. 2011. "Driving into the Sunset: Supporting Cognitive Functioning in Older Drivers." *Journal of Aging Research* 2011(May): 1–6. doi: 10.4061/2011/918782.

Zeeb, Kathrin, Axel Buchner, and Michael Schrauf. 2016. "Is Take-Over Time All that Matters? The Impact of Visual-Cognitive Load on Driver Take-Over Quality after Conditionally Automated Driving." *Accident Analysis & Prevention* 92(July): 230–39. doi: 10.1016/j.aap.2016.04.002.

6 Is the Driver Ready to Receive Just Car Information in the Windshield during Manual and Autonomous Driving?

Élson Marques, Paulo Noriega, and Francisco Rebelo

CONTENTS

6.1 Introduction 82
6.2 Method 84
    6.2.1 Participants 84
    6.2.2 Apparatus 84
    6.2.3 Stimuli 86
    6.2.4 Procedure 86
    6.2.5 Experimental Design 90
6.3 Results 94
    6.3.1 Manual Mode 94
    6.3.2 Activation 97
    6.3.3 Takeover Request 100
    6.3.4 Rear-end Collision 104
6.4 Discussion 105
6.5 Conclusion 110
6.6 Summary 112
References 113

DOI: 10.1201/9780429343513-8

6.1 INTRODUCTION

For the European Commission (2018), road safety is a major societal issue. The number of accidents in Europe has been decreasing since 2001 until the current days according to the European Commission (2018), even so there were more than 26,000 deaths in 2015. In the USA, in 2015, approximately 35,000 people died on road, which is the highest since 2008 (NHTSA, 2016). A naturalistic study conducted in the USA, the SHRP2 study, used online recordings of about 1,000 accidents on road and concluded that 90% of the crashes were due to human error (Dingus et al., 2016).

Autonomous driving (AD) can have a key role concerning the increase of road safety. By 2040, according to the IEEE (2014), it is expected that autonomous vehicles will account for up to 75% of vehicles on the road. It is proposed that the replacement of humans in monotonous driving or day-to-day traffic situations by an automated system could reduce accidents, increase road safety, increase driving comfort and reduce emissions (Gold, Körber, Hohenberger, Lechner, & Bengler, 2015; Kuehn, Hummel, & Bende, 2009).

The success of AD is not only dependent on an accurate and sophisticated automated function but also influenced by the adaptation of the driver and society. One of the important aspects for a good adaptation of the driver is related to the implementation of ergonomics in usability and user experience of the interfaces. The acceptance of new technologies depends on usability and user experience too, so, in that sense, it is very important to have a friendly display that can communicate the same language as that of the end users.

Most people do not associate with the automotive industry, when listening to someone speak about "autonomous vehicle"; they correlate it to full automation (SAE, 2016). However, nowadays the investigation has mainly been focusing on the intermediate levels of automation. To add clarity and consistency to this topic, Society of Automotive Engineers (SAE International) came up with a nomenclature, which is now most widely cited. They defined six levels of automation for automakers, suppliers and policymakers to help classify a system's sophistication.

Looking from the driver's perspective, a new role is associated with the driver from L2 to L5 (SAE, 2016); the supervisor, that is, the car, performs the driving task (steering and managing speed). It is in these levels that we are particularly interested, mainly in L2 and L3 (SAE, 2016).

The main challenges in terms of safety in these levels of automation to human factors are in the transition from manual to autonomous and vice versa. The first transition has the problem of the driver thinking that the AD function is activated after pushing the button, and normally the driver takes hands off the steering wheel quickly, whereas there can be situations where some problem with any actor of the activation process leaves the function not to be activated correctly. The other transition, autonomous to manual mode, is where most of the researchers have focused their attention (Gold et al., 2013; Lorenz et al., 2014). The main challenge is in the driver's taking over control after prolonged autonomous driving. It is very important in such moments that the driver be provided with a clear interface, leaving no doubts about the function's state and what the driver should do.

Manual and Conditional automation (L3) 83

It is equally challenging to provide the best user experience to the end user when AD is activated. It is important to understand whether the user, having activated conditional automation to high speeds (e.g. 120 km/h), prefers information related to road or entertainment.

Recently, many car manufacturers have proposed several development and commercialization plans to equip cars with head-up displays (HUD). This technology has created a new way of interaction between the driver and the vehicle (Phan, Thouvenin, & Frémont, 2016). In the year 2012, 1.2 million vehicles worldwide were equipped with HUD, and the world market will, according to IHS Automotive forecast, expand to 9.1 million vehicles in the year 2020 (Boström & Ramström, 2014). The expected growth of 758% in the next 8 years (IHS iSuppli, 2013) will put a great demand on the car manufacturers to supply new technical solutions and functionalities together with new ways of user-interaction design. However, the HUD brings with it other problems to the driver and a way to solve those problems could be augmented reality (AR). The information projected on the windshield can increase acceptance, trust and awareness of the driver to the autonomous system, once the driver can see the same information that the autonomous car is perceiving.

Because of all the reasons presented above, in recent years, road vehicle automation and AR (HUD plus AR area) have become important and popular topics for research and development in both academic and industrial spheres (Boström & Ramström, 2014; Lorenz, Kerschbaum, & Schumann, 2014; Pauzie & Orfila, 2016).

The main objective of the study was to compare two concepts of displays (AR and instrument cluster (IC)) concerning driver performance and safe driving, both in manual and in AD and transitions. The difference between these concepts is the place where the information was shown.

We focused on six specific objectives:

1. Compare the confidence and the preference of the driver using AR screen and IC during manual mode;
2. Compare the operational area in AR and IC during availability moment, concerning preference, situation awareness, usefulness and driver satisfaction;
3. Compare AR screen and IC during activation concerning preference, situation awareness, usefulness, driver satisfaction and confidence;
4. Compare AR and IC concepts during Give Back (GB, visual-auditory warning) when there is a secondary task (ST, such as watching a movie) in the central console and when there is no ST concerning preference, situation awareness, usefulness, driver satisfaction, driver behaviors and reaction times;
5. Compare operational area and IC during AD is deactivated concerning preference and situation awareness; and
6. Compare AR area and IC during warning for rear-end collision (obstacle) concerning the type of reaction, reaction time and driver's security-lane change.

6.2 METHOD

6.2.1 Participants

There were 29 volunteers (19 men and 10 women) from Centro Technologic Automation of Galicia (CTAG) and external participants (from North of Portugal and Galicia) tested in this study. To participate in this study, participants were required a valid driver's license, normal or corrected-to-normal vision, normal color vision and age between 18 and 65 years. More than 75% of the participants were between 18 and 35 years old, 12% between 36 and 40, and 12% between 46 and 50. More than 50% of the participants had more than 10 years of full driver's license, 28% between 5 and 7 years, 12% between 8 and 10 years, and only 8% between 2 and 4 years. About the traffic accidents, 10 out of 29 participants already had at least one accident, and among them, 3 out of 10 had two accidents, and only one participant had three or more accidents. Out of 29 participants, only one had HUD in their car and 12 had experimented with an autonomous vehicle before (CTAG simulator).

6.2.2 Apparatus

The experiment was conducted in a CTAG driving simulator (Figure 6.1). This simulator is composed of the following subsystems:

- Movement platform with 6 degrees of freedom and 4000 kg payload.
- Visual system comprised three projectors to obtain a 180° cylindrical screen front view and three rearview 7″ LCD displays.
- Acquisition and control systems contained the following elements:

FIGURE 6.1. CTAG simulator.

Manual and Conditional automation (L3)

- four interior cameras.
- one interior microphone.
- an acquisition software developed by CTAG that records all driving and performance measures during the simulation, synchronized with video and audio data.
* SCANeR© II software, which builds a realistic virtual environment. SCANeR© II tool is complemented with EVARISTE and 3DMax that allow the generation of new 3D database and road networks (road geometry, profiles, new buildings, tunnels, etc.) from real environments with the specific requirements for the simulations.
* There is an instrumented vehicle inside the cabin. The only changes done in this commercial vehicle are the replacement of the steering wheel by the active steering wheel system, new IC composed of an LCD screen, the sensors mounted in pedals and gear stick and the mentioned replacement of rearview mirrors. The vehicle has an automatic gearbox.
* The simulator vehicle was adapted to receive the AD function, where the ego vehicle drives by itself when conditional automation was activated.

For this test, the following adaptations have been made:

* The road network selected for the study was placed in a highway, without buildings and curves.
* All the roads used for testing had two lanes in each direction. There was a speed limit of 120 km/h and the traffic density was high during the traffic.
* The speed of the traffic was 45 km/h.

The AR screen was projected with an Acer P5327W projector with a maximum resolution of 1920 × 1200, located above the car, in the simulator screen, so it can simulate an image in windshield (Figure 6.2).

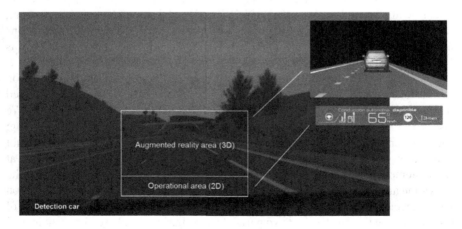

FIGURE 6.2. Vision of the windshield used during the test.

The simulated image in the OA was of 763 × 76 pixels, and in the AR area was of 725 × 322 pixels. The OA was perceived as floating in the air 2 m in front of the car on the driver's side and the AR area was perceived as floating in the air 15 m in front of the car.

6.2.3 Stimuli

The logic (color and content of information) of the AD button and the information in central console was the same for both concepts during all the tests. The difference between concepts was the place where the information was shown (Table 6.1). The "AR concept" used the windscreen to show all the car information and the autonomous function; however, the "IC concept" used the HDD to show the car information and autonomous function, and the operational area (HUD) to show specific information about the car information (current speed and speed limit).

6.2.4 Procedure

Participants were welcomed into the experimental room and were made comfortable. The researcher explained the context of the test as an experimental study on a static simulator. The researcher also explained that the study was part of CTAG research for the autonomous vehicle project (vehicles that can drive alone). After explaining the context to the participant, the researcher explained the general objective of the study. The specific objective of the study was not explained. The aim of the study was to evaluate the human–machine interaction (HMI) of the traffic situation and the human interaction during AD. Participants were then given informed consent, filled out a questionnaire with demographic data and were provided the opportunity to ask any questions. In informed consent it was explained that we collect data using video cameras during all the tests. The researcher also explained that we had a document to guaranty the confidentiality and anonymity of the participants' data and informed that participants' data could be used in our study and possibly presented in scientific dissemination. After the participants signed the confidentiality agreement, they were introduced to the study's structure, where we presented the instructions and how the autonomous system worked. We explained that the AD function was based on a situation with traffic jam, a speed limit and no possibility of lane change. The vehicle drove completely autonomously, accelerating or braking, depending on the vehicle ahead. The participants could make all the activities that they wanted on the touch screen, except for interacting with the mobile phone and sleeping. We also said that there was no risk of accident and, the vehicle managed everything: speed, steering wheel, obstacles; however, if for some reason the vehicle could not manage, it would ask him to take the control. We alerted the participants that at some moments, the system would inform that the AD function was available for the activation and every time such situation occurred, he/she should activate the function with a specific button. After explaining the functions, we explained the procedure of test, informing that the test had two parts: the familiarization phase, for participants' adaptation and understanding the mode and operation of the vehicle, and the test

Manual and Conditional automation (L3) 87

TABLE 6.1.
Schematic depiction of the ten HMI concepts in the different moments

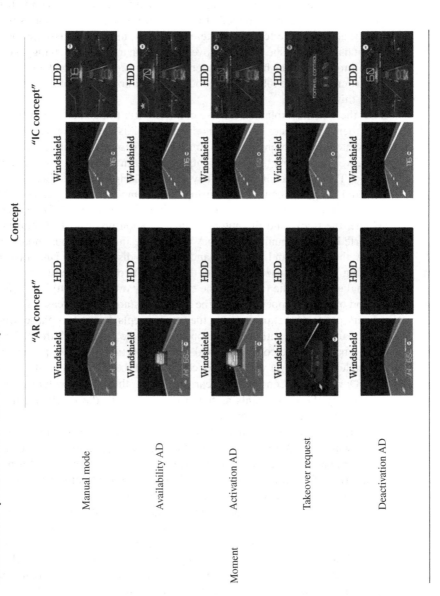

phase, which constituted the main phase. We said that the estimated duration of the test was approximately 1 h. During the instructions, it was also explained that the researcher could ask the participants to do specific STs during manual and autonomous modes. We explained to the participants that there were moments that he/she were in manual mode and others that he/she were going in autonomous mode. We also said that during the test, the system would ask him/her to assume the control. When that happens, he/she must take over and return to manual driving, and for that, he/she had to put his/her hands on steering wheel and foot in brake or accelerator pedal. Participants were told that we would like to know their opinion about driving the car in this study.

About the test, we explained that we could register things that they wanted to tell us about their experiences with autonomous cars. Also about the test, we said that we appreciate their opinion as users to improve the system that they tested. With this objective, we wanted that the participants comment in a loud voice what could be in their mind. This could include what they were thinking, doing, or feeling.

We asked the participants to try during manual mode drive a speed of 120 km/h.

Initially, the participants underwent a 6-min training (Figure 6.3) for adapting to the vehicle, the track, and the autonomous activation process. The familiarization phase started with manual mode. They also drove in autonomous mode to understand how the system and the commands worked, disabling the autonomous function. During the familiarization phase, there were two GBs. The reason for the GB was normal end of traffic (speed of the car in study over 60 km/h) in all cases. The normal end of traffic happened when the car ahead started an increase of speed gradually. Every participant experimented the two concepts during this phase, where half the participants started with the "AR concept" and the others with the "IC concept." We counterbalanced the concepts tested between participants. The participants that started the learning phase with AR concept started the test phase with the AR concept too. During the second AD, in learning phase, the experimenter explained the STs, explaining the participants as how to interact with the touch screen to put a movie running in the central console.

After the preparations, the main test comprised two drives of approximately 10 min each, with a break to answer a questionnaire and prevent phenomena such as fatigue or simulator sickness from affecting the driving performance. This phase had a total duration of 30 min in simulator and the time of each interview was approximately 10 min, but it depends also on the participants.

They drove on both ways, manual and autonomous modes, with the moments in manual mode shorter than autonomous. In the manual-driving condition, participants were entirely responsible for the manipulation of standard longitudinal (accelerator and brake pedals) and lateral (steering wheel) controls. In total, there were four GBs, all them caused by normal end of traffic. Also, there were six moments for the ST during manual mode and two during autonomous mode, and three for interviews and observations during all the tests.

The test phase was composed of two laps. The difference between the participants' group was the concept that was tested first and the order of the ST. We counterbalanced the independent variables, that is, the display and the ST in autonomous

Manual and Conditional automation (L3)

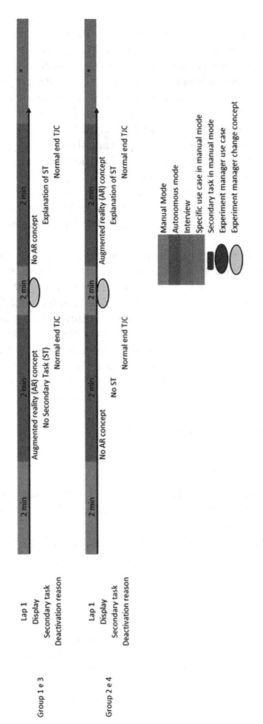

FIGURE 6.3. Familiarization phase.

mode. Group 1 (Figure 6.4) first tested the AR concept and, as a consequence, the first part of the interview was just about the "AR concept." Group 2, on lap 1, tested the "IC concept" first. On lap 2, the groups 1 and 2 tested the other concept that was not tested on lap 1. The second interview was about the second concept tested. The final interview was a comparison between concepts, and the purpose was to know the user preference and the trust between the concepts. The difference between group 1a and 1b, 2a and 2b, 3a and 3b, and 4a and 4b were the order of ST in autonomous mode. The difference between group 1 and 3 was the last use case, a group received the forward collision warning (FCW, group 1) and the other group did not receive the warning (group 3 – baseline "AR concept"). The difference between groups 2 and 4 was the last use case, a group received the FCW (group 2) and the other group did not receive the warning (group 4 – baseline "IC concept"). There were moments that the experimenter asked the participants to do the STs. After the driver puts the movie in the place that the experimenter asked, there was no more interaction with the central console.

In the manual mode in lap 2 the warning was triggered by the driving simulation when the participant's car exceeded a predefined time-to-collision (TTC) threshold of 7 s to the car ahead and the potential collision opponent (Figure 6.5). The obstacle appeared suddenly to make sure the time budget would be the same for all conditions (Radlmayr et al., 2014). The obstacle appeared in a current lane and became visible to the participant that corresponded to a TTC of 7 s = 233 m at 120 km (Lorenz et al., 2014; Radlmayr et al., 2014). Participants could prevent a collision by braking and/or performing a lane change. Obviously, repeating the same scenario several times would eliminate the potential effect of warning, as the participants would remember the type and the position of a recurring hazard; for that, we made a design between subjects.

Finally, we acknowledge their participation.

6.2.5 EXPERIMENTAL DESIGN

The study used a mixed design. The within subject had two factors. First, the HMI concept with two levels ("AR concept" and "IC concept") and, second, the ST with two levels (with and without ST).

The between subject had one factor (type of warning used in a collision avoidance situation) with four levels ((1) visual and auditory warning in AR; (2) no warning in AR condition (baseline to the AR concept); (3) visual and auditory warning in IC; (4) no warning in IC condition (baseline to the IC concept)). Thus, every participant experienced the two concepts of HMI, "AR concept" and "IC concept" and the two levels of ST. Every participant experienced only one of the four warning concepts representing the four possible combinations: (1) visual and auditory warning in AR; (2) no warning in AR condition; (3) visual and auditory warning in IC; (4) no warning in IC condition. The participants were assigned randomly to the groups. The ST during autonomous mode was watching a movie, for the driver to be completely distracted. Zwahlen, Adams and DeBals (1988) pointed out that if a driver's gaze leaves the road for longer than 2 s, then traffic accident risk is significantly increased.

Manual and Conditional automation (L3)

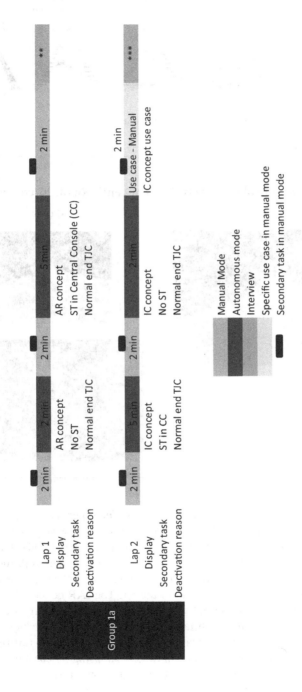

FIGURE 6.4. Test phase, group 1a.

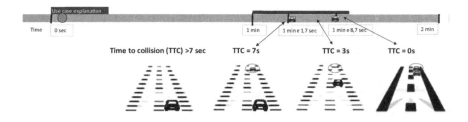

FIGURE 6.5. Use case of possible rear-end collision.

TABLE 6.2.
HMI during rear-end collision. Left: "AR concept"; Right: IC concept"

AR concept

IC concept

This task tried to simulate a real ST during AD. The other ST was during manual mode, which was a mental calculation task (e.g. make successively backward counts of 3, starting with a high odd number to increase mental workload). This task during manual mode had as objective to create driver engagement with the ST. Thus, when in the last stage of the experiment, in manual mode, if the obstacle appears, the driver was confident in the achievement of the ST. The "AR concept" with warning was visual and auditory. The visual warning was displayed in AR area like Tesla Model 3 (2018), directly on the obstacle, with a rectangular 2D shape around the obstacle in red color (Table 6.2). Schall, Rusch, Vecera and Rizzo (2010) found that static cues for hazards had longer reaction times than using no cues; hence, we used a dynamic warning around the obstacle. The auditory warning used was an adaptation of auditory warning of FCW from Tesla Model S (2017), "bip bip bip." The AR

Manual and Conditional automation (L3) 93

concept without warning showed the same information as in manual mode to the AR concept when the obstacle appeared, without visual or auditory obstacle information. The IC concept with warning was also visual and auditory. The visual warning was displayed in IC, in ADAS 3D view, with a rectangular 2D shape around the obstacle ahead in red color. We used the same auditory warning used in "AR concept." The IC concept without warning showed the same information as in manual mode to the IC concept when the obstacle appeared, without visual or auditory obstacle information too.

Concerning dependent variables, we analyzed gaze, driving data as well as subjective measures. During the GB moment, we evaluated the timing and quality aspects (Gold et al., 2013; Kerschbaum, Lorenz, & Bengler, 2014; Lorenz et al., 2014). The timing aspects described the time sequence of drivers' actions after the message appeared. We calculated gaze reaction, road fixation, hands on and take over time. The quality aspects evaluated were the types of reactions and trajectories. Gaze reaction was calculated as the time since the appearance of the GB message until the first saccade from the central console (Lorenz et al., 2014). Road fixation was calculated as the time since the appearance of the GB message until the first glance of the scenery (Lorenz et al., 2014). Hands on was calculated as the time from the appearance of the GB message until the driver has his/her hands on the steering wheel (Lorenz et al., 2014). Take over time was calculated as the time since the appearance of the GB message until the driver put his/her hands on steering wheel and foot in brake or accelerator pedal. The types of reactions were categorized into two groups of interests: brake and steer or accelerator and steer (Gold et al., 2013). Trajectories were calculated for 5 s before the GB message until 5 s after the participant took control.

Other entities in the simulation were recorded by our driving simulator software at a frequency of 10 Hz (Medenica, Kun, Paek, & Palinko, 2011). During the rear-end collision moment, we evaluated the timing and quality aspects. The timing aspects calculated were reaction time, side mirror and indicator. The quality aspects evaluated were the types of reactions and trajectories. Reaction time was calculated as the time since 233 m TTC until the actual driving maneuver began, steering wheel angle > 20 degree or brake pressure > 10% (Gold & Bengler, 2014; Gold et al., 2013). Side mirror was calculated as the time since 233 m TTC until the driver glanced at the side mirror (Lorenz et al., 2014). Indicator time was calculated as the time since 233 m TTC until the driver used the indicator (Lorenz et al., 2014). Types of reactions were categorized into three groups of interests: braking only, braking and steering, or steering only, after receiving the warning (Gold et al., 2013). Trajectories were calculated for 300 m TTC until 100 m after the obstacle. Finally, collision frequencies were observed during the event of possible rear-end collision. After each lap, the researcher asked the participants verbally about the awareness situation through four questions ("How did you know that AD was activated?"; "What did you see?"; "Where did you see the information?"; "How clear was clear for you understand the information? 1–5 Likert scale"). Participants were unaware they would be questioned at the end of the simulation, which allowed them to attend to the environment as they naturally would in normal driving situations (Endsley, 1995a). These questions had some modifications because of the different moments that were used to evaluate the awareness situation: availability, activation, GB and AD deactivated. After each drive, the participants also answered a Van der Laan questionnaire (Van Der Laan, Heino, & De Waard, 1997) as a measure of acceptance of the HMI. This questionnaire was used to evaluate the acceptance of three use cases: availability, activation and GB. When considering the usefulness of a system, such as the AR, people tend to use or not use an application to the extent they believe it will help them perform their job better (Davis, 1989) and performance gains are often dependent on the users' level of willingness to accept and use the system. In the final interview, the participants expressed their preference level by concepts tested through 10-point Likert-type rating scale from "prefer totally a concept" to "prefer totally other concept," showing the images about that specific moment. We evaluate the preference variable in the following use cases: manual mode, availability, activation, GB and deactivation. In the final interview, we also asked the participant verbally in what concept they felt more trust, through 10-point Likert-type rating scale from 1 – "trust totally in this concept" until 10 – "trust totally in other concept," showing the images about the specific moments of both concepts in autonomous and manual modes. We evaluated the trust variable in the following use cases: autonomous and manual mode. We also solicited qualitative verbal comments about the experiment from the participants.

6.3 RESULTS

In this topic, we will present in detail the results of manual mode, transition from manual to autonomous, the transition from autonomous to manual (take over request) and rear-end collision. Before that, we can see in Table 6.3 the summary of results obtained in each moment tested. In terms of preference, "AR concept" was higher and showed significant difference in all the moments when compared with "IC concept." Confidence showed better results with "AR concept" too. The participants were more aware of "AR concept" during availability and activation moment; however, during GB and deactivated moment, the results showed no significant difference between concepts. "AR concept" appeared to be more useful significantly in availability and activation moment. During the GB, the results showed no significant difference. In terms of satisfaction, only during activation, the results showed significant difference between concepts, favoring "AR concept." In terms of reaction time and behaviors, the results showed no significant difference between concepts during the GB. During the rear-end collision, the behavior was safer in "AR concept"; however, the reaction time was better in "IC concept."

6.3.1 MANUAL MODE

In manual mode, we evaluated the preference and confidence between concepts.

Looking at Figure 6.6, more than 75% of the participants (22/29) preferred the AR concept, 3.4% of the participants (1/29) did not prefer anyone, and approximately 24% of the participants (6/29) preferred the IC concept. The mean value that the participants gave to the AR concept was 6.00 points (SD = 2.605 points) and to the

TABLE 6.3.
Summary of results

	Preference	Confidence	Awareness situation	Usefulness	Satisfaction	Timing aspects	Reaction types
Manual mode	AR ($p < 0.05$)	AR 72% vs IC 22%	-	-	-	-	-
Availability	AR ($p < 0.05$)	-	AR ($p < 0.05$)	AR ($p < 0.05$)	No *validate*	-	-
Activation	AR ($p < 0.05$)	-	AR ($p < 0.05$)	AR ($p < 0.05$)	AR ($p < 0.05$)	-	-
Activated		AR 76% vs IC 10%					
GB	AR ($p < 0.05$)	-	No difference ($p > 0.05$)	No difference ($p > 0.05$)	No difference ($p > 0.05$)	No difference ($p > 0.05$)	No difference
☐ Deactivation	AR ($p < 0.05$)	-	No difference ($p > 0.05$)	-	-	-	-
Rear-end collision						IC	AR

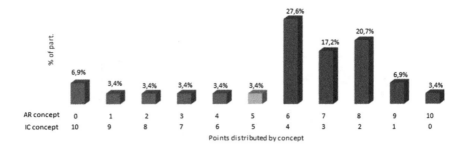

FIGURE 6.6. Percentage of participants according to the points distributed by concept about the preference during the manual mode.

TABLE 6.4.
Descriptive statistics about the preference in specific moment of Manual Mode per concept

	AR Concept			IC concept		
	Mean	Min–Max	N	Mean	Min–Max	N
Availability	6.00 ± 2.605	0–10	25	4.00 = 2.605	0–10	29
Significnat difference			$Z = -2.030, p < 0.021$			

IC concept was 4.00 points (SD = 2.605 points) (Table 6.4). The Wilcoxon analyses showed significant difference when comparing the preference between concepts ($Z = -2.030$, $p = 0.021$) for $\alpha = 0.05$; so it is to be admitted statistically that the participants prefer "AR concept" comparatively to "IC concept" during manual mode.

The participants with a negative response to the AR concept (<5 points preference to the AR concept) gave the reasons that they were accustomed to the IC, and it was easier to drive without lines, and the information in the IC was more visible.

The participants that did not prefer any concept (=5 points preference to each concept) argued that option AR was more effective; however, the IC option had a better design and if he/she had bought a car, he/she would have bought with the IC option because he/she was more used to it. The participants that preferred the AR concept (>5 points preference to the AR concept) argued that the main reasons they were more comfortable with the AR option were because they did not need to take their eyes of the road, and had everything necessary in the windshield. They argued too that they did not miss the IC; they felt disoriented when they tried to look at the IC during the test and that the information on the windshield allowed them to increase the trust, once that appears in their field of view. The lines were seen as more negative than positive by most of the participants.

On the level of confidence, more than 70% of the participants felt more confident in the "AR concept," 7% of the participants did not feel any difference in terms of confidence between concepts, and 21% of the participants felt more confident in "IC concept."

The main reasons presented by participants for feeling more confident in the "AR concept" were that the information appeared at hand. There was no need to change the view of the road; it was easier to adjust the speed, easy to use the information in the windshield, and there were less distractions, because the lines and personal problems when nowadays they try got information of the IC in their own cars.

The main reasons presented by participants to feel more confident in the IC concept were that the AR option impeded them to see the depth of the horizon, gave them more safety because they used the screen (IC) like a confirmation. Because they got used to the IC, the lines seemed like a video game and they preferred to receive less information on road.

The main reasons presented by participants who did not differentiate the confidence between concepts because the HUD is the first place that they looked at when they needed something that appeared in both options.

6.3.2 Activation

During the activation moment, we evaluated the preference, acceptance, awareness situation and confidence, by comparing both concepts.

Looking at Figure 6.7, more than 75% of the participants (22/29) preferred the "AR concept," 10% of the participants (3/29) did not prefer any, and only approximately 14% of the participants (4/29) preferred "IC concept." Table 6.5 shows all the means, maximum and minimum of preference of the two concepts tested. The mean value that the participants gave to "AR concept" was 6.55 points (SD = 2.148 points) and to "IC concept" was 3.45 points (SD = 2.148 points) (Table 6.5). The Wilcoxon analyses showed significant differences when comparing the preference between concepts (Z = −3.111, p < 0.001) for α = 0.05. So, it is to be admitted statistically that the participants prefer "AR concept" in the specific moment of activation comparatively to "IC concept." The participants with negative response to "AR

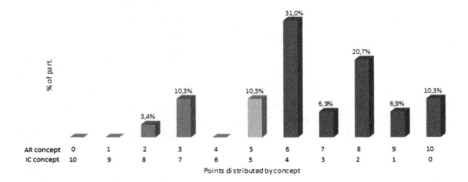

FIGURE 6.7. Percentage of participants according to the points distributed by concept about the preference about AD activation.

TABLE 6.5.
Descriptive statistics about the preference in specific moment of activation per concept

	AR Concept			IC concept		
	Mean	Min–Max	N	Mean	Min–Max	N
Activation	6.55 ± 2.14S	2–10	29	3.45 = 2,148	0–8	29
Significant difference			$Z = -3.111$. $p < 0.001$			

TABLE 6.6.
Distribution of the sample according with the usefulness and satisfaction during the activation moment

	AR Concept			IC Concept			Significate difference	
	Mean	Min–Max	N	Mean	Min–Max	N	Est	p
Usefulness	4.53 ± 0.41	3.7–5	27	4.18 ± 0.73	2.3–5.0	28	−2.862	0.001
Satisfaction	4.60 ± 0.44	3.5–5.0	27	4.36 ± 0.66	2.75–5.0	28	−2.388	0.007

concept" (<5 points preference to "AR concept") gave the main reasons that in IC they could see the car information bigger; the most important was the change to blue color, which was more visible in IC and because the participants do not like to receive information on road. The participants that did not prefer any concept (=5 points preference to each concept) argued the main reason and the most important was the color, which was visible in both concepts. The participants that preferred "AR concept" (>5 points preference to the AR concept) argued as main reasons that the AR concept gave more trust that AD was activated, because they did not look at IC during manual driving, and the driver can see the same that the AD was seeing and because it appeared where the driver was looking in a comfortable way.

Reviewing data analysis showed that there were two participants in the "AR concept" and one in the "IC concept" who mixed the activation moment with other moment by comments. Since all participants should be aware of this moment, we did not consider these three participants for the analysis. Finally, the used sample consisted of 27 participants in the AR concept and 28 participants in the IC concept.

Table 6.6 shows all means, maximum, minimum and standard deviation of acceptance to the two concepts tested in the same moment.

From Table 6.6, we can see that the mean value of the usefulness opinion in both concepts is very positive, close to 4 points, in a scale of 1 until 5. The Wilcoxon analyses showed significant difference when comparing the opinion of usefulness between concepts ($Z = -2.862$, $p = 0.001$) for an $\alpha = 0.05$. So, it is to be admitted statistically that the participants consider AR concept more useful compared to the IC concept in the specific moment of activation.

Relative to satisfaction variable, in Table 6.6, we can to see that the IC concept has a minimum value under the AR concept, but both mean values are good (4.60 points to "AR concept" and 4.36 points to "IC concept"). The Wilcoxon analyses showed significant difference when comparing the opinion of satisfaction between concepts (Est = −2.388, p = 0.007) for an $\alpha = 0.05$. So, it is to be admitted statistically that the participants are more satisfied with the AR concept compared to the IC concept in the specific moment of activation.

About the awareness situation, the first question was how they knew that the AD function was activated, and in "AR concept," what helped the participants more were the change of color, followed by the message "AD activated" (pointed by 56.6% and 37%, respectively). In "IC concept," what was more noticed by the participants to know that AD was activated was the blue color, followed by the change of color (mentioned by 46.4% and 42.9% of participants, respectively).

Then, we asked more specifically what the participant saw in this specific moment and to both concepts; the change of color was the visual information more seen (argued by >50% of participants in both concepts), followed by blue color, which was seen by more than 45% of participants in both concepts. Other items were identified by a few participants, such as the car mark, lines and physical response of the car.

About the place where the participants saw the information, in "AR concept," most of the participants (92.6%) identified the information appeared in AR area, place where the lines and car mark appeared, followed by the OA (pointed by 59.3% of the participants). In "IC concept," most of the participants (92.9%) identified the IC as the place where the AD information appeared and one-third of the participants saw the information in OA during this moment too.

The last question about this moment was how clear was the participant to understand that AD was activated and in both concepts, most of the participants answered positively; however, 93% of the participants in "AR concept" gave the maxim punctuation, against 61% in "IC concept." The Wilcoxon analyses showed significant difference when comparing the clarity to understand the information between concepts ($Z = -2.754$, p = 0.002) for an $\alpha = 0.05$. So, it is to be admitted statistically that "AR concept" showed more clarity to the participants to understand that AD was activated compared to "IC concept."

For the last question, when the participants did not answer "very clear," the experimenter asked the participants the reason for their answer what is that they missed. With regard to the "IC concept," the negative comments were about the missing of some information besides the change of color in the IC, as an icon, that the learning curve was bigger in the "IC concept" compared to the "AR concept," and that the "IC concept" was less indicative. In the "AR concept," the negative comments were because of the size of the letters, the visibility of the icon and the missing of voice messages such as "AD activated."

About the confidence, more than 75% of the participants felt more confident in "AR concept" compared to "IC concept." Of the participants, 14% did not feel difference in terms of confidence between concepts and only 10% of the participants felt more confident in the IC concept.

The main reasons presented by participants for feeling more confident in the "AR concept" were that the information did not distract, the lines gave more perception of color, the information appeared closer to the eyes of the driver, see the same than seeing in AV, and showed that the car was working well. The IC did not call driver's attention and that was not a necessary force to take the eyes of the road to obtain information.

The main reasons presented by participants to feel more confident in the "IC concept" were that the IC was the place where the driver obtained information and the IC was simpler.

The main reasons presented by participants who did not differentiate the confidence between concepts were because was not necessary driver attention in that moment, so the interface (windshield or IC) did not influence the trust and what helped more in this moment was the change of color that was well visible in both concepts.

6.3.3 Takeover Request

During GB moment we evaluated the preference, acceptance, awareness situation, timing aspects and reaction types, comparing both concepts.

Looking at Figure 6.8, more than 85% of the participants (25/29) preferred the AR concept, 7% of the participants (2/29) did not prefer any, and only approximately 7% of the participants (2/29) preferred the IC concept. Table 6.7 shows all the means, maximum and minimum of the preference of the two concepts tested. The mean value that the participants gave to "AR concept" was 7.48 points (SD = 1.703 points) and to "IC concept" was 2.52 points (SD = 1.703 points) (Table 6.7). The Wilcoxon analyses showed significant difference when comparing the preference between concepts ($Z = -4.349$, $p < 0.001$) for an $\alpha = 0.05$. So, it is to be admitted statistically that the participants prefer "AR concept" in the specific moment of GB compared to "IC

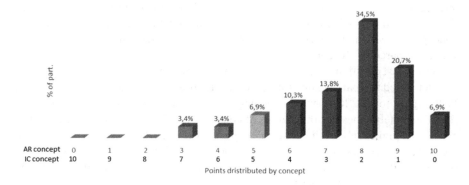

FIGURE 6.8. Percentage of participants according to the points distributed by concept about the preference about GB.

TABLE 6.7.
Descriptive statistics about the preference in specific moment of GB per concept

	AR Concept			IC Concept			Significate difference	
	Mean	Min–Max	N	Mean	Min–Max	N	Est	p
Usefulness	4, S1 ± 0.25	4.0–5.0	29	4.66 ± 0.50	3.4–5.0	27	−1.648	0.051
Satisfaction	440 ± 0.87	1.75–5.0	29	4.32 ± 0.73	2.75–5.0	27	−0.468	0.329

concept." The participants with negative response to the AR concept (<5 points preference to the AR concept) gave the main reason that the information appeared close to the eyes when they wanted to see a movie in the central console. The participants that did not prefer any concept (=5 points preference to each concept) argued as main reason that an icon would be enough with the sound and that the design of IC was better in AR screen; however, in AR only better contrast was missing for it to be perfect. The participants that preferred the AR concept (>5 points preference to the AR concept) argued as main reasons that after listening to the sound, what they did first was to look at the road, so they first saw the information in the AR concept, which the participants considered an important moment, and so the information should appear on the road, that the information was very clear, and that it appeared right where they were looking. When they were alert, what they did first was to look at the road and that if they looked at the IC in this moment, they were losing information on the road and some participants argued that didn't see visual information on the IC.

Review of the data analysis showed that there were two participants in the "IC" concept who mixed the GB moment with other moments by comments. Since all participants should be aware of this moment, we did not consider these two participants for the analysis. Finally, the used sample consisted of 29 participants in the AR concept and 27 participants in the IC concept.

Table 6.8 shows all means, maximum, minimum and standard deviation of acceptance for the two concepts tested in the same moment.

In Table 6.8 we can see that the mean value of the usefulness opinion in both concepts is very positive, close to 4.5 points, on a scale of 1 to 5. The Wilcoxon analyses showed no significant difference when comparing the opinion of usefulness between concepts ($Z = -1.648$, $p = 0.051$) for an $\alpha = 0.05$. Relatively on the satisfaction variable in Table 6.8 we can see that the AR concept has a minimum value under the IC concept, but both mean values are good (4.40 and 4.32 respectively). The Wilcoxon analyses showed no significant difference when comparing the opinion of satisfaction between concepts (Est = -0.468, $p = 0.329$) for an $\alpha = 0.05$.

About the awareness situation, the first question was how they knew that the car was asking the control and in both concepts, what helped the participants more was the sound (noticed by 69% in "AR concept" and 89% in the IC concept). In visual terms, what helped more in the AR concept was the red color (pointed by 76% of

TABLE 6.8.
Distribution of the sample according with the usefulness and satisfaction in the GB moment

	AR Concept			IC Concept			Significant difference	
	Mean	Min–Max	N	Mean	Min–Max	N	Est	p
General	1.94 ± 0.54	1.15–3.58	7A	2.00 ± 0.67	1.06–3.31	21	−0.469	0.329
With ST	2.08 ± 0.63	1.18–3.93	23	2.28 ± 1.01	1.11–4.75	21	−0.574	0.290
Without ST	1.80 ± 0.67	0.80–3.83	23	1.73 ± 0.81	0.89–4.90	21	−0.469	0.329

the participants), followed by the message "Take control" (pointed by 59% of participants). In IC, in visual terms, what helped more was the message "Take control" (pointed by 64% of participants), followed by the red color and icon of SW (pointed by 46% of participants).

Other visual aspects were identified in both concepts such as the change of color, icon of pedal, hand, foot and brake pedal (pointed by less than one-fourth of participants).

About the place where the participants saw the information, to the "AR concept," most of the participants (76%) identified the information that appeared in the AR area, the place where the message actually appeared, followed by the OA (48% of the participants). In the "IC concept," most of participants (93%) identified the IC as the place where the AD information appeared and 4 participants (14%) incorrectly saw the information in the OA during this moment too.

The last question about this moment was how clearly did the participants understand that the AD was activated, and in both concepts most of the participants answered positively. All the participants in the "AR concept" gave the maximum points on the clarity of understanding the information, and in the IC 89% of the participants gave the maximum points. The Wilcoxon analyses showed no significant difference when comparing the clarity of understanding the information between concepts ($Z = -1.732$, $p = 0.125$) for an $\alpha = 0.05$.

As the results presented above show, the results were very positive for both concepts, and so no comments were provided by the participants to improve the clarity of information during GB.

Review of quantitative data analysis showed that six participants in the "AR concept" and eight participants in the "IC concept" didn't get the results well because of technical problems. So, we did not consider these six and eight participants for the analysis. Finally, the used sample consisted of 23 participants in the AR concept and 21 participants in the IC concept.

Table 6.9 shows all means, maximum, minimum and standard deviation of reaction times to the "Hands on" experience of the two concepts tested and the possible influence of the ST. Looking at the hands-on data in general, none of the participants considered doing the ST, and the difference of the "hands on reaction"

Manual and Conditional automation (L3)   103

TABLE 6.9.
Distribution of the sample according with the reaction time of "Hand on" during the GB moment

	AR Concept			IC Concept			Significant difference	
	Mean	Min–Max	N	Mean	Min–Max	N	Est	P
General	2.38 ± 0.79	1.42–4.39	23	2.44 ± 0.71	1.22–3.64	21	−0.608	0.281
With ST	2.61 ± 0.95	1.34–4.89	23	2.85 ± 1.17	1.11–5.13	21	−1.234	0.,114
Without ST	2.13 ± 0.80	0.80–4.14	23	2.02 ± 0.86	1.02–4.90	21	−0.521	0.308

TABLE 6.10.
Distribution of the sample according with the reaction time of "take control" during the GB moment

	AR Concept			IC Concept			Significant difference	
	Mean	Min–Max	N	Mean	Min–Max	N	Est	P
General	2.3 8 ± 0.79	1.42–4.39	23	2.44 ± 0.71	1.22–3.64	21	−0.608	0.281
With ST	2.61 ± 0.95	1.34–4.89	23	2.85 ± 1.17	1.11–5.13	21	−1.234	0.114
Without ST	2.13 ± 0.80	0.80–4.14	23	2.02 ± 0.86	1.02–4.90	21	−0.521	0.308

seems negligibly small, as the means are between 1.94 ("AR concept") and 2.00 ("IC concept") (Table 6.9). Since the data was not normally distributed, we applied the Wilcoxon test, which revealed no significant difference ($Z = -0.469$, $p = 0.329$) (Table 6.9). Just before the GB when the participants were seeing the movie, the "Hands on reaction time" was better in the "AR concept" ($M = 2.08$ s; $SD = 0.63$ s), compared to "IC concept" ($M = 2.28$ s; $SD = 1.01$ s) (Table 6.9). The Wilcoxon analyses showed no significant difference when comparing the "Hands on reaction time" when the participants were distracted with the ST between concepts ($Z = -0.574$, $p = 0.290$) for an $\alpha = 0.05$ (Table 6.9). Just before the GB when the participants were free to look at the road, the "Hands on reaction time" was better in the "IC concept" ($M = 1.73$ s; $SD = 0.81$ s) compared to the "AR concept" ($M = 1.80$ s; $SD = 0.67$ s) (Table 6.9). The Wilcoxon analyses showed no significant difference when comparing the "Hands on reaction time" when the participants were not distracted between concepts ($Z = -0.469$, $p = 0.329$) for an $\alpha = 0.05$ (Table 6.9). The "Hands on" reaction time was significantly worse in both concepts when the driver was doing the ST just before the GB when compared to when the participants were free to look at the road ($Z = -2.972$; $p < 0.001$ when we compared "IC concept" with and without ST and $Z = -1.737$; $p = 0.042$ when we compared "AR concept" with and without the ST).

Table 6.10 shows all means, maximum, minimum and standard deviation of the "Take Control" reaction times of the two concepts tested and the possible influence of the ST. Looking at the "take control time" data in general, without taking into account whether the participants were doing the ST, the difference of the "take control reaction" seemed negligibly small, since the means are between 2.38 s ("AR concept") and 2.44 s ("IC concept"). Since the data was not normally distributed, we applied the Wilcoxon test, which revealed no significant difference ($Z = -0.608$, $p = 0.281$). Just before the GB when the participants were seeing the movie, the "take control reaction time" was better in the "AR concept" ($M = 2.61$ s; $SD = 0.95$ s) compared to the "IC concept" ($M = 2.85$ s; $SD = 1.17$ s) (Table 6.10). The Wilcoxon analyses showed no significant difference when comparing the "take control reaction time" when the participants were distracted with the ST between concepts ($Z = -1.234$, $p = 0.114$) for an $\alpha = 0.05$. Just before the GB when the participants were free to look at the road, the "take control reaction time" was better in the "IC concept" ($M = 2.02$ s; $SD = 0.86$ s) compared to the "AR concept" ($M = 2.13$ s; $SD = 0.80$ s). The Wilcoxon analyses showed no significant difference when comparing the "take control reaction time" when the participants were not distracted between concepts ($Z = -0.469$, $p = 0.329$) for an $\alpha = 0.05$. The time of "Take Control" was significantly worse in both concepts when the driver was doing the ST just before the GB compared to when the participants were free to look at the road ($Z = -2.485$; $p = 0.006$ when we compared IC with and without ST and $Z = -2.585$; $p = 0.004$ when we compared AR with and without the ST).

As the subjects can be differentiated by their reaction they were divided into two groups. In both the concepts, most of the participants "took control" using "steer and accelerator" during an end of a traffic jam. Only one participant used "brake and steer" in both the concepts.

6.3.4 REAR-END COLLISION

During the rear-end collision moment, we evaluated the types and times of reaction. In this collision avoidance emergency, we considered four types of reactions where a hierarchy was established in terms of behavioral safety. We considered the safest behavior to be braking and at the same time turning the steering wheel. The second safest was braking only. The third was turning the steering wheel only. We compared four groups (G1 – "AR + warning condition"; G2 – "IC + warning condition"; G3 – "AR control condition without warning"; G4 – "IC control condition without warning").

For technical reasons with quantitative data collection, one of the participants was eliminated from the sample. Thus, the used sample consisted of 28 participants, 7 in the "AR concept" with warning, 7 in the "IC concept" with warning, 7 in the "AR concept" control condition, and 7 in the "IC concept" control condition.

Figure 6.9 shows the reaction types in each group during rear-end collision.

As can be seen in Figure 6.9, the participants that received warning in the "AR concept" showed a safer behavior with a higher percentage of drivers braking and steering (42.86%) against only 14.29% in the "IC concept" with warning. In the control conditions, the percentage of safer behaviors (braking and steering) were equal; however the "IC concept" control condition was more penalized because the percentage of less safe behaviors (turning the wheel) was 28.57% against 14.9% in the "AR concept" control condition (Figure 6.9).

Manual and Conditional automation (L3)

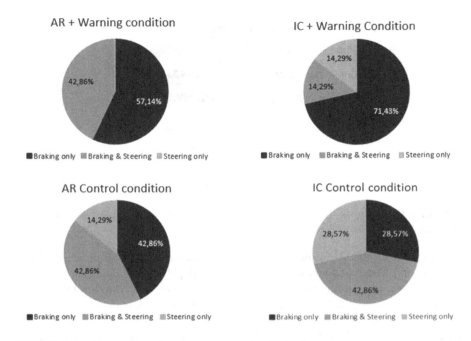

FIGURE 6.9. Types of reaction during rear-end collision by group.

Figure 6.10 shows the reaction times (time since the obstacle is visible until the driver moves the "steering wheel angle >2°" or makes a "difference in brake pressure >10).

Looking at the reaction time, the "IC concept" with warning (M = 2.1 s) showed better results compared to the "AR option" with warning (M = 2.8 s), as well as in the "IC concept" control condition (M = 2.7 s) and the "AR concept" control condition (M = 2.9 s) (Figure 6.10).

6.4 DISCUSSION

Intermediate levels in which the human is expected to monitor the automated driving system may be particularly hazardous because humans are unable to remain vigilant for prolonged periods of time (Casner et al., 2016; D. A. Norman, 2015). In short, the driver during AD level 3 (SAE,2016) is out of loop in some moments. The objective of the current study was to investigate drivers' interactions with a novel AR display and compare with the current displays in the market during manual mode and AD level 3 (SAE, 2016) and the transitions between modes.

We were particularly interested in the driver's user experience, acceptance, awareness situation and confidence with the displays tested. To answer these research questions, a self-report and specific software methodologies were used.

In the manual mode, we evaluated driver preference and confidence. In terms of preference, the results showed a significant difference between concepts, and so we

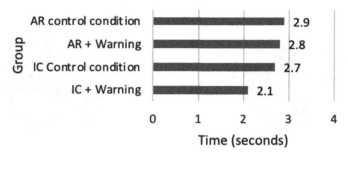

FIGURE 6.10. Reaction time (since the warning appears until the driver: "Steering Wheel angle >2°" or "difference brake pressure > 10%") by group.

can assume that the user experience was clearly better in the "AR concept" (21/29 participants). These findings are in accordance with published studies Park and Park (2019). The main reasons presented by the participants were that the information in Operational Area (OA) was very accessible, that it was less distracting than the information in the IC, and that it appeared right in their field of view. Most of the participants did not distribute the total points equally between the two concepts (10 points for the "AR concept" and 0 points for the "IC concept") because the lines that appeared in the AR area was like a "videogame" and that it was distracting. From a computer graphics perspective, the accuracy of lines wasn't evaluated by the users; however we think that could explain the fact that most of the participants didn't give the maximum points to the "AR concept." Those lines had a delay between the user action in the SW and the real visual information displayed in the windshield, so it seemed that the driver was all the time adjusting his position in the lane. The participants that preferred the "IC concept" (6/29) argued that they were habituated with receiving information in the IC and the lines that appeared on the AR screen. This habit can be changed by the experience with the windshield display, and once the participants used the AR information during the test, they commented that they didn't miss the IC.

In terms of confidence, the results showed a trend toward the "AR concept" again. The main reasons presented by the participants were that they felt more confidence in the "AR concept" during the manual mode in which the localization of information was accessible to the driver, and that the IC had used was now useless, and that it was less distracting to get the information there (AR screen). These results can be explained by the fact that the driver in the "AR concept" didn't need to take his eyes off the road to get car information; on the other side, in the "IC concept" only the current speed and speed limit appeared on the road. The IC used in cars nowadays is negative in terms of safety, since the driver at all times tries get information from the IC and is distracted from the main task. In this test, no quantitative data was obtained to evaluate both concepts; however no specific behaviors were observed.

Manual and Conditional automation (L3)   **107**

In the activation moment, we evaluated the driver preference, acceptance and awareness situation. In terms of preference, the results showed a significant difference between concepts, and so we can assume that the user experience was clearly better in the "AR concept" (22/29 participants). These findings are in accordance with published studies Park and Park (2019), in which the authors argued that the HUD systems have the potential to improve the driver experience. The main reasons presented by these participants were that the user could see the same information that the autonomous vehicle (lines detection and distance car ahead) sees that the information appeared in the field of view of driver, and that it was more accessible. These results can be explained by the fact that the samples tested had low levels of confidence with regard to AD, because this is a new technology under development, so appears the information on the road, adapting to the real conditions, had a positive impact in the user experience. Participants preferred the "IC concept" because in IC they could see the car information bigger, and the most important was the change to blue color, which was more visible in IC, and because the participant did not like to receive information on the road. These negative results of the "AR concept" are because the driver isn't used to receiving the information on the windshield and neither does he see the adaptation of the car information to the real environment (as lines and time gap). The simulator environment could influence the preference for the "AR concept," once the light effects do not affect the information on the windshield. It will be important test and understand the light effect on the user experience of the driver to receive the information in the windshield in real road conditions.

In terms of acceptance, in both variables, the results showed a significant difference between concepts, and so we can assume that the "AR concept" is more useful and left the participants more satisfied compared to the "IC concept." The main reasons presented by the participants that gave a positive response to "AR concept" were that the blue color was more visible, that the icon was more visible, that the car in the "IC concept" didn't recognize the car position and that there was too much information in the "IC concept." The line was perceived negatively by some participants in this moment; however others differed. We believe that the negative feedback on the lines in the autonomous mode was because of their inaccuracy. If we improve the accuracy of the display the information on the AR screen, probably the acceptance of most of the participants will increase. We think that the acceptance was better in the "AR screen" because when the driver was out of loop (example of the ST tested: see a movie), with a simple glance he could check in the windshield that everything was ok; on the other hand, in the IC the information wasn't so clear whether the autonomous vehicle was detecting the lines and the car ahead.

In terms of awareness situation, the results showed a significant difference between concepts. Participants had more awareness in the "AR concept" compared to the "IC concept" because they found the information clearer to understand. The change of color (yellow to blue) helped the participants to know that AD was activated in the "AR concept," followed by the message "AD activated." All the other items were noticed by less than one-fifth of the participants, including the AD icon. The design, position and size of the icon may explain why it was not noticed. In the "IC concept" what helped the participants more were the blue color and the change of color. The message in this concept was noticed by only one-fourth of the sample. These results showed that when the visual information was displayed in the IC, sometimes it wasn't noticed, and so as expected it was clear that the AD was activated in the "AR concept." The human factors team of CTAG used the change of color, the message "AD activated" and the AD icon to help the driver be aware about this moment; however not all the participants noticed that information. Maybe the quantity of new information available to the participants could influence the missing of information during the experimenter interview after each participant tested the concept. On other side, the change of color in this first steps to AD would be enough; however in future it could concern other problems such as a change of color in the display by some internal problem which may be incorrectly associated with activation of AD.

During the take over request, we evaluated driver preference, acceptance, awareness situation, timing aspects and reaction types. In terms of preference, the results showed a significant difference between concepts, and so we can assume that the user experience was better in the "AR concept" (25/29 participants). These findings are in accordance with other studies Park and Park (2019), where the authors argued that the HUD systems have the potential to improve the driver experience. The main reasons presented by these participants were that after listening to the sound, what they did first was to look at the road, and so they first saw the information in the "AR concept," that the participants considered this an important moment, and so the information should appear on the road, that the information was very clear, that it appears right where they were looking, that when you are alert, what you do first is to look at the road, that if you look at the IC in this moment, you're losing information of the road and some participants argued that they didn't see visual information in the IC. These results can be explained by the localization of information, when there is an alert situation (as the GB), the driver has the tendency to look to the road, so if the information appears in windshield as in "AR concept," the driver didn't need "lose" time looking at the IC, and so their preference was more for AR, as verified in this test. The participants that preferred the "IC concept" (2/29) argued as main reason that the information in IC appeared closer to their eyes in the moment of seeing a movie in CC. In fact, the IC was closer to the CC compared to the windshield; however, the normal behavior of the driver when listening to a sound is to look at the road and only after the IC. Maybe these results show that the participant that already knew the function very well and had a good level of trust is already good to use AD.

In terms of acceptance, in both variables, the results showed no significant difference between concepts. These results can be explained by the fact that the sound was the main help to the drivers, and so the visual information was only secondary.

In terms of the awareness situation, the results showed no significant difference between concepts. In both concepts the participants very clearly understood that the car was asking the control. In both concepts, what helped the participants more was the sound. In the "AR concept" the second big help was the red color and in the "IC concept" it was the message "take control." The items of SW, pedal, foot and hands were noticed by some participants too. The GB tested was always the end of traffic jam, and so the participants were already waiting for the car alert to take control

Manual and Conditional automation (L3)   109

and maybe that influenced the driver's awareness positively. The sound wasn't a "bip" as in previous moments (availability and activation), but a female voice saying, "take control" which left the participants without doubts about the action that they should take. One participant in each concept didn't see any visual information, maybe because the reaction time was quick and he had no time to see the information about this moment.

In terms of reaction time, we evaluated the "hands on time" (time since the warning appeared until the driver puts the hands on the SW), the "take over time" (time since the warning appeared until the driver puts the hands on the SW and the foot on the brake/accelerator) and the possible influence of ST in these reaction times. The findings suggest that supporting the driver in his information processing by providing AR information does not have a significant effect on takeover times. These results are in line with previous research (Lorenz et al., 2014).

About the Hands Over Time, the results showed no significant difference between concepts in general. The mean in "AR concept" was 1.94 s and in the IC concept it was 2.00 s. This range of time was in line with the results of previous research (Gold et al., 2013; Lorenz et al., 2014). When the driver was out of loop (seeing a movie), the results showed no significant difference between concepts to HOT too (M = 2.08 s in AR and M = 2.28 s in IC). When the driver's eyes were on the road, the results showed no significant difference too (M = 1.80 s in AR and M = 1.73 in IC). We only tested the GB in the end of traffic jam and that could influence the faster reaction times obtained in this study. The level of engagement in the ST was different among the participants.

About the TOT, the results showed no significant difference between concepts in general. The mean in the "AR concept" was 2.38 s and in the IC concept it was 2.44 s. This range of time is in line with the results of previous research (Gold et al., 2013; Lorenz et al., 2014). When the driver was out of loop (seeing a movie), the results showed no significant difference between the concept to HOT too (M = 2.61 s in AR and M = 2.85 s in IC). When the driver's eyes were on the road, the results showed no significant difference too (M = 2.13 s in AR and M = 2.02 in IC). We only tested the GB in the end of traffic jam and that could influence the faster reaction times obtained in this study. The level of engagement in the ST was different among the participants.

In rear-end collision, we evaluated the types of behavior and reaction times. In terms of types of behavior, the results showed better results to the "AR concept" with a warning compared to the "IC concept" with a warning. The safest behavior, braking and steering, was more frequent in the "AR concept" with the warning condition. In terms of the reaction time, the results showed shorter times to the "IC concept" with the warning, compared to the "AR concept" with the warning. Two possible explanations for these results are: (1) The higher reaction times in the "AR concept" with the warning can be explained by the fact that the participants are not used to receiving warnings in the windshield and were surprised, and so the time to interpret the warning provoked an increase in reaction time. (2) The visual warning projected in the windshield may create a false illusion that the obstacle is far than it is in reality. If we assume that the explanation for these results is the first option, the habituation by the drivers could decrease reaction times when receiving this kind of warning projected in the windshield. For the second option, more design options of a square shape around the obstacle may be tried in future investigations.

6.5 CONCLUSION

Recent announcements that manufacturers will soon sell self-driving cars raise hopes that autonomous vehicles will quickly solve many transportation problems, and several car manufacturers have proposed many developments and commercialization plans about HUD. This technology creates a new way of interaction between the driver and the vehicle (Phan et al., 2016). With an expected growth of 758% in the next eight years (IHS iSuppli, 2013), there will be a great demand for car manufacturers to supply new technical solutions and functionalities together with new ways of user interaction and user design. However, as the HUDs pose other problems to the driver, the AR can be of help to solve those problems. The information projected in the windshield can increase the acceptance, trust and awareness of the driver to the autonomous system, once the driver can see the same information that the autonomous car is seeing. This thesis makes a comparison between two new generations of cars and displays. The objective of the current study was to investigate drivers' interactions with a novel AR display and compare with the current displays in the market during the manual mode of AD level 3 (SAE, 2016) and the transitions between modes.

The results showed that, during the manual mode, in terms of user experience the participants preferred significantly "AR concept" compared to the "IC concept." It appears that all the car information in the operational area was the main reason for this difference. After the experience of receiving the car information in the OA, the participants didn't miss the information in the IC. The lines in the "AR concept" were negative, and more distracting than useful. In terms of confidence, the "AR concept" showed better results. The participants argued that they felt more confident in the "AR concept" because they could keep the eyes on the road while getting the car information and that it appeared in a comfortable zone. These results suggested that the operational area improved the user experience of the driver and that a new expansion of HUD is expected. However, the quantity of information displayed there should be taken into account. New questions can emerge, as to where the fuel and car temperature should appear, without the IC, or should this information always be visible.

The results showed that, during the activation moment, in terms of user experience, the participants preferred significantly the "AR concept" compared to the "IC concept." The main reasons were that they were able to clearly understand that AD was activated, that they could see the same information as the vehicle, and that the information appeared right where the participants were looking. In terms of acceptance (usefulness and satisfaction), the results showed a difference significantly favoring the "AR concept." The information that helped the participants more in this moment was the change of color, followed by the message "AD activated," and blue color, respectively, in the "AR concept." In the "IC concept" what helped the

Manual and Conditional automation (L3)

participants more was the blue color, followed by the change of color. The participants more clearly understood that the AD was activated in the "AR concept," and so we can assume that the "AR concept" enables the participant to be more aware compared to the "IC concept." The icon was noticed by less than one-fourth of the participants and the reason for this is not clear. The participants argued that in AD they felt more confident in the "AR concept" because they could see the same information as the car (lines and time gap). The logic of lines in the AR area during lane change was not evaluated, so future studies should take this into account. The design of the lines and time gap wasn't evaluated; however, a user centered design should be developed.

The results showed that, during the takeover request moment, in terms of user experience, the participants significantly preferred the "AR concept" compared to the "IC concept." The main reasons presented by the participants were that the information appeared right where they were looking, that when you are alert or hear a sound, what you do first is to look at the road and that if you look at the IC in this moment, you're losing information of the road. We can conclude that in important moments, as during this moment, the participants prefer receiving information on the road. In terms of acceptance (usefulness and satisfaction), the results showed no significant difference between concepts. Both concepts had very positive means of acceptance. The information that helped the participants more in this moment was the sound in both concepts. In "AR concept," the second information more noticed was the red color, followed by the message "take control." In the "IC concept" the second information more noticed in this moment was the message "take control," followed by the red color and the icon of SW. The participants clearly understood that the car was asking the driver to take control and the results showed no significant difference between concepts. We can assume that the most important in this moment was the sound in both concepts, where the participants already knew what they should do. In terms of timing aspects, the results showed no significant difference between concepts during "hands on reaction time" and "take over request reaction time." About the reaction types, in both concepts, most of the participants took control using the "accelerator pedal + hands on SW" during the end of traffic tested in a simulator. We can conclude that the display information in the windshield during the "take control" improved user experience; however, in terms of acceptance, awareness situation and reaction times no difference was observed between concepts. During a GB because of end of traffic, the normal behavior of the drivers to resume control was to put the hands on the steering wheel and the foot on the gas pedal.

The wide variety of situations encountered by drivers and the flexibility of displays in what and how they may be able to highlight areas of interest leave many opportunities for additional research into the effects of the AR screen on driver attention, trust and SA. The information displayed on the AR screen can have an important role in the development of the cars of the future (autonomous vehicles), and this should be studied in detail.

The results showed that during rear-end collision, in terms of reaction types, the "AR concept" showed safer behaviors compared to the "IC concept." However, in terms of reaction times, the "IC concept" showed better results. Due to the small sample of the study, we can't draw a conclusion concerning these objective measures.

In terms of future work, in the manual mode, it's important to define the information that should be displayed in the OA and the size and position of that information. With regard to availability, an interesting topic to study is the possible influence of car music when the availability advise is sounded. In this moment too, it would be good to analyze in detail why the icon is not noticed, and to test the possible influence of size, position and form. The size and content of the text should also be studied in detail. About the activation moment, the logic of lines should be tested further, such as when the AV detects an obstacle or when the AV wants change lane. The information that the driver really needs during AD should be studied too. Although the information of GB is clear, the information should be tested in a situation where after the GB an obstacle appears and whether it catches the driver's attention. During deactivation, the icon and text were noticed by only few participants, and so further investigations are needed to understand the reasons for this. Other topics that could be studied are the influence of age and other factors that hinder the driver's acceptance of receiving information from the windshield.

It's important validate these results in real road conditions, as well as in conditions where the participant remains in AD during long periods of time, when situational awareness is worst.

6.6 SUMMARY

We can conclude that the display information in the windshield (AR) improves the user experience in all the moments. The confidence is equal positively influenced by the information on the windshield, both in the manual and autonomous modes, compared to the display information in the IC. In terms of the awareness situation, the display information on the windshield helps the drivers more during the availability and activation of the function. During the GB and deactivation, the location of the information doesn't improve the situational awareness. In terms of acceptance, only during the activation of the function the information in windshield is more acceptable compared to the IC. The display information in windshield during giveback does not improve the reaction times neither does it the change the type of behavior compared to the IC. During rear-end collision, the AR concept showed higher reaction times; however, safe behaviors were more frequent in the windshield than with IC concept.

In general, we conclude that there is a clear advantage of the AR screen compared to the IC and that advantage could increase when people get used to this way of presenting information and also when the technical problems of the lines are solved. These advantages are true for subjective variables and for behavioral reactions in the rear-end collision situation (which is a limitation throughout the study and not just in this use case study). However, concerning reaction times in the rear-end collision situation, there is a disadvantage of AR that has to be further investigated with larger samples and with design changes, such as a square around the obstacle, because it can either cause confusion or make the car appear farther away, creating an evaluation in excess for the time or distance available.

REFERENCES

Boström, A., & Ramström, F. (2014). *Head-Up Display for Enhanced User Experience Department of Applied Information Technology*. (PhD Thesis: Chalmers university of Technology).

Casner, S., Hutchins, E., & Norman, D. (2016). The challenges of partially automated driving. *Communications of the ACM*, 59, 70–77.

Davis, F. D. (1989). Perceived usefulness, perceived ease of use, and user acceptance of information technology. *MIS Quarterly*, 13(3), 319–339.

Dingus, T. A., Guo, F., Lee, S., Antin, J. F., Perez, M., Buchanan-King, M., & Hankey, J. (2016). Driver crash risk factors and prevalence evaluation using naturalistic driving data. *Proceedings of the National Academy of Sciences*, 113(10), 2636–2641.

Endsley, M. R. (1995). Measurement of situation awareness in dynamic systems. *Human Factors: The Journal of the Human Factors and Ergonomics Society*, 37(1), 65–84.

European Commision. (2018). Mobility and transport - Rail market. Retrieved from http://ec.europa.eu/transport/road_safety/specialist/statistics/.

Gold, C., & Bengler, K. (2014). Influence of automated brake application on take-over situations in highly automated driving scenarios. FISITA 2014 World Automotive Congress. *Proceedings of the FISITA 2014 World Automotive Congress*. Maastricht, 02–06 June.

Gold, C., Damböck, D., Lorenz, L., & Bengler, K. (2013). "Take over!" How long does it take to get the driver back into the loop? *Proceedings of the Human Factors and Ergonomics Society Annual Meeting*. 57(1), 1938–1942

Gold, C., Körber, M., Hohenberger, C., Lechner, D., & Bengler, K. (2015). Trust in automation – Before and after the experience of take-over scenarios in a highly automated vehicle. *Procedia Manufacturing*, 3, 3025–3032.

IEEE. (2014). You won't need a driver's license by 2040. IEEE.org, online at http://sites.ieee.org/itss/2014/09/15/you-wont-need-a-drivers-license-by-2040/.

IHS iSuppli. (2013). *Automotive Head-Up Display Market Goes into High Gear*. Retrieved from http://press.ihs.com/press-release/design-supply-chain/automotive-head-up-display-market-goes-high-gear.

Kerschbaum, P., Lorenz, L., & Bengler, K. (2014). Highly automated driving with a decoupled steering wheel. *Proceedings of the Human Factors and Ergonomics Society 58th Annual Meeting*, 1, 1686–1690.

Kuehn, M., Hummel, T., & Bende, J. (2009). Benefit estimation of advanced driver assistance systems for cars derived from real-live accidents. *Proceedings of the 21st International Technical Conference of the Enhanced Safety of Vehicles Conference (EVS)*, Stuttgart, Germany, June 15–18, 1–10.

Lorenz, L., Kerschbaum, P., & Schumann, J. (2014). Designing take over scenarios for automated driving: How does augmented reality support the driver to get back into the loop? *Proceedings of the Human Factors and Ergonomics Society*, 2014 January, 1681–1685.

Medenica, Z., Kun, A. L., Paek, T., & Palinko, O. (2011). Augmented reality vs. street views. *Proceedings of the 13th International Conference on Human Computer Interaction with Mobile Devices and Services - Mobile HCI '11*, 265. Stockholm, August.

NHTSA. (2016). *Traffic Safety Facts: 2015*. U.S. Department of Transportation (August), 1–9.

Norman, D. A. (2015) The human side of automation. In: Meyer, G., Beiker, S. (eds.), *Road Vehicle Automation 2. Lecture Notes in Mobility*. Cham: Springer.

Park, J., & Park, W. (2019). Functional requirements of automotive head-up displays: A systematic review of literature from 1994 to present. *Applied Ergonomics*, 76, 130–146.

Pauzie, A., & Orfila, O. (2016). Methodologies to assess usability and safety of ADAS and automated vehicle. *IFAC-PapersOnLine*, 49(32), 72–77.

Phan, M. T., Thouvenin, I., & Frémont, V. (2016). Enhancing the driver awareness of pedestrian using augmented reality cues. *IEEE Conference on Intelligent Transportation Systems, Proceedings, ITSC*, 1298–1304. Rio Janeiro, November.

Radlmayr, J., Gold, C., Lorenz, L., Farid, M., Bengler, K. (2014). How traffic situations and non-driving related tasks affect the take-over quality in highly automated driving. *Proceedings of the Human Factors and Ergonomics Society Annual Meeting*, 58(1), 2063–2067

SAE. (2016). *Taxonomy and Definitions for Terms Related to Driving Automation Systems for On-Road Motor Vehicles*. SAE International. DOI:10.4271/j3016_201609

Schall, M., Rusch, J. L., Vecera, & Rizzo, M. (2010). Attraction without distraction: Effects of augmented reality cues on driver hazard perception. *Journal of Vision*, 10(7), 236.

Van Der Laan, J. D., Heino, A., & De Waard, D. (1997). A simple procedure for the assessment of acceptance of advanced transport telematics. *Transportation Research Part C: Emerging Technologies*, 5(1), 1–10.

Zwahlen, H. T., Adams, C. C., & DeBals, D. P. (1988). Safety aspects of CRT touch panel controls in automobiles. In: Gale, A. G., Freeman, M. H., Haslegrave, C. M., Smith, P., and Taylor, S. P. (eds.), *Vision in Vehicles II*. Amsterdam: Elsevier, 335–344.

Section 3

Usability and UX in Digital Interface, Game Design, and Digital Media

7 Interface Design and Usability Evaluation of Voice-Based User Interfaces

Martin Maguire

CONTENTS

7.1 Introduction 117
7.2 Study 1 – Survey on the Use of Voice Assistants 119
    7.2.1 Acceptance of Voice-Based Systems 119
    7.2.2 Use of Voice Systems and Purposes of Use 119
    7.2.3 Functions Considered Available but Not Used 120
    7.2.4 Usefulness of Voice-Based Assistant 121
7.3 Study 2 – Use of a Digital Assistant in a Student House 125
    7.3.1 Aim 125
    7.3.2 Method 126
    7.3.3 Results 126
7.4 Study 3 – Comparison of Experienced and Inexperienced Users of a Voice-Based Digital Home Assistant 126
    7.4.1 Aim 126
    7.4.2 Method 127
    7.4.3 Results 127
        7.4.3.1 Results from Inexperienced Users 128
        7.4.3.2 Results from Experienced Users 129
    7.4.4 Discussion 131
    7.4.5 Design Guidelines to Improve User Growth 132
7.5 Conclusion 133
References 133

7.1 INTRODUCTION

Voice user interfaces allow the user to interact with a system using natural speech. Digital assistants such as Apple Siri, Google Assistant and Amazon Alexa are often used in a home setting. Voice interaction has also become a part of many consumer products in the home such as televisions, fridges and lighting. It is also available for drivers to control vehicle systems enabling them to maintain their attention on the road.

DOI: 10.1201/9780429343513-10

There are a number of benefits in designing a system around voice interaction:

- Data entry is possible without needing a keyboard
- It is very useful for tasks when hands and eyes are busy
- It is ideal for people who find typing difficult
- It can give direct access to command structures if statements are processed as natural language
- It is a natural means of interaction as people are good at it.

Voice interaction technology has also developed to the extent that systems can now operate fairly successfully in processing human speech without the user needing to train the system to be able to understand their voice, although errors comprehending human voice input still occur.

It is interesting then that voice interaction still seems not to be used to its full potential in everyday interactions with technology. There are some well-understood reasons for this:

- People may feel self-conscious using voice technology in a public place such as on a train or in an office alongside colleagues
- It may be distracting for others if used in a quiet area
- Speech output is transient in nature, so information becomes hard to receive and remember
- Speech output for reading text is also slower than a human scanning and reading a page of text.

It has also been found that though many "skills" or voice applications have been created for voice-based assistants, only very few have received reviews, thus indicating limited use. It has been said that with voice interfaces becoming more and more common, there is a need to make them more conversational and user-centered.

There are also well-established principles for the design of speech interfaces. Some key principles are:

- *Accommodate conversational speech:* The system should speak in a natural way and adopt human-to-human speech conventions. This acts to increase the interaction flow and comprehension (Cohen et al, 2004, and Yankelovich, et al, 1995).
- *Minimize short-term memory load:* The user's short-term memory load should be kept to a minimum. In the absence of a companion screen display, listed information should be kept short and concise, containing only information necessary to the action being performed. The complexity of concepts the user must understand and the number of things they must learn to use the system must also be kept to a minimum (Damper and Gladstone, 2006, Kim, 2012).

Interface Design & Usability of Voice-Based User Interfaces

- *Provide the ability to control and interrupt:* The system should allow the user to interrupt if routed to a path they do not wish to follow (Yankelovich et al., 1995, Shneiderman and Plaisant, 2010). Example: the user can either interrupt with a new interaction or simply say "stop."

These principles have also been incorporated into a set of heuristics for speech user interactive evaluation (Maguire, 2019).

7.2 STUDY 1 – SURVEY ON THE USE OF VOICE ASSISTANTS

7.2.1 Acceptance of Voice-Based Systems

A general issue relating to voice-based systems is whether consumers will accept them and how they will be used in the future. A key question is what human factor issues are preventing their use at present. A survey of 80 adults was carried out in 2019 to understand their attitudes and experiences of using voice interfaces.

The topics covered in the survey were:

- Use of voice systems and purposes of use
- Functions considered available but not used
- Usefulness of voice-based assistant
- Problems experienced in using voice-based systems
- Expectations of use in the future.

The survey received 80 responses comprising mainly young and middle-aged adults (see Table 7.1).

7.2.2 Use of Voice Systems and Purposes of Use

When asked which voice systems participants had experience of, the results revealed that nine different systems had been used including different types of smart speaker, PC-based systems and an in-car system. Of these instances, 96% were digital assistants (e.g., Apple Siri, Google Assistant, and Amazon Echo), while 4% were more specialist devices such as in-car voice control and automated dictation. This shows how far voice-based digital assistants have come as a mainstream consumer product.

TABLE 7.1.
Survey Age Group Profile

Age group	Number responding	Percentage of sample
18–29	36	45
30–49	29	36
50–75	15	19

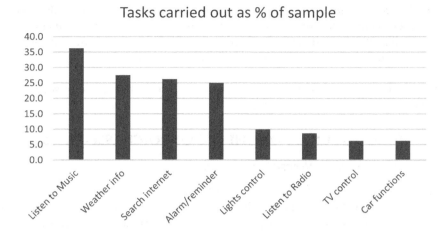

FIGURE 7.1. Frequency of tasks carried out with voice user interfaces.

Participants were asked to list the functions they had used their digital assistant for, including actions performed with other equipment linked to the assistant. As shown in Figure. 7.1, the majority of actions were straightforward tasks that required little or no setup. They included playing music (requiring linking to own streaming service), providing weather information, searching the internet or setting a timer.

Controlling lights which required connection to the external device, i.e., the light, was carried out by 10% of the sample. As the number of devices digitally connected within the home increases, the potential for control of them through the digital assistant will increase although the setup may be a barrier to non-technical users.

Other specific tasks reported included asking to memorize friends' names; asking for directions, phone numbers and train times; receiving messages when driving; controlling AV system; dictating text to be used later; turning on torch; asking to tell a joke; adding items to the shopping list; asking it to speak Cantonese; and general banter with the device. Some tasks might be less frequent because they have not been discovered by users. Over time, the range of things people will think of that they can usefully do may increase. (There were 77 respondents to this question.)

7.2.3 FUNCTIONS CONSIDERED AVAILABLE BUT NOT USED

Participants were also asked to name voice-based facilities that they were aware of, but that they had not used, to get an idea of the potential growth areas for speech. Responses included more complex tasks such as buying online, controlling compatible electronics (lights, heating, CCTV, etc.), playing games, adding events to a calendar, temperature adjustment, home monitoring and helping people with disabilities. Another suggested application area was monitoring the status of elderly relatives through sensors and communicating with them in an emergency. It was thought that voice could also provide access to third-party apps on mobile devices, e.g., messaging or conferencing.

Interface Design & Usability of Voice-Based User Interfaces 121

Some of these activities show the potential for more complex interactions but at the same time, requiring an information structure and feedback, which is perhaps easier to provide via a visual interface – hence the growth of smart speakers with screens. One person thought that the voice interaction could be a channel to give access to all the information that visual interaction had at the moment. This raises the interesting challenge of crafting the voice interface to work in a suitable way without simply matching the visual equivalent. (There were 61 respondents to this question.)

7.2.4 Usefulness of Voice-Based Assistant

Participants also gave a rating of how useful they found their use of a voice-based assistant. A spread of responses was received with around 58% regarding it as "very useful" or "quite useful" (indicating positive support but with reservations), while one-quarter of the respondents found it of little or no use (There were 71 respondents to this question.).

Comments were also elicited to support the rating given. Around 10% of the respondents felt that using speech was useful to perform tasks when the user's hands were not free such as when cooking, doing the laundry, driving or it was simply more convenient, e.g., to turn the lights off or the radio on when in bed, or to change music track while walking along or at a distance from the music player (Figure. 7.2).

While there was a clear view that using voice was quicker than typing for simple tasks such as internet search, playing music or setting a timer, it was felt that this was not very effective for more complex actions such as dictating documents or setting up a playlist. While it was possible to use voice for texting, using a group-messaging app such as WhatsApp was felt to be probably too hard. However, one view was that voice could be a good alternative to typing text, especially if the user is a non-native

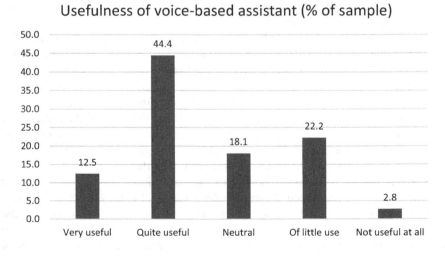

FIGURE 7.2. Rating of usefulness of having voice-based assistant.

speaker. A small number of participants stated that they preferred physical button pressing, e.g., when in a car, to speech interaction, which was felt distracting.

It was said that the real benefit of speech interaction for domestic use was when devices such as lights, heating and TV are correctly connected. This can save time and mean less use of multiple control pads or remote-control devices in the home. As one respondent said, "it needs to be better incorporated into daily life." However, it was also said that many control apps require full access to location, contacts, media, etc. "If permission is not given, then no service will be provided (it's all or nothing)."

There was a desire for voice applications to be able to interact conversationally, but it was felt that often they were not fluent enough at present with the app misunderstanding the speaker too often. It was commented that it was interesting to see children interacting with speech technology and not knowing a world without it. (There were 70 respondents to this question.)

The survey respondents were asked to describe any problems that they had experienced in using the voice-based assistant. By far, the most common problem mentioned was a misinterpretation of the users' input requiring them to repeat it (38%). Related to this, 8.5% of people stated that the system seemed to have problems with their local accent. Other problems that were mentioned by multiple respondents were that their speech system failed to receive inputs correctly due to interference of noise in the environment, e.g., background music, and the system responding when no question was asked. Linking up devices with the smart speaker was also mentioned. Devices can also drop off the network requiring the speaker to be rebooted. Knowing the correct commands to use for particular actions was a problem for a small number of participants, which is a feature of speech systems in contrast to screen-based systems, which can display commands. It was also felt not to be sufficiently accurate with speech-to-text applications especially putting punctuation at the right place. Table 7.2 provides a summary of the problems experienced.

These responses give an indication of the wide range of problems that can occur with speech systems, showing that it is not yet a failsafe means of interaction. (There were 71 respondents to this question.)

Participants were asked to indicate their opinion of whether, in the future, they think most people will be using voice-based assistants for a wide range of purposes. Figure 7.3 shows that there was a strong expectation of the widespread use of the voice-based system with nearly 80% indicating "definitely" or "probably." Only 6.4% thought that it was unlikely to happen. These opinions are likely to be based upon people's current experiences of using voice-based devices and the current level of recognition accuracy, which has improved dramatically over the past few years. As accuracy improves, expectations will also increase and perhaps the belief that voice can be an important means of communication in the future.

Considering age differences both with regard to usefulness ratings of speech interaction and the likelihood of future widespread use, Figure 7.4 shows the results of these questions for the three age groups 18–29, 30–49 and 50–69. The middle-aged group seemed to find voice-based user interfaces (VUIs) slightly less useful than the other two groups, while all three groups were optimistic that voice technology is likely to be an everyday technology that people use in the future.

TABLE 7.2.
Problems Experienced Using Voice User Interface

Category	Problem
Conversational	• Remembering not to say "please" at the end of an input • Waiting for the system to finish a long answer when an input or question was misunderstood • System not easily engaging with the user • System requires responses too quickly before the user has thought them out
Environment	• Problematic when there were multiple receivers in different rooms as the requested music may play in the wrong room • Feeling self-conscious when using the system with others are present
Accuracy and sensitivity	• Not understanding the voice of visitors • Limitation in being able to work in multiple languages at the same time • Needing to speak clearly to be understood • Not understanding a music artist when their name is in non-standard spelling
Performance/Usability	• Knowing the correct commands for particular actions • Slow response times • One system on a phone, when used in a car (in hands-free mode) needed the user to click on the screen to activate it which was not convenient
Security	• One system has a "drop in" feature where it is always enabled for that person. This can cause security issues
Cost	• Add on equipment to operate household devices is expensive

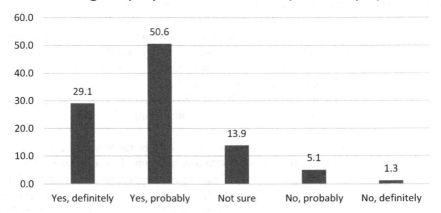

FIGURE 7.3. Expected use of voice-based assistants in the future.

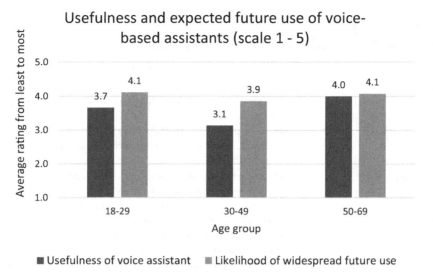

FIGURE 7.4. Comparison of age groups for usefulness of voice assistant and widespread use of voice technology.

Regarding the likeliness of future use, participants were invited to add a comment to qualify their rating. Table 7.3 shows the main themes arising from the comments. The percentage indicates the number of people in the sample who expressed a similar sentiment out of 56 who commented.

It can be seen that both positive and negative comments are listed, some more strongly positive or negative than others.

There was a general acceptance that speech technology is improving. It is becoming useful for certain tasks and will become more accepted, especially by the younger generation. Many felt that it would become integrated into our everyday lives and natural to today's children in the future. However, we may come to depend on it when it is not always the most effective way to complete tasks.

Reservations were expressed – that speech interaction will need to become secure and trustworthy and be able to cope with accents better and different voices if it is to become more accepted. Speech interaction is limited in a public or noisy environment, so other technologies such as eye-tracking, haptics and remote control through mobile devices can be the alternative. It was said that people have built up associations with certain actions being tactile such as interacting with a household device, so voice control may seem unnatural.

It was thought that voice command systems need to be more natural, smarter and responsive to the complexities of human communication. The level of intelligence that comes with voice interaction was an issue with some people who considered it to be the closest thing to a personal assistant while another view was that, while providing useful functions such as efficient internet search, they should stop "pretending to be intelligent." The barriers to using speech are having to set up connections to

Interface Design & Usability of Voice-Based User Interfaces

TABLE 7.3.
Comments Relating to the Potential of Speech Technology for Widespread Future Use

Comment category	Percentage of sample
Becoming more useful, easy and quick	21.6
The technology will keep improving	14.1
It will need to be secure and trustworthy	10.7
It will become incorporated into our lives	9
It is becoming a popular means of interaction	9
I don't have a need for it	7.1
We may become over-reliant on speech	5.3
Can foresee lots of new apps arriving	3.6
I plan to try some of the new speech apps	3.6
It is the closest to being an AI assistant	3.6
It will be natural for future generations	3.6
It needs to be more accurate	3.6
Would use it more if the setup was easier	3.6
It won't be used for everything	3.6
It is impractical in some environments	3.6

other devices and the difficulties of doing this, as well as the expense of purchasing speaker-compatible devices.

Overall, the potential of speech interaction was recognized. It was said that the options available were vast although "we are still not there" yet. Knowledge about system interaction needs to improve before the use of VUI can be fully realized. It was also noted that with the rise of IoT (Internet of Things), users will want more intuitive and convenient ways to communicate with the network rather than, or as well as, touch screens or physical buttons. This analysis was based on 60 people providing a comment.

7.3 STUDY 2 – USE OF A DIGITAL ASSISTANT IN A STUDENT HOUSE

7.3.1 AIM

In this study, a smart speaker was installed into a household to see how much it was used and how attitudes changed after use (Essom, 2018). The aim was to gain an understanding of how users might interact with current voice interaction technology, and hence a field study was conducted by installing an Amazon Echo/Alexa smart speaker in the kitchen of a student house for one week. The participants were four undergraduate students between 18 and 21 years, who were aware of voice-based assistants but stated that they had limited experience of using them. The intention was to see whether the participants used the smart speaker, for what they used it and how they felt about the technology after one week of usage.

7.3.2 Method

At the start of the week, the participants were shown how to interact with the device by using the wake-up command "Alexa" and how to issue different commands such as play music or set a timer. They could interact with Alexa during the following week as they wished. The Alexa software app was available to access the user interactions and speaker responses made during the week.

7.3.3 Results

It was found that only 27 interactions with the speaker were made which included requests to play music or the radio, set a timer (e.g., as a wake-up alarm or for cooking purposes), ask for information or a joke, download a skill or read the news headlines. Possibly more use of the device would have occurred if it had been linked into home devices such as the control of lights or ordering groceries.

Despite the limited use, participants felt that they had interacted with the device effectively. There were comments about the limited accuracy/success rate of interactions:

> Social context – feeling self-conscious when speaking to the assistant with others present and
> Trust – speculation about the microphone being "always-on" and listening in to their conversations.

These results reflect the survey conducted by Milanesi (2016), which showed that people's current use of consumer voice-based assistants may be at a basic level but as more services become reliant on voice assistants and they become integrated into homes, users will become more familiar with them and less self-conscious about using them. A general lesson from the study was that users may be slow or reluctant to use a digital assistant if they are not motivated to use it in some way.

7.4 STUDY 3 – COMPARISON OF EXPERIENCED AND INEXPERIENCED USERS OF A VOICE-BASED DIGITAL HOME ASSISTANT

7.4.1 Aim

A study was carried out to explore the current use of voice-based digital home assistants and how easily they can learn new commands and their perceptions of the system after guided learning (Walker, 2019). Four participants took part in the study. They were chosen to represent different age groups, gender and levels of prior system knowledge and usage as shown in Table 7.4.

Interface Design & Usability of Voice-Based User Interfaces

TABLE 7.4.
Participants Characteristics in Study

Participant number	Gender	Age	Experience with digital assistants
P1	Female	22	Low
P2	Female	49	Low
P3	Male	22	High
P4	Male	52	High

TABLE 7.5.
Rating Statements on a Scale of 1 (Disagree) to 7 (Agree)

Perceived usefulness	Perceived ease of use
It allows me to complete tasks quicker	It is easy to learn new commands
It increases my ability to complete tasks successfully	It is easily controllable
It allows me to be more productive	It is clear and understandable
It allows me to more effectively complete tasks	It is flexible in its use
It makes completing tasks easier	It is easy to become skillful
I find it useful	It is easy to use

7.4.2 Method

The study took place over a period of three weeks. This comprised (1) a first week to create an understanding of each participant's regular use of digital home assistants; (2) a second week where each participant was encouraged to regularly use a selection of new commands that they were shown and (3) a third week where the participants were no longer actively encouraged to use new commands but their usage was tracked to monitor the unprompted use of the newly learned commands.

Participants were asked to complete a short questionnaire in order to gauge the perceived usefulness and ease of use of their system both before and after the study to examine whether the commands learned and experienced gained affected the way the participants perceived their system. The questions used to assess existing perceived usefulness and ease of use were based on the technology acceptance model (Davis, 1989) and adapted for use with the digital assistant. The questions are given in Table 7.5.

7.4.3 Results

Figure 7.5 compares the mean, pre-study and post-study scores for the perceived usefulness and ease of use of the digital assistant to give an insight into the effect of the new commands on perceived usefulness and ease of use (Figure 7.5).

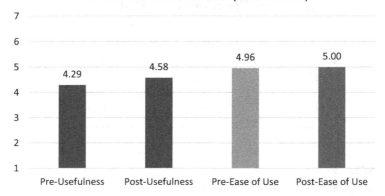

FIGURE 7.5. Comparison of participant ratings pre-study and post-study of voice-based assistant.

It can be seen average *perceived usefulness* score increased from 4.29 (pre) to 4.58 (post), an increase of *0.29* on the 6-point scale. The average *perceived ease of use* score changed from 4.96 (pre) to 5.00 (post), an increase of *0.04*. Participant's perception of usefulness increased by a modest degree after being taught new commands that they could perform with the assistant. Perception of ease of use, which was already reasonably high, only increased marginally. This indicates that being given new commands to learn was seen as helpful and did not seem to decrease ease of use.

7.4.3.1 Results from Inexperienced Users

Participant P1 was a low-usage user with typical commands being "play {song name}'" and "set volume to {volume number}." They regularly used their home assistant only to play music through a linked music account and to alter the volume. P1's changes in rating from the start to the end of the study were shown in Table 7.6.

TABLE 7.6.
Ratings of Usefulness and Ease of Use for Participant 1 before and after the Study

Criteria	Before	After	Change
Usefulness	4.5	4.83	+0.33
Ease of use	5.67	5.17	−0.05

The user admitted to making little to no attempt to learn or complete other tasks using the system. Due to this and the limited usage of the device, their perceived usefulness rating increased only slightly, which can be explained by always using

Interface Design & Usability of Voice-Based User Interfaces

the system for things that they were confident in doing and a lack of desire to learn more. The ease of use score fell slightly based on ratings of "easy to learn new skills" and "easy to become skillful" being lower than pre-study. When questioned as to the lower ease of use score, P1 stated "I had to keep looking up how to complete the task ... I couldn't find it on the app, I had to Google it."

Participant 2 was also an inexperienced user but their usage frequency was lower, only using the system a few times a week with basic commands such as "Alexa, what's the weather like?," and "Alexa, how do you spell {word}." P2 also had a more conversational style of use with questions such as "Alexa, what you up to today?" and "Alexa, what should I do today?" P2's changes in rating from the start to the end of the study are shown in Table 7.7.

TABLE 7.7.
Ratings of Usefulness and Ease of Use for Participant 2 before and after the Study

Criteria	Before	After	Change
Usefulness	4.17	4.50	+0.33
Ease of use	4.00	4.33	+0.33

Although the accompanying app provided with the system often attempted to suggest new commands for users to try, P2 felt that the sheer scope of possibilities hindered their willingness to learn. They compared it to being spoiled for choice as with their TV streaming service.

It was also observed that the questions and commands P2 used were similar to how one would speak to another person (whereas P1 gave commands more in line with that of speaking to machine). The phrases they used were much more diverse, with no single phrase being used regularly. Arguably, P2's pre-study use of the system can be described as the least informed, with the lowest understanding of the potential of the system. One explanation for the large increase of 1.6 in usefulness score could be related to the user discovering the potential of the system, rather than learning new specific commands.

The main hurdle for both P1 and P2 seemed to be discovering what was possible with the speech assistant, rather than an inability to learn new commands. Both users were restricted to commands that they either already knew and were confident in using or limited by their knowledge of the style of conversation adopted with P1 communicating with the assistant as they would with a computer and P2 communicating as they would speak to another human being. This indicated a possible limitation in the way voice systems help users grow in their use of them by not making clear what was an appropriate style to communicate with them.

7.4.3.2 Results from Experienced Users

Participant 3 was a more experienced younger user of the digital assistant. Their changes in rating from the start to the end of the study are shown in Table 7.8.

TABLE 7.8.
Ratings of Usefulness and Ease of Use for Participant 3 before and after the Study

Criteria	Before	After	Change
Usefulness	4.50	4.83	+0.33
Ease of use	5.50	5.83	+0.33

P3 was the most sophisticated user from the group as they owned multiple compatible IoT (Internet of Things) products which they had linked to their digital home assistant system. These included a digital tower fan, Phillips HUE light and Sonos sound system. P3 had invested effort both in the setup of the devices and learning the commands to use them, e.g., "OK Google, turn on the fan," "OK Google, play {song} in the kitchen," "OK Google, turn on the lights."

P3 had the highest score for pre-study usefulness and this can be attributed to a high level of personal investment in the system. The support to extend their use during the study led to a slightly higher usefulness rating. Despite their system being more complex, their ratings for ease of use were as high as P1, who had a simpler setup. After the study, P3's ease of use rating was the highest of all.

It was found that P3 pre-study usage of their system was not very conversational and did not employ the day-to-day abilities that digital home assistant systems are capable of. P3 was asked to use the system as their calendar and general personal assistant for the study, using more administrative and organizational commands such as "OK Google, what do I have planned today?," and "OK Google, can you remind me to call {name} at 6?." The aim was to encourage P3 to integrate the system into the running of their day-to-day life, not simply using it as an audio on/off button. However, it was also observed that after the study, P3 did not continue to use the new commands.

Participant 4 was a more experienced younger user of a digital home assistant. Their changes in rating from the start to the end of the study are shown in Table 7.9.

P4 used their system primarily to organize their life and conduct day-to-day activities. Examples of commands regularly used were "Alexa, call {name}," "Alexa, let {name} know I am on my way," "Alexa, what's in my calendar for this week?"

TABLE 7.9.
Ratings of Usefulness and Ease of Use for Participant 4 before and after the Study

Criteria	Before	After	Change
Usefulness	4.00	4.17	+0.17
Ease of use	4.67	4.67	0

During the study, P4 was encouraged to combine all of these daily actions into a single routine command. They were shown how to create a routine in the accompanying app and then allowed to create one which best suited them. The routine that was created and used was activated by the command: "Alexa, good morning." Using this command would trigger the digital assistant to do the following:

- Talk through the participants' calendar and to-do list for the day
- Read out the headlines from the BBC website
- Play BBC Radio 1.

P4 used this every morning for the duration of the study and then continued to use it in the week following the study, the only participant to consistently continue to use the commands given post-study.

The results of the study showed that encouragement to learn new system commands did not increase or decrease in perceived usefulness and ease of use, in any of the four participants. So either users were satisfied with their current level of system use or the means for growing with the system were not working sufficiently well.

7.4.4 Discussion

The relatively small changes in usefulness and usability scores for the digital assistant show that people's perceptions were relatively stable and were not changed much by the involvement in the study and encouragement to use new features. It could be argued that the participants in this study were close to the maximum score they could achieve in rating the voice-based digital home assistant. This could be due to users having realistic expectations pre-purchase, along with the observation that users quickly fall into a type of use; they are able to quickly learn what they find useful and then do not learn more. This raises the question of how to encourage participants to grow with the home assistant and use it in new innovative ways, and indeed whether this is desirable.

Another interesting finding was that two participants (P2 and P3) were observed to speak to their digital assistant as a machine, while the other two (P1 and P4) interacted with their systems as if speaking to a person. As digital assistants become more "intelligent," and as users become more comfortable interacting with them, perhaps those who interact with them more like people will get more benefits and enjoyment out of them than people who simply interact with them as machines. Designers of these systems may need to encourage a new type of interaction with their systems to allow people to get the most out of them. This raises the question of how "human" the digital home assistant should be and whether it could cause a feeling of awkwardness or discomfort.

As these systems develop, become more intelligent and begin to resemble what we would more classically refer to as artificial intelligence, the way users interact with them is likely to change. At that point, further research should be conducted to determine how comfortable users find a system with a certain level of intelligence. It would theoretically allow the system to better adapt to a user's needs but could cause users discomfort, a phenomenon referred to as the "uncanny valley" (Pollick, 2010). Further development in making interaction more natural is to remove the need for a wake-up command and for the device to become smarter and learn spontaneous scenarios. This could be when the user utters a prominent syllable marking the start of new information. Development in this area could be for the system to recognize when new information is being delivered and when a restart may be intended for incorrect information previously stated.

7.4.5 Design Guidelines to Improve User Growth

In Study 3, it was observed that without external influences users tend to quickly settle into a type of use that they rarely stray from, whether that be using it as an intelligent speaker (P1) or an audio on/off button for compatible devices (P3). These are habits seem to be based on what the individual sees as the most useful that requires the least effort. While P1 would potentially find using the system for things other than playing music useful, going through the process of identifying new commands and integrating them into regular use seems to be regarded as high effort for low perceived usefulness.

The perceived usefulness of a new command that justifies learning it is dictated primarily by the technology and what the market demands, which is out of the control of interface designers. However, what the interface can affect is the amount of effort that is required to acquire the knowledge for these new commands. Currently, the main way these systems encourage users to explore new commands and possibilities is through suggestions on the accompanying app, which often relate to a similar command that was recently used. For example, while P4 used their assistant to remind them to purchase a pack of batteries, when they next went on the app, they could be told that they could use their system to place their order through an online shop and have them delivered at their home.

Based on the results and observation of this study, the following design guidelines are suggested to decrease the perceived effort that a user needs to take to learn new commands, thus increasing the overall use of the system:

(1) Users requirements for the system are collected during system set up so the commands prompted fit into those requirements.
(2) Command prompts are given verbally as part of the system conversation, for example.
User: "Alexa, set an alarm for 9 am."
Assistant: "Alarm set, would you also like me to read out the news?"

By moving the prompts from something that is given after use on a different medium (the accompanying application) to using them in the relevant moment, users are more likely to use the new information.

It is concluded that encouraging users to engage in using new commands to expand their use of speech systems does not seem to be an effective way to achieve greater uptake in their use. Providing guidance to use new commands at the right time and in context could be a more useful means of user learning potential new ways to use the system and adopting them.

7.5 CONCLUSION

The studies report show that voice-based user interaction has a lot of potential for take-up in the home, and when given the chance to experience the capabilities of a digital assistant in the home, come to appreciate the potential that they offer. However, learnability barriers exist and by the nature of speech user interfaces, it is not as straightforward to explore possible functions as it is with visual interfaces. Even after encouraging users by suggesting new commands seemed to have limited effect unless they offered functions that they really felt they needed.

Setting up equipment to use with a digital assistant is a further challenge to the value of digital assistants unless users have experience with these technical aspects.

ENCOURAGING LEARNING OF VOICE-BASED FUNCTIONS

One approach that can be offered to increase the use of the system is to customize the system during setup to provide the functions that the user requires and giving the assistant the capability when acting on a command and to offer further options that the user can take up.

Juang et al. (1998) outline a number of potential future developments including prosody and spontaneity. Spontaneity could be a development to remove the need for a wake-up command and for the device to become smarter and learn spontaneous scenarios. This could be when the user utters a prominent syllable marking the start of new information. Development in this area could be for the system to recognize when new information is being delivered and when a restart may be intended for incorrect information previously stated.

REFERENCES

Cohen, M., Giangola, J., Balogh, J. (2004) *Voice user interface design*. 1st edition. Addison-Wesley, Boston, MA.

Damper, R., Gladstone, K. (2006) Experiences of usability evaluation of the IMAGINE voice-based interaction system. *International Journal of Speech Technology*, 9(1–2), 41–50.

Essom, D. (2018) *The future and effectiveness of speech based automated assistants*. Design Research report, Design School, Loughborough University, Loughborough, UK.

Kim, H.-C. (2012) An experimental study to explore usability problems of interactive voice response systems. In: Pan, J.-S., Chen, S.-M., Nguyen, N.T. (eds.), *Asian Conference on Intelligent In-formation and Database Systems (ACIIDS) Part III*, LNAI 7198, pp. 169–177, 201. Springer-Verlag, Berlin.

Maguire, M. (2019) Development of a heuristic evaluation tool for voice user interfaces. In: *21st HCI International Conference Proceedings, HCII 2019*, Orlando, FL, July 26–31, eds. Aaron Marcus, and Wentao Wang, Part IV, Design, User Experience, and Usability. Practice and Case Studies. Part of the Lecture Notes in Computer Science book series (LNCS, volume 11586), Springer.

Pollick, F. E. (2010) *In search of the uncanny valley*. Lecture Notes of the Institute for Computer Sciences, Social-Informatics and Telecommunications Engineering (LNICST), volume 40. doi: 10.1007/978-3-642-12630-7_8

Shneiderman, B., Plaisant, C. (2010) *Designing the user interface*. 1st edition. Addison-Wesley, Upper Saddle River, NJ.

Walker, S. (2019) *An investigation into interfaces of artificially intelligent systems and their design*. Design Research report, Design School, Loughborough University, Loughborough, UK.

Yankelovich, N., Levow, G., Marx, M. (1995) Designing speech acts. In: *CHI '95 Proceedings of the ACM SIGCHI Conference on Human Factors in Computing Systems*, pp. 369–376. ACM Press/Addison-Wesley Publishing Co., New York.

8 Accessibility Features in Digital Games that Provide a Better User-Experience for Deaf Players
A Proposal for Analysis Methodology

Sheisa Bittencourt, Alan Bittencourt and Regina de Oliveira Heidrich

CONTENTS

8.1	Introduction	136
8.2	Development	137
	8.2.1 Universal Design	137
	8.2.2 Usability	139
	8.2.3 Exclusion of Deaf People in Digital Games	141
	8.2.4 User-Experience	142
8.3	Methodology	143
	8.3.1 Choice of the Analyzed Games	144
8.4	Analyzing the Games	146
	8.4.1 *Marvel's Spider-Man* (2018)	147
	8.4.2 *God of War* (2018)	148
	8.4.3 *The Last of Us Part 2* (2020)	148
	8.4.4 Crossing of the Accessibility Features	149
8.5	Results Evaluation	153
8.6	Conclusion	154
References		155

8.1 INTRODUCTION

Digital games represent an important part of the entertainment industry, both in numbers of avid consumers and in terms of the high amount invested and generated by them. The interest of the industry and the academy also grows at great speed and it is possible to verify the increase of works on this theme, as explained by Westin et al. (2019). Data taken from a study by the North American company NPD Group (2020)* points out that between the years 1999 and 2009, the digital games industry grew over 400% in revenue, a growth that represents more than ten times than the one in the film industry, which has always been seen as a giant in the entertainment industry, which, however, in the same period grew only by 32%.

The interest that has formed around digital games shows that they are also increasingly solidifying as an important form of sociability, an example of this is the use of words related to digital games on the Instagram platform. The word *gamer* has been marked 36.9 million times, while the games *The Last of Us*[†] (2015) and *The Last of Us Part 2*[‡] (2020) appear together with more than 284 thousand mentions. Through the use of hashtags of these words, users can find content and other people who share their interests and thus can practice network sociability. These data were extracted from the Instagram platform itself and are an indication that it is not only the act of playing that is important to the user, but also the sharing of their experiences and perceptions about the games. Because they are frequently cited and referenced, AAA games were chosen as the object of analysis in this work. Aquino et al. (2019) state that this category, AAA, encompasses games that receive high financial and advertising investment, and consequently end up in managing the rules and future aesthetic trends of the games industry. Based on the wide research of AAA games, it is possible to assume that a deaf person who marks these games on their social networks would receive much more interactions with their publication than if they were to post about an independent game made especially for a deaf person, for example. It is important to emphasize that it is not minimizing the importance that educational games have for cognitive development, but this work is limited to games produced by large companies, due to their great reach and, consequently, to the high degree of sociability that these games can generate.

If games have been an important factor in sociability, it is also necessary to think of those who are left out of this expanding universe due to limitations that arise from disability. According to the World Report on Disability (2011), carried out by the UN, there are more than one billion people with disabilities in the world, representing at least one-eighth of the world population, and consequently, a large number of possible consumers of digital games could have been overlooked several times by the game industry. As pointed by Westin et al. (2019), several non-governmental initiatives have pressured various segments of the entertainment industry to also

---

* NPD Group is a North American company specialized in offering data related to market research. NPD Group: Market Research and Consumer Trends (2020) is available at <https://www.npd.com/>. Accessed August 13, 2020.
† *The Last of Us*. Sony Interactive Entertainment, 2015. 1 Electronic game.
‡ *The Last of Us Part 2*. Sony Interactive Entertainment, 2020. 1 Electronic game.

design their product for people with disabilities. This directly moves forward the game industry, which in recent years has sought to correct this mistake. It is increasingly possible to notice that large game producers include accessibility features as a highlight of their releases. It is possible to mention the case of the game *The Last of Us Part 2*, launched in June 2020, which has great prominence in the world press, being even pointed out by the BBC* as the most accessible digital game ever made, with more than 60 options available for accessibility.

Accessibility to digital games is established to remove barriers that cause exclusion of the target audience from the game, maintaining an equivalently challenging user-experience, regardless of whether a player has a disability or not. Therefore, the objective of this article is to analyze how accessibility features, linked to hearing, are related to the principles of Universal Design and Jakob Nielsen's usability heuristics. As a hypothesis, it is expected that this analysis will make it possible to identify strengths that the industry has developed in terms of accessibility features, related to hearing, and to identify opportunities that can be explored in this field in order to reduce the exclusion of people with disabilities in digital games. As the methodology, a comparison will be made between the accessibility features of the three games and the principles of Universal Design and Jakob Nielsen's Heuristics. It is important to note that this article does not intend to assess the efficiency of the accessibility features and, therefore, does not do tests with end users. The intention of this study is to categorize the accessibility features found in the three games analyzed using accessibility and usability principles.

8.2 DEVELOPMENT

This work intends to do an analysis of the accessibility features in digital games, linked to hearing, but for that, it is necessary to specify some concepts that are important for this research. Therefore, this section is divided into four parts: Universal Design, usability, exclusion of deaf people in digital games and user-experience (UX).

8.2.1 UNIVERSAL DESIGN

The Universal Design aims the development of projects that can be used by anyone: children, tall and shorts adults, the elderly, pregnant women, obese, people with disabilities or with reduced mobility. It applies to products or environments, physical or digital. The main concept of Universal Design is that it is not necessary to develop products for a specific group of individuals but to develop them in a way that can be used universally by anyone. According to Bassani et al. (2010), the expression Universal Design was created in 1987 by the American Ron Mace, a wheelchair architect, who needed an artificial respirator. Carletto and Cambiaghi (2008, p. 12) state that "Mace believed that this was the beginning, not of a new science or style,

---

* This BBC article about the accessibility of the game *The Last of Us Part 2* (2020) is available at https://www.bbc.com/news/technology-53093613

but the perception of the need to bring together the things we design and produce, making them usable by all people."*

The concept of Universal Design is not the most suitable for digital games, as a large part of these is not intended to be universally played by anyone. Just like movies and television series, digital games are divided into genres and rating systems of indicative age that end up not only restricting the access of younger players to violent content but making each product developed according to the wishes of their desired target audience. The *Dark Souls*[†] game franchise, for example, is known for its high degree of difficulty and for not having a choice of levels such as easy, normal and hard. Its players seek the extreme challenge, as pointed out by Nuenen (2016). So it is not of great interest for the company, creator of the franchise, to make the game easier, with the purpose of expanding the target audience of its product, not even offering different difficulty options that allow a casual player to be able to enter the franchise, because *Dark Souls* is a difficult game, for people who like difficult games.

The important thing at the moment is to point out that Universal Design was chosen for this research, not because of its conceptual alignment with the way that the games industry develops its products, but because the principles of Universal Design are important for the analysis of the accessibility features made in this research. In the 1990s, a group of architects, product designers, engineers and design researchers was formed to establish the principles of Universal Design. According to Mace et al. (1998, p. 32), "the principles could be applied to evaluate existing designs, guide the design process, and educate designers and consumers about the characteristics of more usable products and environments."

The Universal Design proposes seven principles that seek to supply the needs to organize accessibility into categories regarding possible uses, analysis and even considerations when designing a product (physical or digital) or an environment (public or private). According to Mace et al. (1998), these seven principles are:

1. **Equitable Use:** The design is useful and marketable to people with diverse abilities.
2. **Flexibility in Use:** The design accommodates a wide range of individual preferences and abilities.
3. **Simple and Intuitive Use:** Use of the design is easy to understand, regardless of the user's experience, knowledge, language skills, or current concentration level.
4. **Perceptible Information:** The design communicates necessary information effectively to the user, regardless of ambient conditions or the user's sensory abilities.
5. **Tolerance for Error:** The design minimizes hazards and the adverse consequences of accidental or unintended actions.

---

* Original text in Portuguese Brazil. Translated by the authors.
[†] *Dark Souls*. Namco Bandai Games, 2009. Electronic game.

6. **Low Physical Effort:** The design can be used efficiently and comfortably as well as with fatigue to be a minimum one.
7. **Size and Space for Approach and Use:** Appropriate size and space are provided for approach, reach, manipulation and use regardless of the user's body size, posture or mobility.

It is important to point out that according to the seven principles of Universal Design, creating a product that meets these principles does not necessarily mean developing something that will be used in the same way by everyone. Universal Design recognizes that people create a wide variety of modes of use for the same product and that is exactly what the first two principles, Equitable Use and Flexibility in Use, refer to.

When it is said that a product must be of Equitable Use for people with different capacities, it is proposing that the product should enable the same way of use for all users, be identical when possible, equivalent when not. According to Mace et al. (1998), an example is the door handles that go from the top to the bottom of the door, making it possible for people of different heights or postures to be able to open it. The Flexibility in Use refers to the ability of a product to offer a wide range of preferences regarding its use and a specific example in digital games would be the possibility to reconfigure a game controller functions according to the preferences of each player, making it possible, for example, that all important functions are on a single side of the controller, enabling use by people who have reduced mobility in either hand or even who do not have one of them. These specific cases show that the concept of universality of Universal Design is not about making each product a kind of Swiss Army Knife, with spare features and functions that will not make sense to those who do not need it, but through its flexible use and equivalence, it has the potential to become more specific to each niche of individuals.

Although there are many approaches to Universal Design, and when the gaming industry directs its efforts to offer a large range of accessibility features, it does not intend in any way to exceed the limits of its target audience, because there is a regulatory system of laws which guarantees the control of these limits, in the case of age classification.

8.2.2 Usability

With the advancement of technology, there have been discussions about ways to make digital environments friendlier and easier to use. Bassani et al. (2010) point out that to avoid problems with the use of digital tools, issues related to usability and accessibility must be foreseen from the beginning of the project. The ISO/IEC 25030:20083* defines usability as "the ability of the software product to be understood, learned, used and attractive to the user, when used under specified conditions." Bassani et al. (2010) explain that there are different proposals to evaluate the

---

* This rule provides the requirements and recommendations for specifying software product quality requirements.

usability of a digital product, but that among the different methods, the evaluation by Nielsen's Heuristics stands out.

Jakob Nielsen is the creator of the heuristics, which bear his name, and according to the information displayed on his own website,* it is possible to notice that Nielsen created several usability assessment methods, including the heuristic evaluation. According to Dias (2007), the heuristics were developed in 1990 with the collaboration of Rolf Molich, and in 1994, Nielsen refined the heuristics based on a factor analysis of 249 usability problems, resulting in a revised set of ten heuristics. According to Nielsen (1994), the following are the ten heuristics:

1. **Visibility of system status**: The system should always keep users informed about what is going on, through appropriate feedback within a reasonable time.
2. **Match between system and the real world**: The system should speak the user's language, with words, phrases, and concepts familiar to the user, rather than system-oriented terms. Follow real-world conventions, making information appear in a natural and logical order.
3. **User control and freedom**: Users often choose system functions by mistake and will need a clearly marked "emergency exit" to leave the unwanted state without having to go through an extended dialogue. Support undo and redo.
4. **Consistency and standards**: Users should not have to wonder whether different words, situations, or actions mean the same thing. Follow platform conventions.
5. **Error prevention**: Even better than good error messages is a careful design that prevents a problem from occurring in the first place. Either eliminate error-prone conditions or check for them and present users with a confirmation option before they commit to the action.
6. **Recognition rather than recall**: Minimize the user's memory load by making objects, actions and options visible. The user should not have to remember information from one part of the dialogue to another. Instructions for use of the system should be visible or easily retrievable whenever appropriate.
7. **Flexibility and efficiency of use**: Accelerators – unseen by the novice user – may often speed up the interaction for the expert user such that the system can cater to both inexperienced and experienced users. Allow users to tailor frequent actions.
8. **Aesthetic and minimalist design**: Dialogues should not contain information that is irrelevant or rarely needed. Every extra unit of information in a dialogue competes with the relevant units of information and diminishes their relative visibility.
9. **Help users recognize, diagnose and recover from errors**: Error messages should be expressed in plain language (no codes), precisely indicate the problem and constructively suggest a solution.

---

* Jakob Nielsen's website: https://www.nngroup.com/people/jakob-nielsen/

Accessibility in Digital Games for the Deaf 141

10. **Help and documentation**: Even though it is better if the system can be used without documentation, it may be necessary to provide help and documentation. Any such information should be easy to search, be focused on the user's task, list concrete steps to be carried out and not be too large.

Nielsen's ten heuristics, along with the seven principles of the Universal Design, will be the foundation for the analysis of accessibility features related to hearing in digital games.

8.2.3 Exclusion of Deaf People in Digital Games

The author Collins (2008) says that in the early days of digital games, sound insertion functioned more as a mechanical feedback response to a given action; she offers as an example *Pong** (1972), the first commercially successful digital game that consisted of a kind of digital ping pong where each player should hit a ball with a rectangular platform, throwing it to the other side of the screen. Every time that the player managed to prevent the ball from falling, a sound is played. The sound, in these types of games, had the function of being a complement to the functions that were already evident on screen; they served to reassure the player that the task had been performed as expected. The sound in this scenario was not a restriction for a deaf player, who could be fully capable of playing *Pong*, both against the machine and with other players, without impairing his performance.

Collins (2008) points out that this started to change in the early nineties, when most home computers started to be launched with a CD-ROM player and Sony, in 1995, launched the PlayStation console, which also featured a CD-ROM player for their games. This made it possible for video games to invest much more in the sound of their narratives. The exponential expansion of technology and the popularization of home computers and video game consoles meant that the old technical limitations that prevented the insertion of more complex audio in digital games were overcome, making it possible to develop sophisticated soundtracks and character's speeches, arriving at a point where there are games nowadays in which entire orchestras are able to create the immersion atmosphere as designed by the game's creators. That is the case, for example, in the game *God of War 2*† (2007); the musical team consisted of four composers, three orchestrators, three ensembles (metals, strings and choir), a variety of soloists, a development team and an implementation team (Collins 2008). This, at the time, became common among the major producers of digital games, especially among AAA games.

This technical diversity paved the way for new possibilities, both at artistic and at gameplay level. Many games have started to use sound as elements in their gameplay to achieve the success of a mission, for example. This is the case of the game *God of*

---

* *Pong* – video game released in 1972, named as the first profitable video game in history. Pong is an electronic game from Atari.
† *God of War 2*. Sony Interactive Entertainment, 2007. Electronic game.

War* (2005), where during a challenge in the desert, the player must find and battle with three enemies and although the fights with them are quite challenging, the most difficult part is finding the enemies. This search must be done through different frequencies of sound that are revealed between heavy sandstorms, and even if the enemy is eventually shown briefly, the player's sense of orientation is almost entirely through hearing only. Although there are many positive criticisms regarding the great differential of adding audio as an essential part of the gameplay, an important fact can go unnoticed in this regard: by adding audio as an obstacle to be overcome, it causes the exclusion of an audience that previously was not excluded. If the use of elaborate orchestras can serve for a more refined setting in the story, creating a challenge where the player needs to use his hearing skills to move on, can make a deaf player unable to finish the game and it is exactly in this scenario that accessibility features, linked to hearing, are needed in digital games.

Digital games companies have developed accessibility features that offer ways for deaf players to complete tasks within games using other resources besides hearing. These range from game controller vibrations, to subtitles and visual elements to identify approaching enemies, for example.

8.2.4 User-Experience

This research intends to analyze the accessibility features, linked to hearing, in digital games, so that it is possible to verify whether the experience of this deaf player is really equivalent to what other players, without hearing loss, experience. To do so, at this moment it is necessary to bring the concept of user-experience and the concept of experience itself. Bondía (2002) states that "Experience is what goes through us, what happens to us, what touches us. Not what is going on, not what is happening, or what it touches." In this way, the author establishes that the experience is only relevant if it involves the subject, in fact, that is, if it is important for what the user deems relevant. Borges et al. (2019) state that the main goal of game development is to create a product that is fun to play, presents surprises, provides a challenge to players and promotes social connections. Thus, it is up to the game company to make a game that goes through us, that happens to us and that touches us, being characterized as well as what Bondía (2002) calls experience.

Stull (2018), on the other hand, offers the definition of user-experience, which in a way, is more in line with what is investigated in this research, since the concept of experience itself can have much broader developments. Stull (2018) says that the user-experience includes aspects of cultural anthropology, human-computer interaction, engineering, journalism, psychology and graphic design, and is related to enabling the user to experience a service or product in a satisfactory way, meeting their needs and expectations generated in relation to the product. The user-experience appears as an intersection between the user's expectations, how much he expects from a finished product or service and the company's objectives for that product or service.

---

* *God of War.* Sony Interactive Entertainment, 2005. Electronic game.

Borges et al. (2019) say that user-experience, when referenced in the area of digital games, is commonly used as a synonym for player experience. When we, then, apply the concept of user-experience of Stull (2018) to the digital games, especially with regard to players with disabilities who make use of accessibility features, the user-experience refers to offering a game experience that meets the expectations of this player. But this must be done without simplifying the game or reducing the game's challenge, as this is often part of the experience designed for the user.

8.3 METHODOLOGY

As this research does not plan to apply its results to a specific product, its nature is established as basic, that is, it aims to generate new knowledge useful for the advancement of science without expected practical application. Prodanov and Freitas (2009) say that this type of research involves universal truths and interests.

The way of approaching the problem of this study is qualitative, since it is a selection made from the principle of data collection, fragmenting and extracting what is necessary and always analyzing the value of each information, but without necessarily requiring the use of statistical methods and techniques. Further on, it will be possible to view a table with the accessibility features, in which it is possible to count and list these items, but for that purpose, no specific statistical technique is necessary, which would characterize the work as quantitative.

This work has the purpose of exploratory research, providing, this way, greater familiarity with the problem and allowing the construction of hypotheses. This research adopts bibliographic and documentary procedures, and Prodanov and Freitas (2009) say that the bibliographic procedure is characterized when a work is carried out from material already published, with the objective of putting the researcher in direct contact with material already written on the researched subject. In the case of this research specifically, the search for academic articles was carried out by crossing the words accessibility, digital games and deafness, between the years 2013 and 2020. Seven academic works were found in four different journals through this research. After doing a preliminary reading in their abstracts, it was decided to choose two of them for in-depth research where it was observed how the repetition of concepts was covered and how the theoretical framework was divided. The documentary procedure concerns the use of digital games as objects of analysis. This type of research is based on materials that have not yet received an analytical treatment or that can be reworked according to the research objectives. Prodanov and Freitas (2009) quote as examples the documental procedures of the analysis of "official documents, newspaper reports, letters, contracts, diaries, films, photographs, recordings, etc." (Prodanov and Freitas 2009, p. 56). The authors do not mention digital games in their examples as they are characterized as an audiovisual product, and because of its aesthetic and technical proximity to films and recordings, it was decided that it would be characterized in terms of documentary procedure.

To analyze the accessibility features themselves, different methodological approaches were needed. After choosing the three games to analyze, it would still be necessary to define a procedure for the analysis of these games. Some alternatives

TABLE 8.1.
Stages of the Method of Analysis of Digital Games

(1) Purchase and install all three games.
(2) Play for 60 minutes each game, without activating accessibility features and with audio on.
(3) Access the accessibility features menus and capture screenshots.
(4) List the accessibility features for each game.
(5) Play each game for 60 minutes, with active accessibility features and audio turned off.
(6) Detail the resources found in Step 4, inserting the insights obtained from Step 5.
(7) Cross-check the listed resources with the principles of Universal Design and Jakob Nielsen's Heuristics.

were outlined, but in the end, it was chosen to work as follows: first, it would be necessary to purchase the chosen games. After that, it would be necessary to play at least one hour of each of these, without any accessibility feature activated. After playing them, open the accessibility features menus, one game at a time, capture the image of these screens and then make a list of all the features that are found. After identifying the available accessibility features, they were activated and three games were played for another 60 minutes. With the information on how each of these features operates, it was possible to detail the functioning and objective of each of the features. This procedure uses some approaches to what Kilpp (2010) framing method proposes, which suggests the dissection and division of moving images into groups of information by approximation or similarity, removing fragments from the video and analyzing them in a different way separately, and then, returning to the flow and observing how these elements behave in a fluid environment, but with prior knowledge of how it operates separately.

With the three lists of detailed accessibility features, a table was created where the features linked to hearing could be analyzed through a cross between the principles of Universal Design and Jakob Nielsen's Heuristics. This crossing is an adaptation of a method created by Bassani et al. (2010) to analyze accessibility and usability in distance learning teaching interfaces.

So, after the table is completed and analyzed, it is expected to find strengths that have been practiced by the game industry with regard to accessibility features, linked to hearing impairment, and to also identify possible opportunities to be explored in order to reduce the exclusion rate of deaf people in digital games. A table of this process can be checked below, with the steps of the analysis method that was developed (Table 8.1).

8.3.1 Choice of the Analyzed Games

Since it was defined that a research would be carried out to analyze the accessibility features in digital games, it was summarily defined that this research would be restricted to games launched by major digital game producers, due to the great reach that they have in the current scenario. There have been initiatives to create accessible

Accessibility in Digital Games for the Deaf   145

games made by independent companies and even by research groups at universities for a long time, and many of the features used by the game industry today are due to the pioneering nature of these initiatives that provide a wide range of materials aimed at disabled people. These high numbers can be verified in online stores, such as Google Play Games* or Steam,[†] which have hundreds of games in this line, and many of these games are free.

It is not the intention of this research to say that the big producers of digital games are not all concerned with accessibility, as there has been a general growth in recent years in games that offer accessibility options, including from big producers. Information about this can be found at the Game accessibility guidelines website,[‡] where studies carried out collaboratively between game studios, specialists and academics are exposed, which they intend to produce for the developers a direct reference in ways to avoid the exclusion of players with disabilities and ensure that more digital games can be experienced by as many people as possible. In this website, several games are mentioned and their accessibility features are commented.

In this way, this research initially chose to use the three best-selling games of the year 2020 as the object of analysis. As a source of this data, the NPD Group (2020) was used, which is a company widely cited when referring to digital games, being a research source for newspapers such as *New York Times* and *Metro UK*.

But during the selection, the choice criteria for the games would have to be a little more funneled, compared to what was initially foreseen. Instead of analyzing the three best-selling games of 2020, it was decided that only the exclusive PlayStation 4 games would be analyzed. This choice was made because it was necessary that all the analyses were using the same peripherals that other deaf users would use to play. If it was analyzed as a game that was offered for a different type of console, or a computer, for example, it would have a large number of variables to consider, such as which control the user would be using to perform the tasks, or even if he would be using a keyboard and mouse.

Once it was decided that there was a need to choose a specific console, the choice was for Sony's PlayStation 4, due to the popularity of the console that was pointed out by the NPD Group (2020), as the best-selling console of the decade. Another point that needed to change was the time period of the choice, this had to be changed from "the best-selling games of 2020" to "best-selling games of all time," because the first information was not found in our searches on the internet. So, Table 8.2 was created by the authors and built according to the list released by the NPD Group in July 2020 and published by *Metro UK*,[§] showing the best-selling PlayStation 4 exclusive games of all time, until the time of the survey (Table 8.2).

---

* Google Play Games: the store calls itself as a game manager, where you can purchase paid and free games. Available at https://play.google.com/. Accessed July 15, 2020.
† Steam is a digital game management software and website.
‡ Game accessibility guidelines website: http://gameaccessibilityguidelines.com/
§ The research about the best-selling games can be seen in the site: https://metro.co.uk/2020/07/22/marvels-spider-man-best-selling-ps4-game-ever-us-13023245/?ito=cbshare Accessed September 27, 2020.

TABLE 8.2.
The Five Best-Selling Playstation 4 Exclusive Games in the U.S. until July 2020

1. *Marvel's Spider-Man* (2018)	Release date: September 7, 2018
2. *God of War* (2018)	Release date: April 20, 2018
3. *Horizon Zero Dawn* (2017)	Release date: February 28, 2017
4. *The Last of Us Part 2* (2020)	Release date: June 19, 2020
5. *Final Fantasy 7 Remake* (2020)	Release date: March 2, 2020

Despite that, the list in Table 8.2 presents that the three best-selling exclusive games are *Marvel's Spider-Man*[*] (2018), *God of War*[†] (2018) and *Horizon Zero Dawn*[‡] (2017), and this list has become impractical, because in the second half of 2020 the third game, *Horizon Zero Dawn* (2017), was launched on computers with Microsoft Windows. Thus, this game becomes ineligible for this analysis, because of the reasons previously mentioned. In this way, the third item on the list is eliminated, the three games analyzed according to the announcement made by the *NPD Group* (2020) are *Marvel's Spider-Man* (2018), *God of War* (2018) and *The Last of Us Part 2* (2020).

8.4 ANALYZING THE GAMES

As described in the methodology chapter, to analyze the *Marvel's Spider-Man* (2018), *God of War* (2018) and *The Last of Us Part 2* (2020) games, the games were purchased at first, then downloaded and finally installed. It was stipulated that each game would be played for one hour, without using any accessibility feature and with the audio on. The time of 60 minutes was chosen because it is noticed that this would be the minimum time for checking the accessibility features, since the first minutes are used mainly for tutorials on game mechanics and it would not be possible to accurately analyze the features during these initial moments.

After that time, image captures were made from the accessibility features menus and with these images it was possible to create lists of all the features offered. With this list in hand, the three games were played again, this time with the accessibility features, related to hearing impairment, enabled and the audio turned off. After another 60 minutes of play, it was possible, with these new insights, to create a more detailed list that effectively describes what it serves and how each accessibility feature works effectively. Table 8.3 presents the list of features found in the three games.

Next, all the features found were detailed and how the cross-check was made with the principles of Universal Design and Jakob Nielsen's Heuristics was explained

---

[*] *Marvel's Spider-Man*. Sony Interactive Entertainment, 2018. Electronic game.
[†] *God of War*. Sony Interactive Entertainment, 2018. Electronic game.
[‡] *Horizon Zero Dawn*. Sony Interactive Entertainment, 2018. Electronic game.

Accessibility in Digital Games for the Deaf

TABLE 8.3.
List of Features Found in the Three Games Analyzed

Marvel's Spider-Man (2018)	*God of War* (2018)	*The Last of Us Part 2* (2020)
1. Subtitle	1. Subtitle	1. Subtitle
2. Subtitle background	2. Subtitle name	2. Subtitle name
3. Subtitle size	3. Subtitle background	3. Subtitle direction
4. Combat camera	4. Subtitle size	4. Subtitle background
5. Attack alert	5. Combat camera	5. Subtitle size
6. Vibration	6. Enemy indicator	6. Name color
	7. Vibration	7. Subtitle color
		8. Danger indicator
		9. Damage indicator
		10. Notification of collected item
		11. Frequent dodge alert
		12. Impact marker
		13. Combat vibration alert
		14. Guitar vibration alert

8.4.1 *Marvel's Spider-Man* (2018)

When checking the options menu of the *Marvel's Spider-Man* (2018) game, three accessibility features were found, which are subtitles for dialogues and the possibility of placing a background on it or increasing its size.

Three other features were found in the game, which help to eliminate barriers for deaf people, but these are always active by default and they are as follows: camera positioning assistant, visual alert of enemy attack and a diverse range of game controller vibrations as a return for specific actions in the game.

The features found are explained as follows:

1. **Subtitle:** enables subtitles for character dialogs in most of the game. When enabled, it also shows the name of the character who is speaking next to his caption and defines different colors for different characters, to facilitate differentiation.
2. **Subtitle background:** a background is added behind the subtitles to increase the contrast between the elements and facilitate, in that way, the reading.
3. **Subtitle size:** increases the size of the subtitles that will be displayed on the screen.
4. **Combat camera:** constantly adjusts the camera's position in the game to show nearby enemies. Thus, the player does not need to rely solely on his hearing to identify nearby enemies. This feature is active by default for all players.
5. **Attack alert:** named in the game as "Spider-Sense," this feature visually alerts the player when an attack is about to be launched in the direction of the player's character. This feature is active by default for all players.

6. **Vibration:** several actions in the game, in and out of combat, are confirmed by the vibration of the game controller. These often occur together with sounds and visual cues.

8.4.2 GOD OF WAR (2018)

When browsing through the options menus of the game *God of War* (2018), four accessibility features were found with the purpose of eliminating barriers for deaf people, which are subtitles and the possibility of adding names to it, background or increasing their size.

Three other features were found in the game, which help to eliminate barriers for deaf people, but these are always active by default, in the same way as in the *Marvel's Spider-Man* (2018) game.

The features found are explained below:

1. **Subtitle:** enables subtitles for character dialogs in most of the game.
2. **Subtitle background:** a background is added behind the captions.
3. **Subtitle size:** makes it possible to increase the subtitle size.
4. **Subtitle name:** shows the name of the character who is speaking next to the subtitles.
5. **Combat camera:** automatically adjusts the position of the camera in the game to show enemies that are attacking. This feature is active by default for all players.
6. **Enemy indicator:** this feature visually shows, through arrows, in which direction the enemies are. But only the enemies that are not being seen on the screen. Also exhibits different colors of arrows to differentiate enemies between who are carrying out physical attacks and who attack with projectiles. This feature is active by default for all players.
7. **Vibration:** several attacks, actions and interactions with objects are accompanied by vibrations of the control, facilitating their identification. These often occur together with sounds and visual cues.

8.4.3 THE LAST OF US PART 2 (2020)

When checking all the options on the menus of the game *The Last of Us Part 2* (2020), 13 accessibility features were found.

Another feature was found in the game, which helps to eliminate barriers for deaf people, but it is always active by default, which are indicators of damage.

The features found are explained below:

1. **Subtitle:** enables subtitles for character dialogs in and out of combat.
2. **Subtitle name:** shows the name of the character who is speaking next to the subtitles.
3. **Subtitle direction:** an arrow is visible next to the subtitles to show the direction of the speaking character.

Accessibility in Digital Games for the Deaf   149

4. **Subtitle background:** a background is added behind the subtitles.
5. **Subtitle size:** makes it possible to increase or decrease the subtitle size.
6. **Name color:** brings several color options for the character's names displayed next to the subtitles.
7. **Subtitle color:** allows the user to change the colors of the subtitles text.
8. **Danger indicator:** visual indicators appear on the screen warning that an enemy is about to see the player's position and in which direction the enemy is. When this option is off, the danger and direction indicators are transmitted to the player via audio only.
9. **Damage indicator:** visual indicators appear on the screen showing in which direction the enemy attacked and caused damage to the player's character.
10. **Notification of collected item:** visual notifications appear on the screen when collecting an item. When this option is off, only one sound is heard when collecting an item.
11. **Frequent dodge alert**: the dodge button appears on the screen whenever an enemy initiates a body attack. Players without hearing impairment can identify the enemy's attack by the increase in sound it makes when approaching the player's character.
12. **Impact marker**: the visual marker that is in the middle of the screen, and is used as a crosshair to fire firearms, changes the color to red, for a short period, to show that an enemy was defeated with one shot.
13. **Vibration alert in combat:** enables game controller vibration alerts (1) when an enemy initiates a body attack, (2) when the character is aiming at an enemy and (3) when a shot hits an enemy. When this option is turned off, part of this information is only transmitted to the player by sound. An example of this is when the player hits a shot at the enemy, and when the enemy screams, the sound of screams makes the player to understand that enemy is hit.
14. **Guitar vibration alert:** game controller vibrations occur when hitting a guitar note.

8.4.4 Crossing of the Accessibility Features

Step 7 of the game analysis procedure consists of crossing the accessibility features found in the three games analyzed, the principles of Universal Design and Jakob Nielsen's Heuristics. As can be seen in Table 8.4, the principles of Universal Design are represented in each of the columns in the table, and the Jabok Nielsen's Heuristics in each of the lines. The 27 accessibility features identified were placed at the cross between these concepts, so that it is possible to identify patterns and opportunities between the features related to accessibility for deaf people. In order to see the results at once in the same table, acronyms of the games are placed in parentheses next to each of the accessibility features: **SM** for *Marvel's Spider-Man* (2018), **GW** for *God of War* (2018) and **LU** for *The Last of Us Part 2* (2020) (Table 8.4).

TABLE 8.4.
Crossing of the Accessibility Features in the Games

Universal Design – Columns Nielsen's Heuristics – Lines	Equitable Use (18)	Flexibility in Use (8)	Simple and Intuitive Use (4)	Perceptible Information (22)	Tolerance for Error (0)	Low Physical Effort (9)	Size and Space for Approach and Use (0)
Visibility of the current state of the system (14)	(SM, GW e LU) Subtitle (GW e LU) Subtitle Name (LU) Subtitle direction (SM e GW) Combat camera (GW) Enemy indicator (LU) Danger indicator (LU) Damage indicator	(LU) Name color (LU) Subtitle color	(SM e GW) Combat camera (SM) Attack alert	(GW e LU) Subtitle Name (LU) Subtitle direction (LU) Name color (LU) Subtitle color (SM) Attack alert (GW) Enemy indicator (LU) Danger indicator (LU) Damage indicator	—	(LU) Subtitle direction (GW) Enemy indicator (LU) Danger indicator (LU) Damage indicator	—
Correlation between the system and the real world (4)	(SM e GW) Vibration (LU) Combat vibration alert (LU) Guitar vibration alert	—	—	(SM e GW) Vibration (LU) Combat vibration alert (LU) Guitar vibration alert	—	(SM e GW) Vibration (LU) Combat vibration alert	—
User control and freedom (0)	—	—	—	—	—	—	—
Consistency and standards (0)	—	—	—	—	—	—	—

(Continued)

Accessibility in Digital Games for the Deaf

TABLE 8.4. (CONTINUED)
Crossing of the Accessibility Features in the Games

Universal Design – Columns Nielsen's Heuristics – Lines	Equitable Use (18)	Flexibility in Use (8)	Simple and Intuitive Use (4)	Perceptible Information (22)	Tolerance for Error (0)	Low Physical Effort (9)	Size and Space for Approach and Use (0)
Error prevention (17)	(SM e GW) Combat camera (GW) Enemy indicator (LU) Danger indicator (SM e GW) Vibration (LU) Notification of collected item (LU) Frequent dodge alert (LU) Combat vibration alert (LU) Impact marker	(SM, GW e LU) Subtitle background (SM, GW e LU) Subtitle size	(SM e GW) Combat camera (SM) Attack alert (LU) Frequent dodge alert	(SM, GW e LU) Subtitle background (SM, GW e LU) Subtitle size (SM) Attack alert (GW) Enemy indicator (LU) Danger indicator (SM e GW) Vibration (LU) Notification of collected item (LU) Frequent dodge alert (LU) Combat vibration alert (LU) Impact marker	–	(GW) Enemy indicator (LU) Danger indicator (SM e GW) Vibration (LU) Notification of collected item (LU) Combat vibration alert (LU) Impact marker	–
Recognition instead of memorization (1)	(LU) Frequent dodge alert	–	(LU) Frequent dodge alert	(LU) Frequent dodge alert	–	–	–
Flexibility and efficiency of use (0)	–	–	–	–	–	–	–

(Continued)

151

TABLE 8.4. (CONTINUED)
Crossing of the Accessibility Features in the Games

Universal Design – Columns Nielsen's Heuristics – Lines	Equitable Use (18)	Flexibility in Use (8)	Simple and Intuitive Use (4)	Perceptible Information (22)	Tolerance for Error (0)	Low Physical Effort (9)	Size and Space for Approach and Use (0)
Aesthetic and minimalist design (0)	–	–	–	–	–	–	–
Support for users in collecting, diagnosing and recovering errors (0)	–	–	–	–	–	–	–
Help and documentation information (0)	–	–	–	–	–	–	–

8.5 RESULTS EVALUATION

When analyzing the completed Table 8.4, it is noticeable that some rows and columns have more accessibility features than others, and some do not have at least one item in their extension. It is possible to realize that when it comes to Universal Design, the principle of *Equitable Use* is the one that has received the most attention from AAA games. Of the 27 identified features, 18 of them are intended to match the dynamics and mechanics based on sound in the analyzed games. These resources serve to bring more than one sensory dimension (sight, hearing and touch) to information that would often only be transmitted by one of these dimensions.

Still talking about Universal Design, it is possible to realize that the principle of *Perceptible Information* is also widely used with regard to accessibility features, being identified in 22 of the 27 features found. This happens when there is important information to be transmitted during the game. In these moments, this information is transmitted in multiple ways, such as being passed through image, sound and vibration, from the accessibility feature. This fits perfectly with one of the guidelines of this principle of Universal Design, guideline number 4a, which according to Mace et al. (1998, p. 34), says "Use different modes (pictorial, verbal, tactile) for redundant presentation of essential information."

Another principle of Universal Design that was repeated in several features was *Low physical effort*, with an impact on 9 of the 27 accessibility features. These features mean that the player does not need to move the character excessively or rotate the camera constantly, just to check information, which would be passed to the player only by audio, if these features were not activated. This, in a way, minimizes the fatigue that the player could experience, if the images on his screen were constantly moving and rotating.

The principle of *Flexibility in Use* was related to 8 of the 27 resources, and all of them are related to making subtitles more flexible. In a general way, 14 of the 27 resources, more than 50% of the features are linked to subtitles. This shows that the use of subtitles receives a lot of attention from game developers. Emilia Schatz and Matthew Gallant, the developers of *The Last of Us Part 2* (2020), confirm this by showing that among the 12 most used accessibility features in the game, 7 of them are related to subtitles (#Gaconf 2020).

As for Jakob Nielsen's Heuristics, one of the most recurring was *Visibility of the current state of the system*, being found 14 times in the 27 accessibility features analyzed. In general, these features were used with the function of visually showing what is happening with the player's character and around him. Part of this information would be transmitted only by audio to the player, but with the feature activated, they are transmitted in a visual and sonorous way.

Still on Jakob Nielsen's Heuristics, it is possible to note that *Prevention of errors* is also found with great recurrence, in 17 of the 27 cases found. It is noted that the features related to this item are mainly related to danger alerts and confirmations of essential actions in the game that are transmitted mainly by audio to the player when these features are not activated.

8.6 CONCLUSION

The hypothesis of this work was that a detailed analysis of the accessibility features and the position in which they appear in the crossing table would serve to identify strengths that the industry has developed so far in terms of accessibility features. It was also expected to identify the opportunities to be explored in this field, in order to find alternatives to provide a better user-experience for deaf players. And so, in that way, also identify ways in order to decrease the exclusion rate in digital games Thus, in general, it can be said that the initial hypothesis was confirmed, and by filling in the table it was possible to see empty areas that show opportunities for the game industry in terms of accessibility and usability for deaf players.

It is possible to, for example, mention that a strong point developed by the game industry is the features that refer to *Perceptible Information*, according to the principle of Universal Design. When information is important for the game to progress, it appears in different ways, with sound, messages on the screen and through the tactile vibration feature of the game controller. There is great concern in the industry to make the gaming experience comparable at the most diverse levels, and this is already related to another principle of Universal Design, which is the *Equitable Use*. Important information is passed on in different ways to the user, and redundantly, to ensure its transmission to the player and to contemplate different cognitive abilities.

On the other hand, there is a gap in the table, which can be interpreted as an opportunity to be explored, would be the lack of resources that refer to *Flexible Use*, appearing only in 8, of the 27 features. At the #GAconf (2020),* international conference on the accessibility of digital games, during one of the panels, disabled players unanimously said that they prefer to have access to a list with the greatest number of accessibility features possible, as this makes it possible to activate features that meet their accessibility needs. This claim places the *Flexible Use*, a major priority for players with disabilities and yet resources with this characteristic do not receive as much attention from the industry when compared to other principles of Universal Design. The situation becomes even more problematic when it is realized that all eight *Flexible Use* features mentioned are related to subtitles, that is, of all the resources that the deaf player can access, only the subtitles can be changed more flexibly according to their own needs. A scenario where the deaf player has the possibility of greater flexibility in accessibility features, in terms of size, frequency, contrast, colors, intensity, among others, seems to be an alternative that could solve many problems of accessibility and usability.

As future research proposals, this project intends to study in greater depth the empty spaces in this intersection table through agile design methodologies, such as design thinking and design sprint. Further, with the partnership of deaf players, it is planned to find new ways to offer the players a better user-experience through solutions that can fill the existing gaps related to accessibility features in order to reduce the exclusion of deaf players.

---

* #GACONF: Game Accessibility Conference, 6, 2020, Online. Available at <https://www.gaconf.com/gaconf-online-2020/>. Accessed September 19, 2020.

REFERENCES

Aquino, A. C. G., Obregon, R. F. A. and Couto, H. D. 2019. Reflexões acerca do realismo e da representação visual em games tendências de mercado e jogos AAA. In: *A produção do conhecimento na engenharia da computação*, ed. Martins, E. R., 324–327. Ponta Grossa: Atena Editora.

Bassani, P. B. S., Behar, P. A., Bittencourt, A. S., Heidrich, R. O. and Ortiz, E. 2010. Usabilidade e acessibilidade no desenvolvimento de interfaces para ambientes de educação à distância. *Renole*, v. 8 n. 1, 1–10. <https://seer.ufrgs.br/renote/article/view/15180/8947> (Accessed July 10, 2020).

Bondía, J. L. 2002. Notas sobre a experiência e o saber de experiência. *Revista Brasileira de Educação*, n. 19(April), 20–28. <http://www.scielo.br/scielo.php?script=sci_arttext&pid=S1413-24782002000100003&lng=pt&nrm=iso> (Accessed September 24, 2020).

Borges, B., Sidarta, I., De Souza, A. M., Coelho, B. and Darin, T. 2019. Experiência do Usuário em Jogos Digitais: Uma Catalogação de Instrumentos de Avaliação. Paper presented at the *Workshop about interaction and research of users in the development of games*, Vitória.

Carletto, A. C. and Cambiaghi, S. 2008. *Desenho universal: um conceito para todos*. São Paulo: Mara Gabrili.

Collins, Karen. 2008. *Game Sound: An Introduction to the History, Theory, and Practice of Video Game Music and Sound Design*. Cambridge, Massachusetts: The MIT Press.

Dias, C. 2007. *Usabilidade na web: criando portais mais acessíveis*. 2ª ed. Rio de Janeiro: Alta Books.

Kilpp, Suzana. 2010. *A traição das imagens: espelhos, câmeras e imagens especulares em reality shows*. Porto Alegre: Entremeios.

Mace, R. L., Mueller, J. L. and Story, M. F. 1998. *The universal design file: designing for people of all ages and abilities*. <https://projects.ncsu.edu/ncsu/design/cud/pubs_p/pudfiletoc.htm> (Accessed July 10, 2020).

Nielsen, J. 1994. Heuristic evaluation. In: *Usability inspection methods*, eds. Nielsen, J., and Mack, R. L. New York: John Wiley & Sons, 25–64.

Nuenen, T. V. 2016. Playing the panopticon: Procedural surveillance in Dark Souls. *Games and Culture*, v. 11, n. 5, 510–527. <https://www.researchgate.net/publication/273495440_Playing_the_Panopticon_Procedural_Surveillance_in_Dark_Souls> (Accessed November 10, 2020).

Prodanov, C. C. and Freitas, E. C. 2009. *Metodologia do Trabalho Científico: Métodos e Técnicas da Pesquisa e do Trabalho Acadêmico*. Novo Hamburgo: Feevale.

Stull, Edward. 2018. *UX Fundamentals for Non-UX Professionals: User Experience Principles for Managers, Writers, Designers, and Developers*. Berkeley: Apress.

Westin, T., Brusk, J. and Engström, H. 2019. Activities to support sustainable inclusive game design processes. *EAI Endorsed Transactions on Creative Technologies*, v. 6, n. 20 (July), 1–9. <https://eudl.eu/pdf/10.4108/eai.30-7-2019.162948> (Accessed November 10, 2020).

World Health Organization. 2011. World report on disability. <https://www.who.int/disabilities/world_report/2011/report/en/>. (Accessed July 23, 2020).

9 Game On
Using Virtual Reality to Explore the User-Experience in Sports Media

Ragan Wilson, Nina Ferreri and Christopher B. Mayhorn

CONTENTS

9.1 Examining the Sports Media Experience through Usability within a Hedonic System 159
9.2 Theories of Arousal and Affect Underlying User-Experiences with Hedonic Systems 159
9.3 Technology Manipulations to Enhance Arousal 160
9.4 Understanding the Impact of User Characteristics 161
9.5 Viewing Context Matters 162
9.6 Expanding Sports Media Viewing Beyond 2D: The Emergence of Virtual Reality 163
9.7 Pilot Study: Sports Viewing in HMD within a Virtual Environment Is Different from 2D 164
9.8 Conclusion 166
References 167

On March 5th, 1733, a written article detailing a prizefight in England was published in the *Boston Gazette*, and became one of the earliest examples of sports media. It was imported from Europe, heavily delayed in its' distribution to the well-educated and wealthy colonials in North America (Bryant and Holt 2006). For the next one hundred years, media coverage of sports would remain a pastime reserved solely for the wealthy until newspapers began targeting urban, middle-class audiences. Given the much more inclusive modern landscape of television, writing and radio, it is difficult to conceptualize how early sports media could have ever been so delayed or restricted to such small audiences. However, the history of sports media is a history of innovation to reach new audiences (Bryant and Holt 2006).

Sports media coverage most notably began with newspaper columns by sports journalists in the 1830s (Bryant and Holt 2006), which specifically started to cover sports like horse racing. With the invention of radio technology, fans began listening to sporting events as they occurred in real time, often with editorialized "play-by-plays" from an announcer. Such innovation in event coverage introduced a social element when fans gathered in their homes around the family radio set to listen to events (Bittner and Bittner 1977). Later, technological innovation replaced the radio with television as the most popular media type for experiencing sporting events. Television combined the audio experience of radio, but now people could see what was happening on the screen (Bryant and Holt 2006). With that screen, fans could now see the action live, with replays and close-ups, without ever leaving their living rooms.

In much the same manner that radio and television revolutionized how we experience sporting events, the Internet changed it even further. With readily available information at their fingertips, sports fans could suddenly find information about their favorite teams and view clips of events involving that team on several digital media outlets (Bryant and Holt 2006). The advent of online streaming services like Hulu + Live TV™ (Hulu© n.d) and ESPN+™ (ESPN© n.d) further encouraged sports viewers to subscribe to their services partially for information about their favorite sports and sports teams. Professional sports organizations have developed their own platforms for sharing information such as the Major League Baseball (MLB) At Bat™ application (MLB© 2019) and the National Hockey League (NHL)® application (BAMTECH LLC 2019; Newman 2017). These streaming services not only imitate what television broadcasts do but are able to expand the functionality beyond traditional television by giving the user more information about their sports team at the press of a button (Hulu n.d; ESPN n.d; Newman 2017). While the age of streaming sports provides information access via monitors, televisions or mobile devices, streamed sports remains limited to 2D presentation. With the recent arrival of virtual reality (VR) technology, sports fans can now experience an even more-realistic, 3D viewing of events via head-mounted displays (HMDs).

While not currently in widespread use, HMD sports media coverage is expanding with companies such as NextVR© (NextVR, 2018). These companies use specialized, custom-created equipment (Brown 2019) to record sporting events typically in a format that combines 180-degree sports footage with another 180 degrees of computerized backdrops (NextVR, 2018). This creates a 360-degree experience that can be displayed on HMDs such as the Oculus Rift™ (Facebook Technologies 2016) and HTC Vive™ (HTC Corporation 2016) such as what NextVR produces (NextVR 2018). As they pursue the development of an immersive experience for users within sports entertainment, NextVR has partnered with sports leagues and broadcasters alike (e.g., Fox Sports, NBA, NHL) to provide an HMD viewing experience via their headset-mobile application (NBA n.d; NextVR 2018; NHL Public Relations 2019) allowing users to experience being close to the action in the headset, while remaining at home.

With sports media platforms evolving at such a rapid rate, user-experience research needs to catch up and examine not only whether there are differences between watching sports using different technologies, such as on a 2D television screen vs. an HMD, but also whether or not those differences are interpreted as good or bad. A related issue is whether these sports experiences offered via new technology enhances user satisfaction and enjoyment. To address these empirical questions, the remainder of the chapter is organized to summarize previous research that resulted in the classification of sports media as a hedonic system that influences user arousal, analysis of user characteristics and then a consideration of viewing context. Using this approach, we later describe results from a pilot study that should inform our conclusions regarding the future of user-experience testing within the domain of technology-facilitated sports media viewing.

9.1 EXAMINING THE SPORTS MEDIA EXPERIENCE THROUGH USABILITY WITHIN A HEDONIC SYSTEM

Even though the sports media experience occurs within a highly specialized context, the foundation of user-experience and usability at its most basic level involves the interaction between a user and the technology. The user operates this technology to display the content that they want, which in this case is a sports game. The combination of previously examined general usability and user-experience theory with sports media viewing theories can provide more insight into these relationships between the user, the technology and the content that the technology is displaying, thus providing a base for more specific investigations of these highly specialized situations, as detailed in the next sections.

By its very nature, sports media viewing typically involves a hedonic entertainment source. This understanding is notable given differences between hedonic (pleasure-focused) and utilitarian (productivity-focused) user acceptance models where hedonic systems differ in the design and end goals from utilitarian-focused systems (Van der Heijden 2004). Designers of hedonic systems often prioritize the user's experiences of pleasure by making design decisions such as the addition of colored displays that encourage prolonged use, often the end goal. By contrast, utilitarian systems prioritize design decisions that minimize potential distractions to users and emphasize efficiency in leading the user towards their task goal. These systems are focused on the end goal of getting a user to complete a task in as efficient and timely a manner as possible. Thus, results from the assessment of more hedonic systems demonstrate that the behavioral intention to use such systems is strongly related to perceived enjoyment and perceived ease of use, whereas perceived enjoyment was less important to utilitarian systems (Van der Heijden 2004).

9.2 THEORIES OF AROUSAL AND AFFECT UNDERLYING USER-EXPERIENCES WITH HEDONIC SYSTEMS

Because hedonic systems are designed to elicit pleasure and enjoyment, an understanding of how emotional arousal is influenced by the choice of entertainment media is needed. According to the excitation transfer theory developed by Zillmann (2008), arousal is derived from the physiological activation of the sympathetic nervous system by exposure to stimuli from different sources that combine to intensify a user-experience. Within this theoretical framework, exposure to stimuli from one source (e.g., a game viewed on the television) results in residual excitation that can be transferred to another source (e.g., switching the channel to another game). Thus, successive emotional changes might be intensified by previous exposures. Consider a sports viewer who switches channels from a hedonically negative event where she was disappointed in the poor performance of a favored team to another game where the competition is heightened. Residual excitation from the first game might intensify the subsequent euphoric experience of viewing a highly competitive match where the outcome is less certain. In effect, the user might experience intense enjoyment due to the suspenseful conclusion of the second game.

A related concept is that the enjoyment of sports entertainment is derived from affective disposition theory (Zillmann and Cantor 1976; Raney 2017) where users make judgments about their dispositions towards characters (i.e., players or teams). This affects the user's emotional reaction to the characters' actions and struggles, which then affects the user's enjoyment. If this theory (Zillmann and Cantor 1976; Raney 2017) is applied to a sports context, users should experience heightened enjoyment when "liked" individuals or teams experience positive outcomes, such as a win, or when "disliked" individuals or teams experience a negative outcome, such as a defeat. There is a sports-specific, theoretical application of this concept known as the disposition theory of sports spectatorship (see Zillmann, Bryant, and Sapolsky 1989; Peterson and Raney, 2008) where viewers develop affiliation to a particular sports team that can be from extremely positive to extremely negative on a continuum. That then impacts the viewer's experience and enjoyment of the content. In effect, sports spectator enjoyment is a byproduct of the emotional valence and outcome. As will be discussed below, this emotional valence might be further related to fanship, a user characteristic associated with enjoyment. Users high in fanship or fans in this case being users who display intense interest and emotional investment into certain activities, like watching sports (Cummins 2009).

9.3 TECHNOLOGY MANIPULATIONS TO ENHANCE AROUSAL

Although the primary emphasis of the current work will address how emerging technologies might enhance sports viewing, it is apparent that television continues to be a popular technology used to watch sports (Kim, Cheong, and Kim 2016; Peterson and Raney, 2008). Thus, it is no surprise that much of the controlled, laboratory-based research on sports media has focused on viewers' reactions to sports games displayed on television screens. These efforts included research such as studies by Gan, Tuggle, Mitrook, Coussement, and Zillmann (1997) and Peterson and Raney (2008) that have explored suspense and enjoyment as part of the sports user-experience with events on 2D screens. Some of these previous research used methods reminiscent of traditional usability and user-experience studies, most notably the works of Cummins (2009), and Cummins, Keene, and Nutting (2012).

In the first of two related studies, Cummins (2009) focused on the user-experiences of presence defined as a feeling of "being there," emotional arousal and enjoyment as experienced via subjective camera angles. He also examined how user characteristics, such as fanship, could impact that user-experience. This study was a 2 (camera angle) × 3 (sports fanship) × 3 (viewing order) study with both within-subjects (camera angle) and between subjects (sports fanship and viewing order) variables. Subjective camera angle focused on the first-person view of the action from the athlete's perspective, such as focusing on what the driver sees while driving in NASCAR (Cummins 2009; Narducci 2007) while an objective camera angle described a more third-person, sideline perspective akin to traditional camera angles of football (Cummins 2009; Cummins, Keene, and Nutting, 2012). Participants viewed 16 plays that were split by camera angle into two separate groups, subjective camera and objective camera. The videos were paused in between plays so that measures of sports fanship, presence and enjoyment could be collected. The technology-focused results of this study indicated that the use of an overhead subjective camera enhanced both spatial presence and engagement for the users. Other results involving fanship and gender along with camera angle and presence variables will be described in Section 9.4.

In a follow-up study, Cummins, Keene, and Nutting (2012) re-visited many of the same variables by examining differences in presence, arousal and enjoyment by the camera angle. They also examined if clips with more excitement would lead to greater presence and arousal than clips with less excitement. Finally, they examined a potential interaction between camera angle and perceived excitement by assessing whether exciting clips played using subjective camera angles would demonstrate more arousal and excitement than clips played with objective angles. This study used a 2 (camera angle) × 2 (potential for excitement of clip) × 4 (viewing order) experimental design using 14 clips taken from a televised football game between Florida State University and the University of Miami. As with the previous study, stimuli video clips were classified into two groups based on their camera angle as either subjective or objective. Arousal was measured using skin conductance throughout the plays, with the participants completing additional self-report measures in between plays to assess presence and enjoyment. Results indicated that camera angle was related to the reported presence and skin conductance as a physiological measure of arousal. Clip excitement did elicit more subjective presence and arousal along with larger, although not more frequent, measured skin conductance arousal. An interaction effect revealed that less exciting clips presented with a subjective camera angle were self-rated as more arousing whereas more exciting clips presented with a subjective camera angle were self-rated as less arousing.

9.4 UNDERSTANDING THE IMPACT OF USER CHARACTERISTICS

Although technology-based manipulations can influence the enjoyment of sports media as demonstrated above, user characteristics such as gender must also be considered. Previous research indicates that gender can influence viewing preferences and motivations for fanship, as well as be a predictor for overall enjoyment of sports-watching experiences (Dietz-Uhler, Harrick, End, and Jacquemotte 2000; Gan et al. 1997; Peterson and Raney 2008; Sargent, Zillmann, and Weaver III 1998). With regard to spectating preferences, men appear to be fonder of confrontational types of sports such as ice hockey, football and basketball while women tend to gravitate towards competitive but non-aggressive sports such as gymnastics, figure skating and diving (Sargent, Zillmann, and Weaver III 1998). Additionally, potential gender differences in motivations for being a sports fan were explored by Dietz-Uhler, Harrick, End, and Jacquemotte (2000). Men tended to be motivated by previous experience of playing sports and getting sports information while women tended to be motivated because they attend games and watch with family and friends. However, the results of other research that has considered gender as a possible predictor of enjoyment have been mixed (Gan et al. 1997; Peterson and Raney 2008).

Perhaps more promising than simple gender differences is recent literature on fanship that has examined differences between sports fans and non-fans (Cummins 2009; Cummins, Gong, and Kim 2016). A sports fan can be defined as someone with a strong emotional attachment towards sports (Gantz, Wang, Paul and Potter 2006; Cummins 2009). People who identify as sports fans have their own cognitive and experiential reactions to sports media (Cummins 2009; Cummins, Gong, and Kim 2016; Potter and Keene 2012). For example, viewing behavior can be different, even at the eye level, between fans and non-fans. Cummins, Gong, and Kim (2016) investigated how fanship and other user characteristics could impact attention to information graphics in televised sports using a 2 (sports clip excerpt) × 4 (infographic type) × 2 (fanship, sports statistics and fantasy sports knowledge) quasi-experimental design. Independent variables were fanship, sports statistics and fantasy sports knowledge, while the dependent variables were fixation frequency, gaze duration and observation frequency. Regarding the part of the study that dealt exclusively with fanship and selective attention, results partially supported that selective attention for sports information graphics was different between people who were high and low in sports fandom. There were significant group differences in gaze duration but not with fixation frequency. Along with cognitive reactions, user perceptions of media can be different for sports fans. As an example, sports fans tend to feel more presence and enjoyment while watching sports compared to non-fans, as shown by Cummins (2009). In the study mentioned earlier on subjective camera angles, Cummins (2009) found people higher in sports fanship overall enjoyed watching the sports clips more and experienced more spatial presence and engagement than non-sports fans.

9.5 VIEWING CONTEXT MATTERS

While technology manipulations such as camera angle and user characteristics such as gender and fanship are important considerations in sports media enjoyment, other research has demonstrated that the context of viewing is also important. For instance, watching a game in a usual at-home setting is much different from experiencing the game "in person" at a sports arena. Although home television experiences have been thoroughly investigated (Kim, Cheong, and Kim 2016; Lombard and Ditton 1997), relatively few studies have documented the viewing experiences outside the home. An examination of the literature revealed that these previous studies have ranged from more qualitative studies of viewer behavior in places such as sports bars (Eastman and Land 1997) to actually comparing differing viewing experiences of watching a sports game in a theater setting versus at home (Kim, Cheong, and Kim 2016).

An early effort in exploring viewing context was published by Eastman and Land (1997) who described a naturalistic observation of sports-watching behavior outside the home setting in sports bars. Trained observers were sent to six sports bars on weekends and afternoons where the average period of observation was reported to be 2.3 hours. The written and audiotaped notes about behaviors and conversations in the bars during those times were supplemented with additional information from open-ended interviews with staff at the bars. From those observations, the researchers reported that there were several reasons to explain why people choose to watch sports outside their homes, including active community participation, conversational and other social opportunities, easier access to restricted sporting events and as a diversionary activity for non-fans while waiting to eat and/or drink. Importantly, results from this study indicated that different viewing contexts could elicit different experiences for users.

In another study, the role of viewing context was investigated by examining how the holistic experience of viewing sports media in a theater varies from the viewing experiences with individual components in a home (Kim, Cheong, and Kim 2016). They examined differences in presence, suspense and enjoyment in home viewing versus theater viewing of sports media events. The study focused on examining a series of relationships between presence, game attractiveness, suspense and enjoyment that predicted that presence components including realism, immersion and physiological response along with game attractiveness contribute to perceived suspense and enjoyment. A total of 240 participants were recruited and randomly assigned to watch one of two FIFA World Cup games either at home or at a theater before completing a questionnaire that assessed presence, game attractiveness, suspense and enjoyment of the experience within one day of watching the game.

Results indicated that the three measures of presence (i.e., realism, immersion and physiological response) along with suspense and enjoyment were significantly higher in the theater condition than in the home-viewing condition. They also found significant relationships between presence, game attractiveness, suspense and enjoyment, with the presence components (except physiological response) contributing to game attractiveness to the suspense which then contributed to enjoyment.

9.6 EXPANDING SPORTS MEDIA VIEWING BEYOND 2D: THE EMERGENCE OF VIRTUAL REALITY

As the summary of previous sports media findings suggests, much of the extant research has been limited to studying user-experience within a 2D, television context. Because technology continues to evolve at a rapid pace, this work needs to be updated and expanded as emerging technologies such as virtual reality systems begin to appear as an alternate means to present sports media. Not only do technological components like head tracking within virtual reality result in higher levels of user perceptions of presence (Cummings and Bailenson 2015) but the creation of an entire virtual scene around a viewer can easily mimic a context similar to a sports arena or even a home setting. Therefore, if virtual reality is going to create a possibly more pleasurable experience, future research designs studying it and sports media should be informed by what we know of hedonic systems (Van Der Heijden 2004).

The hedonic framework from marketing literature states that a user's adoption of a product is driven by his/her own imaginal (fantasy, role project and escapism) and emotional (enjoyment, emotional involvement and arousal) responses (Hirschman 1983; Hirschman and Holbrook 1982; Lacher and Mizerski 1994; Holsapple and Wu 2007). Given this concept, Holsapple and Wu (2007) adapted the hedonic framework to create a new framework for examining virtual technologies that described emotional responses as consisting of both arousal and emotional involvement. Arousal occurs when outside stimulation leads to a heightened mental and emotional state whereas emotional involvement describes how emotionally involved a user is in the task that they are doing (Holsapple and Wu 2007). Interestingly, these same factors directly determine the quality of user-experience with sports media as described in excitation transfer theory (Zillmann 2008) and the disposition theory of sports spectatorship (Zillmann, Bryant, and Sapolsky, 1989).

Due to the potential for virtual reality technology to enhance sports-viewing experiences through presence and suspense via the arousal and emotional involvement mechanisms described above, the next section details the first step into this process, examining how the role of video format presentation could be different between HMD (360-degree) and 2D (monitor) presentations.

9.7 PILOT STUDY: SPORTS VIEWING IN HMD WITHIN A VIRTUAL ENVIRONMENT IS DIFFERENT FROM 2D

Expanding on previous research published by Kim, Cheong, and Kim (2016) that directly compared sports-viewing experience by context (theater vs. home), Wilson and Mayhorn (2019) explored the reported differences in presence, suspense and enjoyment across two new contexts: HMD (360-degree) video vs. 2D (monitor). As shown in Figure 9.1, this pilot study was also designed to extend Kim et al.'s (2016) observed relationships between presence, game attractiveness, suspense and enjoyment to inform the user-experience within the virtual reality context of HMD.

To examine these differences and potential relationships in Wilson and Mayhorn (2019), two hypotheses were developed, based on the Kim et al. (2016) hypotheses, which are summarized as (1) There will be significant, positive relationships between presence components and suspense, game attractiveness and suspense, and suspense and enjoyment and (2) that HMD (360-degree) would outperform 2D (monitor) conditions on presence, suspense and enjoyment.

The video used in this pilot study was a five-minute video clip of the North Carolina State University Wolfpack football team scoring a touchdown. This moment happened during the first quarter of the North Carolina State Wolfpack vs. Florida State Seminoles game on November 3rd, 2018. This video clip was displayed in either HMD (360-degree) if played on the Oculus Rift (Facebook Technologies, 2016) or a

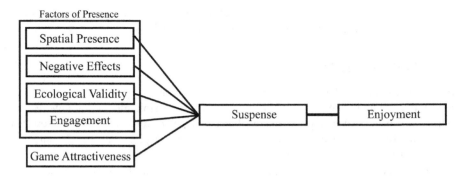

FIGURE 9.1. Proposed relationships between presence, game attractiveness, suspense and enjoyment (Based on relationships from Wilson and Mayhorn, "Examining the role of media in sports media viewing." Santa Monica, CA: Human Factors and Ergonomics Society 2019).

monitor (2D) in the Movies or TV application on Microsoft Windows. Both versions of the videos were installed on an MSI gaming computer running Windows 10 OS (Wilson and Mayhorn 2019).

Beyond the preliminary sample described in Wilson and Mayhorn (2019), 83 participants provided informed consent and then completed a demographics survey adapted from Lessiter, Freeman, Keogh, and Davidoff (2001)'s ITC Sense of Presence Inventory. Afterward, they were randomly assigned to watch the video in either the HMD (360-degree) or the monitor (2D) condition. After the video, they completed the following surveys in counterbalanced order: an attention maintenance survey that queried participants about events in the video, an adapted version of the ITC Sense of Presence Inventory (Lessiter et al. 2001), game attractiveness items (Kim, Cheong, and Kim 2016) including a measure of team disposition adapted from Knobloch-Westerwick et al. (2009). Lastly, items intended to measure suspense in the media clip (Oliver and Bartsch, 2010) and enjoyment (Gan et al. 1997; Peterson and Raney 2008) were also administered in this set before the participant was debriefed and dismissed (Wilson and Mayhorn 2020).

Although the Kim et al. (2016) model was not a good match as indicated via confirmatory factor analysis, the proposed relationships were partially supported. In retrospect, this poor match is not very surprising given that it was derived from 2D environments such as theaters rather than HMDs within virtual environments. Most importantly, results revealed differences between the two video format conditions regarding presence, suspense and enjoyment. From the presence constructs measured in the ITC Sense of Presence Inventory (Lessiter et al. 2001), spatial presence and engagement were significantly heightened in the HMD (360-degree) condition compared to the 2D (monitor) condition. The same differences between the groups were also found for suspense and enjoyment, with participants in the HMD (360-degree) condition experiencing more suspense and enjoyment than those in the 2D (monitor) condition. In short, this pilot study demonstrates that even if the model previously described by Kim et al. (2016) was not a good fit to describe the users' experience of sports media, measurable differences between users' experiences of presence, suspense and enjoyment between HMD (360-degree) and 2D (monitor) viewing of sports media were observed (Wilson & Mayhorn 2020).

9.8 CONCLUSION

Sports media has evolved from delayed announcements imported from overseas to a very limited audience to a quick, almost always accessible media that anyone anywhere can access at any point during the day (Bryant and Holt 2006). However, sports media research has tended to focus on the 2D presentation of content as illustrated in the work by Cummins (2009). While it is evident that sports media presentations in different viewing contexts result in measurably different user-experiences (Eastman and Land 1997; Kim, Cheong and Kim 2016), there has been little exploration of context as a technological variable. With the advent of virtual reality displays that use HMD, the need to explore how these new modes of sports media presentation influence user-experience as both a technology and a context is obvious. Interested companies such as NextVR (2018) driven by the commercial benefits of providing "next-generation" sports media viewing technology often implement their system designs without the benefit of peer-reviewed research findings. Likewise, the user-experience literature needs to further explore how virtual reality sports-viewing experiences address some of the known relationships between arousal, emotion, usability and enjoyment. These efforts might be further informed by how the properties of technologies (Van der Heijden 2004) as described in the hedonic framework (Holsapple and Wu 2007) have the potential to shape the user-experience of sports viewing.

The pilot study from Wilson and Mayhorn (2020) highlights an early attempt to explore differences in the user-experience of participants watching a sports game either in a traditional 2D technology or with the HMD of a VR system. The results from this study indicate that there are differences between the two conditions, although the relationships between the components of presence, game attractiveness, suspense and enjoyment did not follow previous literature (Kim, Cheong, and Kim 2016; Peterson and Raney 2008). The partial support for the hypothesized results may be due to several limitations present in the study such as a lack of a measure for fanship, which is important to perceptions of enjoyment of sports media (Cummins 2009). Also, the video clip that participants viewed was relatively short in length including five minutes before a touchdown was scored. The clip might not have been long enough and also the content itself might not have been considered suspenseful by the participants as one team was engaged in a "drive" to score and the other team was solely engaged in defensive action. Likewise, there is a variety of other factors that could be manipulated to enhance user perceptions of presence within the virtual environment. For instance, a current project in our laboratory is investigating whether the addition of multisensory feedback (i.e., haptic vibrations to mimic crowd noise) enhances arousal and enjoyment of collegiate sporting events. Given these potential limitations and future directions, we anticipate heightened research interest in this area to further explore what factors influence sports media viewing using HMDs within virtual environments.

One deliverable from this chapter is the realization that the combination of rigorous methodologies from both sports media and user-experience research can and should be implemented to explore this emergent topic. Moreover, it is our hope that other researchers and practitioners have gleaned an understanding from this manuscript that the evolution of sports media to include virtual reality (and future technologies) does not occur in a vacuum. Context and user characteristics matter. Because VR sports viewing represents a hedonic system, this research arena represents a unique contextual niche for usability professionals because the main focus is not on operational efficiency and productivity but emotional enjoyment and pleasure. As we have discussed, there is a wealth of literature that describes how user characteristics and technology interact with such outcome variables to influence the quality of the user-experience in very predictable directions that promote usability as well as technology adoption. Rather than haphazardly designing systems as the technology becomes available and perceived demand for services increases, this growing area of technology-mediated presentation represents a challenge and an opportunity for human factors professionals. Armed with empirical research findings from a variety of areas that address seemingly diverse topics such as emotional arousal, usability and technology design, human factors researchers and usability practitioners from allied disciplines such as computer science and industrial systems engineering can purposely collaborate to develop VR systems that provide both a usable and emotionally arousing sports-viewing experience to users. Game On!

REFERENCES

BAMTECH LLC. 2019. NHL. Apple Application Store, 11.2.1 (2019). Accessed 10 December 2019.

Bittner, J. R. and D. A. Bittner. 1977. *Radio journalism*. Englewood Cliffs, NJ: Prentice-Hall.

Brown, L. 2019. Top 10 professional 360 degree cameras. *Wondershare Filmora*. November 6, 2019. Accessed 10 December 2019. https://filmora.wondershare.com/virtual-reality/top-10-professional-360-degree-cameras.html.

Bryant, J. and A. M. Holt. 2006. A historical overview of sports and media in the United States. In *Handbook of sports media*, eds. A. A. Raney and J. Bryant, pp. 21–43. Mahwah, NJ: Lawrence Erlbaum Associates, Inc.

Cummings, J. and J. N. Bailenson. 2015. How immersive is enough? A meta-analysis of the effect of immersive technology on user presence. *Media Psychology* 19, no. 2 (May): 272–309. doi: 10.1080/15213269.2015.1015740.

Cummins, R. G. 2009. The effects of subjective camera and fanship on viewers' experience of presence and perception of play in sports telecasts. *Journal of Applied Communication Research* 37, no. 4 (October): 374–396. doi: 10.1080/00909880903233192.

Cummins, R. G., Gong, Z., and H. S. Kim. 2016. Individual differences in selective attention to information graphics in televised sports. *Communication & Sport* 4, no. 1 (January): 102–120. doi: 10.1177/2167479513517491.

Cummins, R. G., Keene, J. R., and B. H. Nutting. 2012. The impact of subjective camera in sports on arousal and enjoyment. *Mass Communication and Society*, 15, no. 1 (January): 74–97. doi: 10.1080/15205436.2011.558805.

Dietz-Uhler, B., Harrick, E. A., End, C., and L. Jacquemotte. 2000. Sex differences in sport fan behavior and reasons for being a sport fan. *Journal of Sport Behavior*, 23, no. 3 (September): 219–231. https://search.proquest.com/docview/215880803?

Eastman, S. T. and A. M. Land. 1997. The best of both worlds: Sports fans find good seats at the bar. *Journal of Sport and Social Issues* 21, no. 2 (May): 156–178.

ESPN. n.d. "ESPN+." ESPN. Accessed 10 December 2019. https://plus.espn.com/.

Facebook Technologies. 2016. Oculus rift [Hardware]. https://www.oculus.com.

Gan, S.-L., Tuggle, C. A., Mitrook, M. A., Coussement, S. H., and D. Zillmann. 1997. The thrill of a close game: Who enjoys it and who doesn't? *Journal of Sports and Social Issues* 21, no. 1 (February): 53–64. doi: 10.1177/019372397021001004.

Gantz, W., Wang, Z., Paul, B., and R. F. Potter. 2006. Sports versus all comers: Comparing TV sports fans with fans of other programming genres. *Journal of Broadcasting & Electronic Media* 50, no. 1 (June): 95–118. doi: 10.1207/s15506878jobem5001_6.

Hirschman, E. C. 1983. Predictors of self-projection, fantasy fulfillment, and escapism. *The Journal of Social Psychology* 120, no. 1 (May): 63–76. doi: 10.1080/00224545.1983.9712011.

Hirschman, E. C., and M. B. Holbrook. 1982. Hedonic consumption: Emerging concepts, methods and propositions. *Journal of Marketing* 46, no. 3 (July): 92–101. doi: 10.1177/002224298204600314.

Holsapple, C. W. and J. Wu. 2007. User acceptance of virtual worlds: The hedonic framework. *ACM SIGMIS Database* 38, no. 4 (November): 86–89. doi: 10.1145/1314234.1314250.

HTC Corporation. 2016. HTC VIVE [Hardware]. https://www.vive.com/us/.

Hulu. n.d. Watch sports on Hulu Live TV. Hulu Live TV. Hulu. Accessed 14 December 2019. https://www.hulu.com/live-sports.

Kim, K., Cheong, Y., and H. Kim. 2016. The influences of sports viewing conditions on enjoyment from watching televised sports: An analysis of the FIFA World Cup audiences in theater vs. home. *Journal of Broadcasting & Electronic Media* 60, no. 3 (September): 389–409. doi: 10.1080/08838151.2016.1203320.

Knobloch-Westerwick, S., David, P., Eastin, M., Tamborini, R., and D. Greenwood. 2009. Sports spectators' suspense: Affect and uncertainty in sports entertainment. *Journal of Communication* 59, no. 4 (December): 750–767. doi: 10.1111/j.1460-2466.2009.01456.x.

Lacher, K. T. and R. Mizerski. 1994. An exploratory study of the responses and relationships involved in the evaluation of, and in the intention to purchase new rock music. *Journal of Consumer Research* 21, no. 2 (September): 366–380. doi: 10.1086/209404

Lessiter, J., Freeman, J., Keogh, E., and J. Davidoff. 2001. A cross-media presence questionnaire: The ITC-sense of presence inventory. *Presence: Teleoperators & Virtual Environments* 10, no. 3 (June): 282–297. doi: 10.1162/105474601300343612.

Lombard, M. and T. B. Ditton. 1997. At the heart of it all: The concept of presence. *Journal of Computer-Mediated Communication* 3, no 2. (September): JCMC321. doi: 10.1111/j.1083-6101.1997.tb00072.x.

MLB. 2019. MLB at Bat. Apple Application Store, 12.9.0. Accessed 10 December 2019.

Narducci, M. 2007. On sports media: New views of NASCAR races. *Philadelphia Inquirer*, February 9, 2007. https://www.inquirer.com/philly/columnists/20070209_On_Sports_Media___New:views_of_NASCAR_races.html.

NBA. n.d. "XR." Sit Courtside in VR & in a digital lounge in MR. *NBA*. Accessed 10 December 2019. https://www.nba.com/xr.

Newman, M. 2017. NHL.TV offers live games on more than 400 devices. *News Headlines*. NHL, January 31, 2017. https://www.nhl.com/news/nhltv-games-schedule-features-for-2016-17/c-282066602.

NextVR. 2018. NextVR. Oculus Rift Store. 2.0.2.0 (2018). Accessed 10 December 2019. https://www.oculus.com/experiences/rift/1651137804962098.

NHL Public Relations. 2019. NHL partnering with NextVR for virtual reality at marquee events. *News Headlines*. January 31, 2019. Accessed 10 December 2019. https://www.nhl.com/news/nhl-partnering-with-nextvr-for-virtual-reality-at-marquee-events/c-304339710.

Oliver, M. B. and A. Bartsch. 2010. Appreciation as audience response: Exploring entertainment gratifications beyond hedonism. *Human Communication Research* 36, no. 1 (January): 53–81. doi: 10.1111/j.1468-2958.2009.01368.x.

Potter, R. F. and J. R. Keene. 2012. The effect of sports fan identification on the cognitive processing of sports news. *International Journal of Sport Communication* 5, no. 3: 348–367. https://journals.humankinetics.com/view/journals/ijsc/5/3/article-p348.xml.

Peterson, E. and A. A. Raney. 2008. Reconceptualizing and reexamining suspense as a predictor of mediated sports enjoyment. *Journal of Broadcasting & Electronic Media* 52, no. 4. (December): 544–562. doi: 10.1080/08838150802437263.

Raney, A. A. 2017. Affective disposition theory. *The International Encyclopedia of Media Effects* 2017: 1–11. doi: 10.1002/9781118783764.wbieme0081.

Sargent, S. L., Zillmann, D., and J. B. Weaver III. 1998. The gender gap in the enjoyment of televised sports. *Journal of Sport and Social Issues* 22, no. 1 (February): 46–64. doi: 10.1177/019372398022001005.

Van der Heijden, H. 2004. User acceptance of hedonic information systems. *MIS Quarterly* 28, no. 4 (December): 695–704. https://www.jstor.org/stable/25148660.

Wilson, R. and C. B. Mayhorn. 2019. Examining the role of media in sports media viewing. In *Proceedings of the Human Factors and Ergonomics Society 63rd Annual Meeting*, vol. 63, no. 1 (November): 1978–1982. Los Angeles, CA: SAGE Publications. doi: 10.1177/1071181319631424.

Wilson, R. and C. B. Mayhorn. 2020. On the field: Examining differences in video format in sports media viewing. In *Proceedings of the Human factors and Ergonomics Society Annual Meeting*, vol. 64, no. 1 (December): 781–785. Los Angeles, CA: SAGE Publications. doi: 10.1177/1071181320641181

Zillmann, D. 2008. Excitation transfer theory. *The International Encyclopedia of Communication*. doi: 10.1002/9781405186407.wbiece049.

Zillmann, D., Bryant, J., and B. Sapolsky. 1989. Enjoyment from sports spectatorship. In *Sports, Games, and Play: Social and Psychological Viewpoints*, 2nd ed., ed. Goldstein, J. H., 241–278. Hillsdale, NJ: Lawrence Erlbaum Associates, Inc.

Zillmann, D. and J. R. Cantor. 1976. A disposition theory of humor and mirth. In *Humor and Laughter: Theory, Research, and Application*, eds. A. Chapman and H. Foot, 93–115. London: John Wiley & Sons, Ltd.

Section 4

Usability and UX in Fashion Design

10 Expropriating Bodies
Immateriality and Emotion in Costume Design

Alexandra Cabral

CONTENTS

- 10.1 Introduction 173
- 10.2 Costume Design and the Scope of Fashion as Art 174
- 10.3 Costumes and the Embodied Experience in Immersion 177
- 10.4 Usability and User-Experience in the Personal and Social Construction of Meaning 179
- 10.5 Immateriality and Emotion in a Practice-led Research 182
  - 10.5.1 Costume Design for *Garb'urlesco* (2019) 183
  - 10.5.2 Costume Design for *Cinderella* (2018) 185
- 10.6 Conclusion 188
- References 189

10.1 INTRODUCTION

Fashion, since it became part of the contemporary art context, has been promoting cross-bordering approaches – the same that characterize the art practices – in the fashion design context itself. Therefore, presentations of collections in catwalk shows have been shifting to performative practices driven by the conceptual intentions of the designers, some held in the cultural spaces of contemporary art. The cross-contamination occurring among the contemporary arts has also allowed the fashion audience to engage in a more immersive way into the works of fashion designers. On the other hand, under the influence of other forms of art, fashion proposals became ever more conceptual, questioning so many preconceptions on wearing and communicating, since "clothing, seen as 'anti-fashion', gained new ways to express one's thoughts, under this artistic universe grounded by a wave of experiments" (Cabral, 2010:44). Quinn gives the example of Chalayan: "Chalayan builds bridges between the visual, the ideological, the invisible and the tangible, his designs challenge preconceived notions of what clothing can mean" (Quinn, 2005 apud Bugg, 2014:33).

Similarly, performance art is a multidisciplinary practice, also interweaving the performing and the visual arts, sharing with fashion the body as a means of expression – as so does costume design. Costumes, necessarily attached to the fictional narratives, are able to respond, in a fluid manner, to diverse practices in numerous site-specific contexts, with costume design taking references in fashion design practices and fashion semiotic references.

Nonetheless, while, in the fashion design context, usability is commonly related to understanding appropriate fitting, comfort, performance and suitability to a certain social context, including "enjoyment and satisfaction regarding the needs of the subject" (Neves, Brigatto and Paschoarelli, 2015:6135) and with usability as the "reach by which a product can be used by certain users to achieve specific goals with effectiveness, efficiency and satisfaction in a certain context of use." (ISO 9241-11 apud Tavares et al., 2019:583-584), in the costume design performative context, the emphasis given to the user-experience carries a revision of these definitions.

Tavares et al. (2019:586) mention Gomes Filho's reading system of ergonomic factors in design, from which we can point out comfort, reach wrap, posture, application of force and materials. Nevertheless, those factors are reframed in costume design, with comfort under the scope of discomfort as a tool for acting, muscular fatigue in reach wrap as a goal in specific gestures, posture as intentionally disrupted in certain circumstances, difficulty in handling garments as an acting resource – and materials following all those requirements. Costume design can give a new sense to usability while serving a narrative imposition that the body will correspond to. As the body is no longer the primary criterion, within the impositions of a construction of a performative body, ergonomic factors meet somatics in an esoteric way, shaping cognitive and physical ergonomic approaches at the will of performance practice.

10.2 COSTUME DESIGN AND THE SCOPE OF FASHION AS ART

At the turn of the millennium, fashion was exhibited in museums of contemporary art, either alongside other artworks (as in the Florence Biennale, 1996) or by itself, conveying the inspiration behind the fashion collections and thus exposing the designers as artists, in refreshing new kinds of exhibitions (e.g., Chalayan at the Design Museum in London, 2009). These, also attractive to the museums, gave fashion recognition as a form of art and were interesting to a diverse audience that understood fashion in a different context.

It's still an ever-growing interest, since, according to Armani, "in our digital era there is undoubtedly growing curiosity about what is concrete, literal and real" (Pinnock, 2019) and there's a shift in the way brands communicate, as Theyskens points out "there's also the reaction of people, the pictures taken, the social media around it, from a communication standpoint, it's extremely interesting" (ibidem). Fashion brands have also been creating their own museums, such as the Fondazione Prada (1993), Chanel Mobile Art Container (2008), Prada Transformer (2009), Armani/Silos Museum (2015) or Le Musée Yves Saint Laurent Paris (2017), passing the subliminal message that fashion is not only culture but is in the sphere of art.

The symbology present in wearing clothes embedded fashion as art with meanings, the same way it has permeated costumes in performance art. Gained from the socio-cultural context, it is filtered when serving the performative body and contributes to the way the spectators read costumes, while subjects capable of creating meaning in real-time during the performative communicative exchange, which comprises an "aesthetic and emotional discourse as a linguistic construction" (Cabral and Carvalho, 2013:275). There is a suggestion of a mixture of personal interpretation, wearing as a socio-cultural experience, and common sense knowledge present in Bugg's research on clothing the body as performance that seeks "to engage viewers and wearers on an emotional and experiential level by connecting to cognitive understanding and memory" (Bugg, 2014:29). Bugg proposes that the plasticity of clothes and the ability of the body to perform has placed fashion and performance in a hybrid territory, allowing conceptual fashion and costume design to cross-border, putting "emphasis on clothing the body as a visual and physical communication strategy and in relation to research in the fields of performance, costume design, fashion design and fashion communication" (ibidem).

Conceptual fashion tends, generally, to be more on the side of art than of design, for the creative appropriations of meaning that frequently objectify the body, thus apparently more related to costumes, although any garment can be read as a costume, regardless of being more or less conceptual. Even though we try to categorize the pieces in any of the fields – fashion, conceptual fashion or costumes – the categorization is in the eyes of the spectators, so the effort is somehow in vain. The context of performance (the narrative and *mise-en-scène*) and the perspective it is looked at as such, in contrast with the usual fashion shows/presentations, might influence spectators in whether seeing "conceptual fashion" or "costumes", since the garments can actually be the same exact pieces. The performance becomes the filter of fashion because depending on the conceptual framework, there can be allowed any kind of fashion expressions.

Kawakubo seems to pose this discussion by placing similar clothes on the runway and on stage (in *Scenario* by Merce Cunningham), in *Body Meets Dress and Dress Meets Body* (1997). On the runway, spectators find the attire deforming the bodies to the extent that they question whether the designer is launching a new fashion that season or proposing costumes; on stage, spectators wonder on how the performer is telling stories by wearing garments that have a fashionable sense, despite the volumes that deform the body. That is, the observer is reading the same object but through different lenses.

Conceptual fashion is seen by Hazel Clark as related to conceptual art, known by its capability of making "statements that posed questions but that rarely provided clear answers" (Clark, 2012:67), and so does performance art nowadays, especially due to its multidisciplinary and cross-bordering way of exposing facts of life, so in tune with the hypermodern ways of communicating globally, while "a vehicle that could carry a complex layering of iconographic information (...) [that] might give an artist instant visibility" (Goldberg, 2011:227). Clark also mentions that "conceptual fashion design that has emerged since the 1980's (...) had the intellectual ground prepared for it (...) by conceptual art, which in turn had its precedents in the art of the early twentieth century" (Clark, 2012:67), which also means that conceptual approaches to fashion already existed before.

The twentieth century is full of examples that take fashion as a means of expression that questions the body as an object, either in the visual or the performing arts, besides other artworks that explore the body in context – providing fashion design with more than symbolic materials to work with, today. Fashion and body were used to address either the dressed body towards a conceptual exploration of an objectified body or the body as a support of works, resulting in fashionable conceptions that could be seen as costumes.

For example, Schlemmer, in *Triadic Ballet* (1927), placed the body as a pivot of exploring relationships between music, dance and costumes in a defined space, for which costumes had a particular role, since they redefined the body shapes and the possibilities of movements. Either contemplative or interactive, other approaches and artworks enclosed a self-explanatory facet to a certain extent, inspired by the body. Their creators counted on our capability of abstraction in recalling previous experiences: through plastic conceptions that considered the living energy of the body, as the Boccioni's sculptures – such as *Unique Forms of Continuity in Space* (1913) – showed with their implied movement, as a way "to obtain a pictorial emotion" (Boccioni, 1912 apud Ruhrberg et al., 2005:434); or through the constructions of new fictional realities in unsettling reconfigurations of the body as in Man Ray's *Coat Stand* (1929), an articulated body-inspired mask with a stand, photographed on a naked body. The dressed body, also considered in a broad way, was implicit in Robert Rauschenberg's chair costume for the *Antic Meet* (1958) show by Merce Cunningham, where the body was customized with a chair worn like a backpack, an approach that paradoxically gave the body new static limbs that unbalanced it.

The inclusion of the spectator as an intervenient in the work of art in the performative approaches of the 1960s and 1970s encompassed practices that explored the bodily expression in space, both within a conceptual framework and directly in a social context, under the scope of semiotics – to which the spectator, while a participant, played a significant role as a promoter of social critical discourse. This was brought up also when fashion started to make its statements in the contemporary art scene at the turn of the new century as spectators were exposed to the concepts behind the garments in order to get more involved, which empowered them again. The premises and the results of these approaches have been inspiring contemporary fashion and costume design too.

The twenty-first century brought new possibilities of immersion to either spectators and performers, mainly due to new technologies that could be worn, emphasizing that "a new dialogue between body and dress is unfolding right now" (Quinn, 2012:12). Ying Gao proposes, in *(no)where, (now)here* (2013), the invasion of the performer's private space, when the costume changes in shape before the presence of the spectator, turning their relationship literally interdependent. Also, and contrarily to the chair costume by Rauschenberg that limited the body physically, in *Nama Project* # 2 (2013), Textile Resistance empowered the user's body physically and virtually, through an extracorporeal dimension given by a textile with motion sensors that allowed the user to control sound fractals projected on a screen. There, a double experience occurred when the tactile and kinesthetic sensations were expanded into a virtual immersion in graphics and sound. Earlier on, in 1980, an even more visceral approach was proposed by Stelarc, who "wore" technology to expand the body, in its physical, performative and immersive capabilities, namely with a prosthetic

Third Hand, combined with *Laser Eyes* and *Amplified Body*, making use of biomedical sensors that record the activity produced by muscles, brainwaves and heartbeat, reconverting it into sound information and light projection in the action space of his performances.

The dynamics of meaning in dressing thus implies the plasticity of the body in such a way that it is possible, in costume design, to have the attire taking advantage, as the storyteller, in visual narratives (Cabral and Figueiredo, 2019); to get clothes to be used without the body; to get the naked body to express without clothes what once clothes helped the dressed body to express. The body in reminiscence – and the user – can pervade in costumes, since "a garment deprived of its bearer keeps containing it" (Cabral, 2019:11) and "clothing even when devoid of the body still retains the ability to perform" (Pantouvaki, 2014 apud Bugg, 2014:32).

Moreover, these intricate interchanges allow clothes even to overtake the body and become its surrogate, as a new plastic entity, being added layers of meaning to the ones previously attained during the wearing experience. We acknowledge the impact of performance on the body as it "transforms it into pure signifying material, placing Man in the same level of signs" (Santos, 2008:43) and of fashion as an artistic practice, through its own techniques and creative constructions, in proposing new signs trough the form of embodied clothing.

10.3 COSTUMES AND THE EMBODIED EXPERIENCE IN IMMERSION

Monks (2010) exposes how costumes can become identity shapers, from the moment they are being put on. Interestingly, the author struggles more to establish differences between the character embodied and the actor preserved as a subject, than to propose similarities. In fact, we can read that "the attempt fails. We cannot distinguish between the performers and what they wear" (ibidem:19). Moreover, she admits that there is a "strange permeability between actors' bodies and their clothes" (ibidem:14) and that "our sense of who they are is determined entirely by their relation to costume" (ibidem:15). This indicates that the user-experience, in performance art, can be very much indissociable from the spectator's immersion, since the reading is done by the audience.

The actor and director Tiago da Cruz considers that costumes should not distract the spectators and, if they do, that might mean that they do not work in the narrative context, but he admits that "some actors can only actually find the character after trying the costume on" (Da Cruz, 2014:300) and that while they don't have access to the final costumes, they rehearse in "clothes with similar characteristic from the ones of the costume" (ibidem:300). Tiago da Cruz also sees the actor's work as a result of a transformation from the within, because he admits that "we can have nothing and having nothing as our costume" (ibidem:301), showing his interest in Grotowski's concept of a Poor Theater (Grotowski, 1969), which explores the abilities of the body to express along with a mental focus needed to build a character.

The performer Sally E. Dean does not ignore those capabilities in a performer, but she has a broader vision on what dressing a costume implies, considering the spectrum of its influence from the inner self point of view to a sense of collectiveness allowed by it. She registers the term "somatic costumes" as a trademark, as she experiences the acting process as enabled by costumes, the same way she recognizes, in turn, that costume design should be "started with a somatic instigator of some kind (...) that brings awareness to (...) parts of the body, so any (...) can lead to story, narrative" (Dean, 2015:291), like in any artistic process. This means that, for her, it is important to consider the "moving live body in the costume design process and also in its presentation, somehow" (ibidem: 295). We, therefore, can perceive an impossibility in untying bodies and costumes, as we also cannot detach costumes from their context of creation, which is where the bodies move. Moreover, the notion of a collective self is very much present in costumes that give access to the space of action, because that space also implies certain performative postures before it. Dean even considers a spiritual approach that turns the environment into a body in itself, recognizing the idea of the body as transversal to human and non-human elements that are present in a certain performative situation (ibidem:292).

Indeed, a costume can "generate performance through its sole existence" (Pantouvaki, 2014 apud Bugg, 2014:39), as we have seen in Sally E. Dean's work, but so does its inexistence. A naked body alone, or wearing props instead of clothes, proposes the quest on figuring out what it means, in contrast to what if it wore clothes in turn. Props or accessories are read as clothes, for example in the works of Rebecca Horn where the body is dressed metaphorically, in contexts that seem to test the vulnerabilities of the human body, where the artifacts impose movements and allow immersions in specific environments. *Finger Gloves* (1970) and *Unicorn* (1972) extend the body extremities with their prosthetic-like accessories and lead the user to embody outer-body entities that surround it.

In this context, also the theater director Oscar Wagenmans considers the action field as part of a performer's body, when he proposes avatar performances in the Second Life platform, such as *Wear To Move* (2014). Those performances place the spectators as the users, when they are invited to participate by wearing virtual costumes and "are able to both project themselves in an imagined space and perceive their presence in a dream-like reality" (Cabral and Figueiredo, 2014:2433). Apart from embodiment being seen as an immersive tool in the happening that places the spectator literally "in the performer's shoes", the way costumes are combined in/with the action field and scenario is done to question the general tendency in projecting personal and daily life routines into the creation of virtual fictional spaces and situations. Wagenmans explains that "sometimes it gets so intense, the virtuality, that people want reality in there" (Wagenmans, 2015:305), which is conflicting with the idea of becoming someone or something else in virtual fictional situations. Costumes are then used to dress the bodies and also the spaces, leading to a questioning of the idea of the body, namely why it should be seen as humanoid and not as something else, if that becomes totally arbitrary in a virtual situation. "I challenge people not to wear clothes when they can wear buildings, they can wear paintings" (ibidem:308), he says, which will lead participants to embody a new conscience.

Expropriating Bodies 179

This meets our opinion that by wearing a costume we are inhabiting a space (Cabral and Figueiredo, 2014:2431) and that is essential for the spectator's immersion, but we could also say that the wearer is inhabiting an idea – in some way, it is becoming part of it. Thus, a body that is transformed into something else, enables the immersion into an unexpected reality proposed by the transformation itself. That demonstrates that the abstraction of embodiment into other objects, situations and abstract visual languages is to be considered and this might be why Bugg reflects on the new practices of scenography when she says that "the rules of scenography and performance making are clearly being reassessed and developed" (Bugg, 2014:35).

There are also contexts that test not only the performers but also the spectators in their ability of evasion through costumes, as well as of dematerialization, such as the embodiment into virtual realities or those portrayed in cinematic fictions. This emphasizes the costumes' abilities in allowing us to drift between physical and imagined realities, as if we could (or perhaps we can) live simultaneously in both real and virtual setting, leading Bugg to recognize the "layering of the real and virtual" (ibidem) that needs to be taken into account in performance art. That is explicit in the movie *Avatar* (2009) by James Cameron, where costumes are portals for new dimensions, since "the avatars bridge reality and fiction in the main characters' lives as well as it does in ours, allowing us to feel the characters' experiences as our own" (Cabral and Figueiredo, 2015:143). Tangibility and intangibility become parallel and complementary concepts in the fluid immersion of embodiment. Therefore, materiality in fashion, when challenged by the performative contextualization by being turned into a costume, is proposed as a fragment of a narrative, enabling the construction of a sensation, feeling or undetermined underlying message. The more abstract it is, though, the more it might become a new conceptual construction of immaterial connotation.

This can explain our feelings of evasion when we get immersed in works like Nick Cave's *Sound Suits* (since 1992), where he proposes a musical experience through a fashion materialization in a profusion of colors, textures and shapes on the body. Ironically, the physicality of the media is only a teaser for the spectator to build a correlation between the sound and a fashion gadget that could be read as a musical instrument. It is the usability of the fashion item that then turns the wearer into a sound, meaning that he/she can pass through the experience of putting something on just to be able to become intangible. This reminds us of Rancière, when he comments that "the veil is the music, because it is the artifice through which a body extends itself to engender forms into which it disappears" (Rancière, 2013:97), although, in this context, we can risk saying that the veil is the costume.

10.4 USABILITY AND USER-EXPERIENCE IN THE PERSONAL AND SOCIAL CONSTRUCTION OF MEANING

Is the user the performer only, in the performative experience, when fruition is part of it? Costumes promote immersive states because they "play a decisive role in triggering the spectators' whish of dematerialization for an embodiment in the characters of their choice" (Cabral and Figueiredo, 2015:146), the same way they "become gateways into new imaginary realities felt like their own" (ibidem). Yohji Yamamoto allows spectators to test their expectations, by letting them try the clothes on, which they once had imagined themselves wearing, in his retrospective exhibition at the MoMu in Antwerp (2006). By doing that, the designer found a way of placing spectators as part of the collection concept, for the reason that body and clothes are read as one and only. The spectator's projection and validation are commonly paired, because of the socio-cultural convergences in wearing clothes and memories of implicit tactile feelings, lingering on their skins throughout their lifetime.

Bugg mentions that models, in a fashion show, do not necessarily have an embodied experience by wearing the attire, as the body only "becomes a canvas or site for the communication" (Bugg, 2014:4), opposed to the embodiment experimented by the performers on stage – which is indeed perceived by the spectator as emotions and metaphors in a more open context of communication, generally attributed to art fruition. But costumes can either have resemblance to clothes or be more abstract and we will give a few examples of how both forms allow performance to change the way we read the body and/or the situation it is involved in.

Maria Vieira, in her exhibition Muros de Abrigo/Shelter Walls (2011), reinforced the role of the spectator as part of the work of art, in some installations where she used textiles. In one of them, *Pronomes/Pronouns* (2001), she seemed to suggest that clothes can be portals to interpretative dimensions, when she suspended thick wool capes resembling the Azores' ones (capote-e-capelo) – the region where she lived – as if they were tents, accompanied by an audio recording of murmured pronouns. Although they could not be entered, spectators felt invited to seek protection under those shells of warmth and the sound of spoken pronouns reinforced how impersonal and isolated a person could become by wearing them. Nevertheless, spectators were impelled to seek the individuality of each absent body in every suspended piece, which generated performance through an imminent intromission. Like costumes, the capes enabled the spectator's experience of belonging to or generating an environment, ironically of presence, in a conceptual place of absence. The bodily projection inside the capes as if spectators were in the wearer's shoes gave them new lenses to feel the socio-cultural connotation in the traditional clothes and better read the installation, at the same time it turned them into less passive observers.

The New York Textile Month 4 exposes the artwork *Masked Feelings* (2019) by Elodie Blanchard as capable of inciting "self-discovery" (Blanchard, 2019:3) through masks with anthropomorphic references. The sculptures, that explore how the spectators can be placed out of their comfort zone, make use of a material textile exploration with techniques one can recall and different scales and proportions suggesting ambiguous feelings and relationships that can be formed within the "self" and with the "other," "encouraging interaction, [and] self-reflection" (ibidem), although they are exhibited in walls as contemplative pieces. *Calçada-Mar/Coblestone-Sea*, in turn, a sculptural wearable piece we produced in 2009, had an intriguing facet by being more evasive in meaning than the masks by Elodie Blanchard or *Muros de Abrigo* by Ana Vieira, due to its abstract configuration. It unlocked diverse options in communication and interpretation that followed the possibilities of being worn in different ways (Figure 10.1).

Expropriating Bodies 181

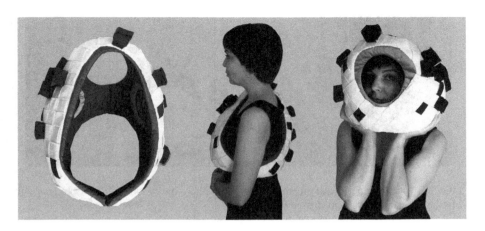

FIGURE 10.1. Open possibilities of wearing in *Calçada-Mar/Coblestone-Sea* (2009), by Alexandra Cabral.

The emotional territory in the design of this piece was explored through a visual resemblance in shapes and colors with the Portuguese cobblestone and the sea, suggested and contextualized by the title, beforehand. Nonetheless, its wearing versatility allowed a broader exploration of the performative experience, giving the performer and the spectator more freedom to construct meaning from what the body could express along with it. In fact, the piece was exhibited so as to allow the spectator to try it on. With this and the previous examples, we see that familiar or traditional elements promote the construction of meaning in the user's experience and the spectator's reading, while the shapes and sizes of the pieces add new ways of wearing, performing and interpreting. And, again, by giving the spectator the possibility of wearing costumes, the emotional experience in the performative practice is reinforced, as it is reinforced the plasticity role of costumes in the construction of fictional narratives through their materiality.

Usability and user-experience were both tested during a scenic space and costumes' course of the degree in Performative Arts of ESTAL (*Escola Superior de Tecnologias e Artes de Lisboa*) that we lectured, in 2018. Students were given costumes to perform with, resembling Martha Graham's *Lamentation* (1930) tube costume, Rebecca Horn's *Finger Gloves* (1972) tube extensions, Franz Erhard Walther's *Just Before Dawn* (1967) textile band joining a group of walkers and Vahid Tehrani's *Cloth Dance* (2012) top-to-toe cover on a moving body. The experiment took its own course when students understood the ability of costumes in promoting new forms of interaction among them. Costumes allowed certain movements that implied a collective collaboration, adding to the functionality the authors of the works had initially proposed. This implied that the conceptual frameworks were surpassed to give space to new meanings and fantasies arising from the experience. The Rebecca Horn's *Finger Gloves* simulation, made with straws, led students to want to be touched as if they were the walls being scratched, and the Martha Graham's tubes' interpretation

FIGURE 10.2. Experiments in interaction inspired by the works of Rebecca Horn and Martha Graham. Workshop class at ESTAL (2018).

proposed a human amalgam, when students tried to hold hands to become interconnected (Figure 10.2). The Franz Erhard Walther's fabric band trial promoted a movement experience that led students to feel part of a "horse gallop," losing individual controls in order to support a collective flow. In the Vahid Tehrani's *Cloth Dance* trial, since the fabric was printed with a motif (contrarily to its original plain color), it helped a better camouflage of the human body inside, that could be read as "something" moving, instead of "someone" moving. This case was particularly interesting because the reaction of the wearer was of "fear" and "weirdness," before the urge of the others to "attack" the moving mass she became part of – they felt "curiosity" and "anxiety" on not knowing what was inside; that is, the user was no longer a person, but a thing.

This leads us to our definition of ergonomy. Since costume design comprises an implicit art practice, we see ergonomy as a scientific discipline applied to costume design that considers a user's physical and cognitive processes, as well as the "immersion of performers and spectators in the works, taking the body as their own simulacrum. It comprehends the new correlations originated by the body customization and its potential in expropriating itself into other physical or virtual entities and by its placement into post-human realities" (Cabral, 2019:281).

10.5 IMMATERIALITY AND EMOTION IN A PRACTICE-LED RESEARCH

The case studies *Garb'urlesco* (2019) and *Cinderella* (2018), costumes designed by the present author, focus either on traditions or/and appeal to the collective imagery, according to their themes. The cases expose, by their dynamic or interactive facets, user-centered design approaches in specific moments of the narratives. Those facets also promoted the revision of the narratives by the users themselves, who developed their own ways of using and embodying them, the characters or the story. The creative process encompassed experimental practices that placed textiles as means of expression and creation, with the body as a mediator of the fictional discourse.

10.5.1 Costume Design for *Garb'urlesco* (2019)

The project *Garb'urlesco* was a multidisciplinary performance promoted by Contemporaneous (*Associação para a Promoção da Arte Contemporânea*), which we designed costumes for, in the realm of our research on costume design based at CIAUD-Research Center (University of Lisbon). The show included music, dance, theater – and costumes, as storytellers and the only scenic elements besides the musical instruments.

Concept-wise, the project was inspired by the intense touristic activity in the Algarve (Portugal) and its impact, which decharacterized places in their landscape and routines, compromising the habits of locals and the preservation of identity. It was then relevant "the broadcasting of material and immaterial heritage, (...) [as an] urgent attitude for the preservation of the cultural values" (Cabral, 2012:1). Costume design was, therefore, influenced by typical elements from architecture to artifacts and garments, such as the pottery from Porches, traditional chimneys, tapestry, the traditional cape (*bioco*), wicker baskets, traditional foldable chairs. The cast included musicians acting as the Four Elements in Nature (Earth, Fire, Water and Air) and a Guarding Angel, and actors/dancers performing as a Tourist and a Local Inhabitant. The first ones signified the harmony or disharmony with touristic activities and the Angel, the divine protection in a Christian community.

The costumes for the Tourist and the Inhabitant needed to be conceptually complementary, non-literal in their aesthetics and representative of their inner selves and how they perceived each other. A wealthy tourist is commonly looked at as a divinity or savior, justifying the proposal of a golden figure. The costume had a layer of muscles placed on ordinary clothing, to emphasize them as improbable in a well-fed tourist with a prominent belly. In contrast, the Local Inhabitant looked more modest: in silver and pale hues of beige, seemed subservient to the tourist. There were references to the full traditional skirt, though, printed with the Porches' pottery motifs. The traditional foldable wood chairs were also present in a vest, assembled with silver parts that folded the same way (Figure 10.3). The costumes were both based on symbolic meanings evoking socio-cultural references as they were the gear that helped actors perform. The "chair" vest, when folded, became a tray used to serve the Tourist or mimicked the chair itself. The performer's body would thus become the tray and the chair. The outfit turned the character subservient to the narrative as it disclosed the hierarchy among characters, due to the signifying parts present in it (but absent in the fashion vocabulary), translated in wearable paraphernalia suitable to a surrealistic approach.

The Four Elements of Nature's costumes, apart from being inspired by the Elements themselves, were also related to the sounds and expression of the musical instruments that represented them: the cello as Wind, the flute as Water, the harpsichord as Earth and the violin as Fire. Poetic and cognitive relationships were created towards a user-experience that linked the sound produced by the musicians (through their instruments) and their costumes, promoting an embodiment into the natural element they represented. For example, the cello itself was embraced by the costume, circumscribed by an involving textile floor piece resembling a cloud (Air

FIGURE 10.3. The wealthy Tourist and the subservient Local Inhabitant: costumes for *Garb'urlesco* (2019), designed by Alexandra Cabral.

element), and the textile elements looking like water cascades were suspended from the wrists of the flute player (Water element). It is therefore easy to understand that the immersion of the musicians into their Elements was facilitated by the silhouettes, colors and textile surfaces of their costumes. Some motifs were also inspired in the Algarve, such as the prints resembling typical chimneys in the Fire costume or the layers of the ceramic soils in the Earth costume.

Costumes allowed musicians to play the instruments freely, but they were designed to generate ergonomic dependencies among them and the actors. They could simply correspond to pre-established moments of the narrative or provide new possibilities of use, due to detachable parts that enabled mimicry dialogs, contributing to improvisation and revision of the staging – they could become props in new discourses. This led their usability to be reshaped along with the user-experience. The Earth's costume detachable upper skirt, inspired by the traditional tapestry, was handled by the Inhabitant, who acted as if she was sewing it (Figure 10.4); the Inhabitant's *bioco* cape, divided in panels connected by pressure buttons, allowed the Elements of Nature to undress her, leaving her vulnerable to the touristic invasion (Figure 10.5). In the past, the *bioco* used to hide a woman's identity, but now she was left exposed. The scene was read as a ritual and introduced the performance itself, in a mixture of mystery and supposition, being also a maneuver to help the musicians cross the scenic space and reach their instruments.

The presentation included a preamble in which spectators had to cross a hall with chairs, where the musicians were randomly placed, making them sit next to each other while waiting for the music show to begin. As a still nature, the performers' costumes had a role beyond wearability, because they characterized the space as contemplative for being the focal points of interest in a bland décor. Covering their users head-to-toe, also due to their masks, their relative abstraction was captivating and intriguing in terms of what concepts they represented. Again, costumes addressed immaterial aspects that placed the body as secondary, on the behalf of a creation of a performative moment. Nevertheless, the body supported the costumes and completed their meanings, since it was still perceived underneath them.

Expropriating Bodies

FIGURE 10.4. The Local Inhabitant borrowing a detachable part of the Earth's costume as a prop: costumes for *Garb'urlesco* (2019), designed by Alexandra Cabral.

FIGURE 10.5. The Elements of Nature undressing the Local Inhabitant's *bioco:* costumes for *Garb'urlesco* (2019), designed by Alexandra Cabral.

10.5.2 Costume Design for *Cinderella* (2018)

The case study *Cinderella* was part of our PhD research in Design, carried out at the University of Lisbon, a research titled *Design of Dynamic and Interactive Costumes for Performance Art* and tutored by Professor Carlos Manuel Figueiredo and Professor Manuela Cristina Figueiredo. The goal was to understand how dynamism and interaction in costumes could contribute to better ergonomic performance, communication and fruition in performance art. The study revealed the challenges and benefits in terms of usability and user-experience, namely their contribution to "immersion and emotional involvement in the plot, by the spectators (..), besides understanding if they would help out and bring advantages in the acting of the performers involved" (Cabral, 2019:173).

The elements of fashion that were present in the narrative of *Cinderella* by Charles Perrault, such as the crystal shoe or the dress for the ball, were important to help the spectators get situated in the scenes in a re-written narrative, the same way they helped actors get immersed in their performance and extrapolate new ways of acting. Originality and surprise met a limited calendar, budget and cast. Thus, the design process implied several trial and error tests, based on usability. The adapted narrative allowed a more personal embodiment, allowing us to test the prototypes in the very beginning and before the actors did, which also placed us in the role of the spectators. It was our job to understand if the proposals met our storyboard along with the desired impact in the action space we created, namely "a fluid manipulation of the textiles with visual impact form far; prints with a readable scale and with dynamic features working on the body and under intermittent lights" (Cabral, 2019:185).

The Cinderella's shoe places Cinderella as the only wearer, but many other girls could wear the same shoe size, creating an inconsistency in Perrault's story. In our proposal, the shoe fitted a standard 38 European size, but as it had LEDs, they would only light on when Cinderella wore them, due to conductive materials in her costume. Thus, other girls could wear the shoe, but with no effect. The transformation in the shoes contributed to creating a moment of fantasy, simultaneously turning the narrative more credible. The impressions of the actress on the invasive intensity of the LEDs, influenced her acting: "often, the actress referred to feel as she was «taking off»" (Cabral, 2019:218).

The evil aspect of the Stepmother and her two Daughters' costumes was due to their being quite alike in their grim personalities. As the most known ancient version of *Cinderella* is Greek, we took Hydra of Lerna as an inspiration: the monster serpent of nine heads was killed by Hercules, a hero that could be symbolized by the Prince, rescuing Cinderella from the others. The costume covered the actress head-to-toe, almost annulling her, facilitating the embodiment into the characters themselves, reinforced by the texture reminding the skin of a serpent. The costume fitted the user's figure, but it was not designed to facilitate movements or avoid discomfort, the actress had to adapt to it. The head accessory, resembling a helmet with three heads, was an extra weight and size she had to handle, along with the fact it compromised her earing. The costume both forced and promoted her to keep a certain posture and balance. It respected the psychological characterization of the characters, the same way it created a new synesthetic experience for the user, who needed extra rehearsals with it. The stretching fabrics implied specific movements, but the actress also had to give life to the three different characters, with her voice. The electroluminescent wires had to change in color, together with the different voices and were visible along the arms, which created a new stylized character in the dark (Figure 10.6). The performer's body was therefore reshaped into a floating figure moving along with the music, testing the possibilities of an emotional bond with the spectators and promoting the creation of new imagery in their heads about the creatures they met, turning a real action into an idea of animation, in which, according to the theater/tv director Maria João Rocha, "the body gains an extra-human dimension, due to an element in the costume" (Cabral, 2019:231).

The Cinderella's Bird Friend costume allowed the actor to perform a bird in a humanoid figure, shifting to a bird-like cartoon image that the fluorescent prints helped to achieve in the low-key environment, where the actor's feet seemed absent. The juxtaposition of a garden décor with similar textiles helped the character merge in it and reinforce that animation-like impression of the acting. As in the Stepmother/ Daughters proposal, this could probably be helpful in chroma-key filming, allowing the actors to perform somewhat more immersed in the narrative than when they can only project a result. The costume helps the embodiment in the character and the

Expropriating Bodies

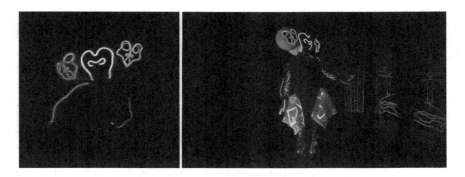

FIGURE 10.6. The presence of the Stepmother and Daughters under different lighting: *Cinderella* (2018), by Alexandra Cabral. Video credits: Multimedia Center, Lisbon School of Architecture.

FIGURE 10.7. The feeling of love in the first dance of Cinderella and the Prince: *Cinderella* (2018), by Alexandra Cabral. Video credits: Multimedia Center, Lisbon School of Architecture.

action it belongs to, simultaneously allowing the spectators a better immersion and belief in the scene.

As for Cinderella and the Prince, Perrault proposes that they meet in the ball and we kept that moment of the story, since we could use it to emphasize the narrative realm of costumes, relating them to the feeling of love and promoting mimicry. Thus, their costumes seemed to gain their own light when the characters met, which also helped circumscribe the action space (Figure 10.7). The dynamism in costumes allowed the actors to perceive each other differently because they, indeed, looked different. When the characters met again in the last scene to dance once more (the moment we added to the story), they wore colorful costumes. The euphoria of love inspired the colors and the prints that seemed to change along with the music, following the dancing bodies; bodies seemed to intermittently disappear and appear, giving the illusion of changing along with the environment (Figure 10.8). Costumes allowed characters to mimic the dream-like reality proposed by the performative moment, in a post-human approach. The shifting circumstances between predictability and unpredictability refrained the actors to control the scene, in such a way

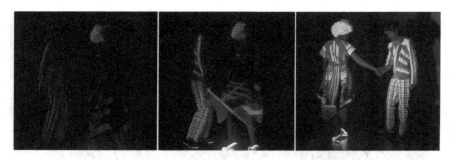

FIGURE 10.8. Cinderella and Prince's bouncing bodies in space, in euphory: *Cinderella* (2018), by Alexandra Cabral. Video credits: Multimedia Center, Lisbon School of Architecture.

that they were used, after all, as visual elements of an intangible musical and emotive message.

10.6 CONCLUSION

Costumes question the role of the performer at the service of the performance, since his/her body can be subservient to a performative narrative the same way it can impose itself in the performing process. This is more than a simplistic view, because it can be applied to an extensive field of practice: from the costume designer's point of view, being simultaneously a director and a performer in the process of creation which can, in itself, include the narrative; from the performer's perspective, who tests the costumes and promotes new framings and approaches either to costume design and performance creation; from the spectator's standpoint, who embodies the costumes and thus the situations they are involved in, adding complexity and value to the performance itself.

The body as a site for communication and creation of meaning has thoroughly been discussed in theory and practice, but we propose a body that is surprisingly and momentarily left behind after the embodiment in a performative situation, emphasizing the immediate post-moment of a costume's role on the scene. That is when the expectations on the performative experience are tested and new unplanned possibilities can arise. This puts to the test the spectators' ability of abstraction and creation of their subjectivity in reading the performance. This means that, although the body of the performer is used to create a connection with the costume and the narrative, based on semiotics, the embodiment lets the performer to be set aside as secondary. It also happens when readings of fashion, present on costumes, are blurred with more abstract or cross-fertilized practices that deconstruct the idea of dress, of wearing and of moving a dressed body in context, since they promote extrapolations into new constructions of meaning.

Physical attributes of the wearer, formal characteristics of costumes, all are blended in an emotional experience that, in the limit, leaves little space to either costume or body. This is easier to understand if we consider that although costumes and body are commonly addressed in visual interpretations, intangible inputs in performance such as sounds and lights can revise those readings, attaching the materiality of the performer (read as body and costume) with immaterial elements, transposing him/her to the edges of reality/unreality where symbolic meanings are placed at the service of perception and understanding. Hence, the correlation between performer and costumes is unlocked, because it goes beyond the somatic experience into a certain fictional reality, it becomes an immersion into a mindset where performance art is placed. Therefore, the approach becomes less body focused than it apparently seemed to be, although it relies on an experience based on the five senses, recognized by the spectator: tactual, given by a costume's materiality filtered by the memories of wearing; visual, given by the projection of their body in costume and of it in space and circumstance; auditive, merging the previous with sounds; and olfactive and tasteful, being some references given to the audience or to the performer to sense.

This perspective of analysis is not focused on the introspective usage of the costume by the wearer alone. The design premises that can arise, either for costume design or performance art, can't ignore the fact that performance is in the spectator's eye. The performer's user-experience is dubious because they are at the service of the performative fiction, and even if they might feel part of a new immersive state, they will still be asked to convince the spectators either they are or aren't. Costumes can then be enablers of the bodily experience into fiction, but the framing by the other elements of the performance space (such as sound and lighting) can add significance to the meaning created on set, which gives usability a broader sense.

REFERENCES

Blanchard, E. 2019. Masked Feelings. *New York Textile Month 4: Talking Textiles*: 3–11.
Bugg, J. 2014. Emotion and Memory: Clothing the Body as Performance. In: *Presence and Absence: The Performing Body*, eds. Anderson, A. and Pantouvaki, S., 29–52. Oxford: Inter-Disciplinary Press.
Cabral, A. 2010. *Fashion and Contemporary Work of Art: Approaches, Practices and Cross-Contaminations in the Work of Joana Vasconcelos*. M.A. thesis. Universidade de Lisboa.
Cabral, A. 2012. Design de Moda e Património Lusófono. In: *Anais do 8º Colóquio de Moda*. Rio de Janeiro: ABEPEM.
Cabral, A. 2019. *Design of Dynamic and Interactice Costumes for Performance Art*. Ph.D thesis. Universidade de Lisboa.
Cabral, A. and Carvalho, C. 2013. Design de Figurinos Inteligentes e Criação de Discursos Simbólicos. In: *Actas da UD'13, 2º Encontro Nacional de Investigação Doutoral em Design*, 271–279. Porto: Universidade do Porto.
Cabral, A. and Figueiredo, C. M. 2014. Costume Design: Ergonomics in Performance Art. In: *Proceedings of the 5th AHFE: International Conference on Applied Human Factors and Ergonomics*, 2430–2438, eds. T. Ahram, W. Karwowski and T. Marek, Kraków.
Cabral, A. and Figueiredo, C. M. 2015. O papel somático do figurino na imersão em mundos cinemáticos virtuais: entre as memórias do espectador e o seu desejo de desmaterialização. In: *AVANCA/CINEMA - Internacional Conference*, 141–148. Avanca: Cine-Clube de Avanca.

Cabral, A. and Figueiredo, C. M. 2019. Performative Approaches in Designing Costumes: Ergonomics in Immersion and Storytelling. In: *Proceedings of the 10th AHFE International Conference on Applied Human Factors and Ergonomics*. Washington, DC: Springer.

Clark, H. 2012. Conceptual Fashion. In: *Fashion and Art*, eds. Geczy, A. and Karaminas, V., 67–75. London and New York: Berg.

Da Cruz, T. 2014. Interviewed by Alexandra Cabral. In: *Design de Figurinos Dinâmicos e Interactivos para a Performance Artística*. PhD thesis, 297–302. Universidade de Lisboa.

Dean, S. E. 2015. Interviewed by Alexandra Cabral. In: *Design de Figurinos Dinâmicos e Interactivos para a Performance Artística*. PhD thesis, 287–295. Universidade de Lisboa.

Goldberg, R. 2011. *Performance Art*. 3rd ed. London and New York: Thames & Hudson.

Grotowski, J. 1969, November 27. Interview by Jean-Marie Drot. "Jerzy Grotowski on the Notion of Poor Theatre". Video. http://fresques.ina.fr/europe-des-cultures-en/fiche-media/ Europe00064/ jerzy-grotowski-on-the-notion-of-poor-theatre.

Monks, A. 2010. *The Actor in Costume*. Hampshire and New York: Palgrave & MacMillan.

Neves, E., Brigatto, A. and Paschoarelli, L. 2015. Fashion and Ergonomic Design: Aspects that Influence the Perception of Clothing Usability. In: *Proceedings of the 6th AHFE International Conference on Applied Human Factors and Ergonomics*, 6133–6139. Las Vegas: Elsevier BV.

Pinnock, O. 2019. The Growing Popularity of Fashion Exhibitions. *Forbes* (March 14th). https://www.forbes.com/sites/oliviapinnock/2019/03/14/the-growing-popularity-of-fashion-exhibitions/#7b5f24debbb1.

Quinn, B. 2012. *Fashion Futures*. London and New York: Merrell.

Rancière, J. 2013. *Aisthesis: Scenes from the Aesthetic Regime of Art*. London and New York: Verso.

Rhurberg, K., Schneckenburger, M., Fricke, C. and Honnef, K. 2005. *Arte do Século XX*. Vol. I. Koln: Tashen GmbH.

Santos, D. R. 2008. *Anything Goes? Uma discussão sobre a necessidade de uma orientação ética na arte contemporânea*. M.A. dissertation. Universidade de Aveiro.

Tavares, A., Soares, M., Marçal, M., Albuquerque, L., Neves, A., Silva, J., Pimentel, S. and Filho, J. 2019. The Use of the Virtual Fashion Tester: A Usability Study. In: *Design, User Experience, and Usability. Practice and Case Studies*, HCII 2019, *Lecture Notes in Computer Science*, vol. 11586, eds. Marcus, A. and Wang, W. Cham: Springer.

Wagenmans, O. 2015. Interviewed by Alexandra Cabral. In: *Design de Figurinos Dinâmicos e Interactivos para a Performance Artística*. PhD thesis, 303–312. Universidade de Lisboa.

11 UX, Design, Sustainable Development and Online Selling and Buying of Women's Clothes

Carolina Bozzi, Marco Neves and Claudia Mont'Alvão

CONTENTS

11.1 Introduction	191
11.1.1 Corporate Social Responsibility (CSR)	192
11.1.2 Consumer Socialization	192
11.2 Selling and Buying of Clothes Online: Main Issues	194
11.2.1 The Serial Returners	194
11.2.2 Easy Returns	195
11.2.3 Size Inconsistencies Stimulate the Serial Returner Behavior	196
11.2.4 Impact on the Reverse Logistics System	199
11.3 The Analysis	200
11.4 Market and User Context	202
11.4.1 In Brazil	202
11.4.2 In Europe	203
11.4.3 In Portugal	203
11.5 The Project	204
11.5.1 Benchmark	204
11.5.2 Ideation	209
11.6 Conclusion	213
References	213
Annexures	217

11.1 INTRODUCTION

This chapter intends to present how Design can contribute to improving the apparel e-commerce user-experience while aiming at a more sustainable process to help fulfill some of the Sustainable Development objectives.

The United Nations defines Sustainable Development as "Development that meets the needs of the present without compromising the ability of future generations to meet their own needs" (United Nations, 1987, p. 43). This definition emerged during the World Commission on Environment and Development, created by the United Nations to discuss and propose ways to harmonize economic development and environmental conservation. To this end, the 17 Sustainable Development Goals were set in 2015 as part of the 2030 agenda. The goals build on the progress and learnings from the 8 Millennium Development Goals which took place between 2000 and 2015 (United Nations, n/d).

Regarding e-commerce and Sustainable Development, the most relevant goals and their sub-themes were selected as shown in Table 11.1.

11.1.1 CORPORATE SOCIAL RESPONSIBILITY (CSR)

To reach the Sustainable Development goals, it is paramount for companies, as well as consumers, to adopt a new mindset and take responsibility for their actions and consequences. The idea behind corporate social responsibility is that "business and society are interwoven rather than distinct entities; therefore, society has certain expectations for appropriate business behavior and outcomes" (Wood, 1991, p. 695).

Companies should not be responsible for solving all social problems. They should, however, be responsible for the problems they cause and help to solve social problems and issues related to their operations and business interests. An automaker, for example, is rightly responsible for helping to solve vehicle safety and air pollution issues, and that company can be reasonably involved with driver education programs and public transport policies (Wood, 1991). As should an e-commerce company be responsible for minimizing its logistics footprint by optimizing its processes and seeking ways to make it more efficient.

The conception of a company's responsibility to society is closely related to the precepts of Sustainable Development. Objectives 8 and 12 are linked to sustainable consumption and waste reduction issues, respectively, regarding responsible consumption and production. This term implies the importance of raising awareness about the impacts that consumer decisions may have on the environment, on their health and society (Giesler and Veresiu, 2014). Additionally, ethics in online marketing is essential, as consumers should trust product descriptions and presentations to facilitate their decision-making process.

11.1.2 CONSUMER SOCIALIZATION

Size tables vary from company to company and differ according to the country, consequently presenting the consumer with a series of different options of sizes to choose from. Due to this complex system of size tables, it is essential to focus efforts not only on providing consumers with rich and complete product descriptions but also on assisting them to understand this information and providing means for them to use it (Mason, De Klerk, Sommervile and Ashdown, 2008). Without the apprehension of information, consumer socialization is not possible. This concept was developed by Ward (1974, p. 2) and is defined as "the process by which young people

TABLE 11.1.
Selected Sustainable Development Goals and Sub-themes. Source: sustainabledevelopment.un.org. Accessed on: September 16, 2019

Goal	Description	Relevant sub-theme
Goal 8: Decent work and economic growth	Promote sustained, inclusive and sustainable economic growth, full and productive employment and decent work for all	8.4 Improve progressively, through 2030, global resource efficiency in consumption and production and endeavor to decouple economic growth from environmental degradation, in accordance with the 10-year framework of programs on sustainable consumption and production, with developed countries taking the lead.
Goal 12: Responsible consumption and production	Ensure sustainable consumption and production patterns	12.1 Implement the 10-year framework of programs on sustainable consumption and production, all countries taking action, with developed countries taking the lead, taking into account the development and capabilities of developing countries.
		12.2 By 2030, achieve the sustainable management and efficient use of natural resources.
		12.4 By 2020, achieve the environmentally sound management of chemicals and all wastes throughout their life cycle, in accordance with agreed international frameworks, and significantly reduce their release to air, water and soil in order to minimize their adverse impacts on human health and the environment.
		12.5 By 2030, substantially reduce waste generation through prevention, reduction, recycling and reuse
Goal 16: Peace, justice and strong institutions	Promote peaceful and inclusive societies for sustainable development, provide access to justice for all and build effective, accountable and inclusive institutions at all levels	16.6 Develop effective, accountable and transparent institutions at all levels.
		16.10 Ensure public access to information and protect fundamental freedoms, in accordance with national legislation and international agreements.

acquire skills and knowledge relevant to their role in the market as consumers." The initial concept was established regarding young people and adolescents as they form the set of future consumers, at the time, the concern was how to direct advertisements aimed at this audience and how they would affect them. Recent studies apply this concept to adults to know how they acquire information to qualify and position themselves as consumers throughout their lives and not only when they are young (Mason et al., 2008; Moreira, Casotti and Campos, 2018).

These issues are intensified when commercializing products with non-digital attributes, as clothes. Non-digital attributes are difficult to communicate digitally without losing information (Lal and Sarvary, 1999; Bell, Gallino and Moreno, 2017). Thus, people still find it necessary to have a multi-sensory contact with clothes, and product descriptions can be difficult to elaborate due to the aforementioned attributes, such as fit and feel. Additionally, the lack of size standardization practiced by the industry may lead the consumer, in certain situations, to buy products that do not meet their needs or expectations. This situation may cause frustration and increase the number of returns and/or product disposal.

Searching for information is a recognized behavior that is important for the consumer decision-making process (Kotler, 2000; Katawetawaraks and Wang, 2011; Solomon, 2011). By this means, consumers acquire the necessary information about a product or service to make the best purchasing decision (Flavián, Guerra and Orús, 2009). Therefore, consumers should have tools that facilitate the search, acquisition, processing and comparison of information related to the products and services they want to buy (Flavián, Guerra and Orús, 2009).

Furthermore, in the effort to create a more conscious and "green consumer," company leaders should "partner up to help consumers make more sustainable decisions but also encourage consumers to see themselves no longer as protected citizens who have something to gain from stricter environmental laws but as environmental stewards for whom every action – from installing a green thermostat to buying energy-saving light bulbs – is an investment in their own and the planet's future" (Giesler and Veresiu, 2014, p. 848).

11.2 SELLING AND BUYING OF CLOTHES ONLINE: MAIN ISSUES

11.2.1 THE SERIAL RETURNERS

One of the problems is the difficulty to communicate non-digital attributes of clothing may generate a high number of returns, also known as serial returns. Subsequently, serial returners are people who buy several units of the same product in different sizes and colors to try them on and return those that do not meet their expectations or their needs. Regarding clothing, this behavior is mainly generated because the purchased products do not fit as expected, or because the color is not the same as in the photographs, or even, the fabric quality or type is different from what the consumer had envisioned (Hope, 2018). One of the ways consumers find to mitigate this uncertainty concerning fit is to buy a product of the size they believe is correct, a larger and a smaller one, and choose a range of colors. After receiving the items, they will check which fit is correct and return the remainder (Ricker, 2017; Banjo, 2013).

The quality of information presented online to consumers, whether visual or textual, is fundamental to the formation of a mental image of the product (Yoo and Kim, 2014). Effective product presentation is not only attractive to consumers, but it also supports their decision-making process because of the lack of direct contact with the product. In a digital environment where consumers are not able to fully inspect a product before a purchase, product presentation is a key factor to achieve effective cognitive responses, subsequently impacting the shopping experience and its results. The role that online product presentation plays becomes even more vital for products such as clothing that require a multi-sensory experience as part of the consumer decision-making process (Kim and Lennon, 2008; Yoo and Kim, 2014).

11.2.2 Easy Returns

According to Kotler and Keller (2016), the degree of perceived risk varies depending on the amount of money invested, the uncertainty about product attributes and the level of self-confidence. The authors explain that consumers develop mechanisms to reduce uncertainty and the negative consequences of risk, such as delaying decisions, gathering information from friends and choosing brands that are available in their country, and offer guarantees. It is important to understand what factors cause consumers to feel at risk and provide information and support to reduce it (Kotler and Keller, 2016).

When buying online, this perception of risk may be intensified due to the difficulty of performing a thorough inspection of the product before purchase, especially regarding experience products. Therefore, many companies offer easy return policies, thus encouraging serial returners (Hope, 2018). Alternatively, retailers should improve their product presentations to help consumers to make better and more informed decisions.

As a rule, physical stores are not required to offer returns for repentance. However, when buying online there are laws to protect consumers and allow them to return products based on the assumption that purchase decisions were based on a brief description and product photos (Código de Defesa do Consumidor [CDC], 2017; The European Parliament and the Council of the European Union, 2011; Hope, 2018). These policies reduce the perceived risk when buying online and stimulate the business.

In Brazil, according to the Consumer Protection Code (Código de Defesa do Consumidor – CDC), products purchased on the internet may be exchanged within seven days if the customer is not satisfied and/or if the product is not as advertised.

The consumer may withdraw from the contract, within seven days of its signature or upon receipt of the product or service, whenever the contracting of supply of products and services occurs outside the commercial establishment, that is, on the internet, on the phone, or catalogs* (CDC, 2017).

Under the European Union (EU) rules, a dealer must repair, replace, reduce the price or give a refund if the goods purchased are defective or do not look or function as advertised. If you have purchased a product or service online or outside a physical store (by phone, mail, door-to-door seller), you also have the right to cancel

---

* Free translation from: O consumidor pode desistir do contrato, no prazo de sete dias a contar de sua assinatura ou do ato de recebimento do produto ou serviço, sempre que a contratação de fornecimento de produtos e serviços ocorrer fora do estabelecimento comercial, isto é, por meio da internet, de telefone ou de catálogos de vendas (Código de Defesa do Consumidor [CDC], 2017).

and return your order within 14 days, for any reason and without justification (The European Parliament and the Council of the European Union, 2011).

In September 2019, in Europe, a campaign was released to raise awareness about consumers' rights with a focus on digital transactions. One of the pillars is the guaranteed returns, and the example depicted in the publicity banners is a woman who is unsatisfied with a dress that does not fit well (Figures 11.1 and 11.2). This campaign is clear evidence of people's dissatisfaction and fear of buying clothes online.

11.2.3 Size Inconsistencies Stimulate the Serial Returner Behavior

Buying clothing online is stimulating serial returners. A study revealed that in the UK half of what consumers spend is reimbursed by retailers (Barclaycard, 2018). Consumers assume that items bought online will not fit and order multiple sizes and colors of the same product, as mentioned previously. According to the report, two out of five consumers declared returning items of clothing because they did not fit as expected. The main reason could be the variation of size measurements across and within brands. Consumers are frequently confused about which size they should order.

FIGURE 11.1 Campaign to raise awareness about consumers' rights when buying products and services online, Portuguese version. Source: Europe (n.d.)

Online Selling and Buying of Women's Clothes

FIGURE 11.2 Campaign to raise awareness about consumers' rights when buying products and services online, English version. Source: Europe (n.d.)

To minimize this problem, consumers demand that retailers attempt to improve the references used for patternmaking, whether by standardization or using new technologies that help them visualize how the product will fit. On the other hand, retailers are looking for solutions to fight the growing return rates. A way to do this could be by presenting consumers with more accurate product information such as the exact measurements of each product (Barclaycard, 2018).

Maggie Alderson summarizes the situation in the article "Size Matters":

"What dress size are you? I'm a size 4. Well, that's what it says in the label of my Donna Karan jacket, so I must be. But on my Zimmermann dress it says 14. I've got Collette Dinnigan garments in small, medium and large. My jeans are 30, my Agnes B. skirts are 38 and my Giorgio Armani jacket is 44. I'm a size 10 trouser at David Lawrence, but a size 12 jacket. My Easton Pearson new best dress is a 10, but I can't even fit into a size 14 Alannah Hill. So what size am I really? Damned if I know" (Alderson, 1999: 28, cited in: Kennedy, 2009).

The general non-compliance of the clothing industry with a size standard is mainly responsible for this confusion. Size designations are identified by an ad hoc (scalar) one-size scale, for example, 12, as there is no direct relationship to body measurements, they can be easily manipulated or altered (Ashdown, 1998).

Size designations have evolved and are no longer serve their original purpose of helping consumers find clothes that best fit them. Canada, for example, uses anthropometric data from other regions to develop its measurement systems (Faust, 2014). A measurement system is a table of numbers representing the dimensions used to classify the body shapes for each size group in that system (Petrova, 2007).

It has often been noticed that size designations have evolved so as not to serve their intended purpose of helping consumers find clothes that best suit them. Some countries, such as Canada, use anthropometric data from other regions to define their measurement systems (Faust, 2014). A measurement system is a table of numbers representing the value of each of the dimensions used to classify the bodies found in the population for each size group in that system (Petrova, 2007). When databases from other populations and groups are used as a reference, consumers' expectations are frequently not met because these products were not designed for them. Moreover, many anthropometric databases are outdated and no longer correspond to the current reality; they do not reflect the changes in the population's measurements due to the growing number of obese and the greater mix of races leading to a more diverse population (Petrova, 2007).

To produce ready-to-wear garments that fit their target audience, manufacturers use various measurement systems. Resourcing to different references may confuse manufacturers, retailers and consumers (Brown and Rice, 2001). A system that effectively communicates fit and measurement enabling consumers to identify the appropriate size is a critical factor in creating an effective global measurement system so that consumers can easily find clothes that fit them (Aldrich, 2007).

According to Faust (2014), the benefits of a standard size designation are:

- Helping consumers find and choose the right garment for their size and shape;
- Reducing time spent shopping;
- Improving the shopping experience;
- Increasing satisfaction;
- Reducing returns/exchanges; and
- Eliminating their confusion by understanding the information on the size label.

Aldrich (2007) explains that when consumers select a ready-to-wear product, a size label offers information about the product's dimensions before they try it on. The label is a means of communication between the manufacturer and the consumer to enable efficient purchasing decisions. If the information on the size label is unreliable or incomplete, consumers will have to try on different sizes to find one that fits. Consumers end up wasting their time in the fitting room, and this often results in an unsatisfactory experience.

The author also discusses the different existing size systems from various countries in different parts of the world. Most of these systems are not easy to understand if a consumer is unfamiliar with them and can also cause some difficulties in the global apparel trade (Aldrich, 2007). With the increase in the international clothing trade, it is difficult to communicate the size dimensions of imported and exported garments. In order to develop an international sizing system to alleviate this confusion about size codes within and between countries, the International Organization for Standardization (ISO) (1991) established an anthropometric size labeling system based on the communication of major body dimensions. However, it is not mandatory, and many companies do not comply with the standard.

Petrova (2007) clarifies how confusing size labels and tags might be. Manufacturers frequently copy previously developed size tables or use size specifications that they have developed based on the knowledge of their current and former customers. She explains that sizing systems are often created and adjusted through trial and error, with feedback from brief consumer surveys and the analysis of sales and returned merchandise reports. Changes in dimensions are made gradually, often without altering the size designation. As a result, sizes can vary from company to company, so products that have different sizes are labeled as the same and products of the same size may be labeled differently.

Sizing systems can be extremely confusing and these difficulties are more evident when buying clothes online because consumers are not able to fully assess products before buying. They must make purchase decisions based on visual and textual information. Product presentations are crucial for successful online sales and therefore the efficiency of this communication must be one of the major concerns to provide a positive and satisfying user-experience and to decrease the return and exchange rates.

11.2.4 Impact on the Reverse Logistics System

Although e-commerce "saves" multiple trips to stores and eliminates the costs to maintain brick-and-mortar structures, the burden of delivery can be heavy. Instead of transporting large quantities of products to stores and being able to plan optimal routes, e-commerce involves transporting small packages to several different places that are unknown until the time of purchase. The same is true for the stage of reverse logistics.

Table 11.2 summarizes some of the main differences between traditional and e-commerce logistics. Novaes (2007) clarifies that the main difference is product flow, which is no longer processed in boxes but storage units. Consumers' expectation of product delivery is also another factor that changes, as they expect almost immediate delivery, making logistics planning much more dynamic than traditional logistics. Another difference cited by Novaes (2007) is the unpredictability of demand, in the traditional logistics it is pushed, what is delivered is pre-determined. A push system allows companies to have predictability in their supply chains since they know what will come and when (Lander, 2019). Whereas under a pull system, products enter the supply chain when customer demand justifies it (Lander, 2019).

When a product is returned by a consumer, to be exchanged or returned, the reverse logistics process is initiated. The Council of Supply Chain Management Professionals – CSCMP (2013, p. 168) defines reverse logistics as "A specialized segment of logistics focused on the movement and management of products and

TABLE 11.2.
Traditional Logistics vs. E-commerce Logistics. Based on: McCullough (1999). Cited in: Araujo et al. (2013)

Operation	Traditional logistics	E-commerce logistics
Delivery type	Large volume	Small volume
Consumer	Known	Unknown
Demand type	Pushed	Pulled
Stock order flow	Unidirectional	Bidirectional
Destiny	Concentrated	Highly scattered
Demand	Stable, consistent	Highly unstable, fragmented
Responsibility	One vehicle	Through whole supply chain

resources after the sale and after delivery to the customer. Includes product returns for repair and/or credit."

There are a few different types of reverse logistics and for this study only the post-sales type is being considered.

Leite (2009, p. 18) defines it as:

The specific area of activity dealing with the equation and operationalization of the physical flow and the corresponding logistic information of unused or underused post-sale goods, which, for different reasons, return to the different links in the distribution chain, which form part of the reverse channels through which these products flow.

One of the reasons raised by Leite (2009) for the increased concern about reverse logistics by virtual retailers is the pressure of legislation, which allows consumers to reject the product for several reasons, thus characterizing a channel subject to high return rates. Regarding retail chains, end-consumers return newly purchased or unconsumed products for various reasons, among them are products that are in disagreement with the consumer's expectations (Rogers and Tibben-Lembke, 2001; Leite, 2009; Beuren, Gausmann and Diedrich, 2015).

As mentioned previously, products bought online in Brazil (or at any out-of-store channel) can be returned within 7 days, if the consumer is not satisfied or if the product is not as advertised. In the European Union, consumers have 14 days to return or exchange a product. Returned products usually end up in companies' warehouses mostly due to violations in packages and because of small defects caused by transportation (Araujo, Matsuoka, Ung, Hilsdorf and Sampaio, 2013).

11.3 THE ANALYSIS

Issues involving measurement systems in the apparel industry are far from being solved, however, it is possible to improve how this information is communicated to the consumers. Tufte (1990 p. 50) argues, "It is not how much information there is, but rather how effectively it is arranged that it does not matter how much information there is; but how efficiently it is organized." The way in which information is presented is important because it has an effect on the meaning-evocation process, it not only affects written information but will also influence other ways of representing concepts (Zwaga, et al. 2004).

Bearing in mind the difficulty faced by consumers in determining their sizes and how it reflects on the returns and exchanges, it was relevant to analyze the measurement tables provided by clothing e-commerce intended to support the consumer's decision-making. Based on this analysis a study was carried out to develop an alternative to aid this step of the consumer journey. The existing solutions served as a starting point to create an alternative method. An interactive system named iTMI, the Interactive Measuring Table (*Tabela de Medidas Interactiva*), was created.

The primary objective of the iTMI is to:

- Improve the user's interaction with size tables so that it is more effective, efficient and satisfying and, consequently, reduce the number of returns.

The secondary objectives are to:

- Present information more clearly to support the user's purchase journey and their decision-making process;
- Optimize the table according to the device being used; and
- Allow users to compare different size/measurement tables.

This project was created using a mobile-first approach as the importance of smartphones in the user's shopping journey is continuously increasing. According to Criteo's Global Commerce Review (2018), 39% (North America) to 51% (Asia Pacific) of online purchases happen through a mobile device. Due to their small screens, poor internet connections, among other issues, developing projects that prioritize mobile devices has resulted in good usability. Initiatives like mobile-first and mobile-only, highlight the importance of mobile devices. Another reference to how relevant they are becoming is Google's new algorithms that prioritize mobile-friendly websites (Zhang, 2018).

The mobile-first approach considers that it is much more difficult to understand complex information on a small display. Lack of space means fewer visible options, so users explore the interface more. They scroll to other parts of the content rather than simply viewing it at a glance, as a result, more of their short-term memory is required to understand the information (Nielsen and Budiu, 2014). Due to this, a simple interface is preferred to avoid cognitive overload. This happens because, as the user scrolls, more information needs to be stored in their working (reduced capacity) memory because what they no longer see needs to be remembered for the information to make sense. The cognitive load increases as the screen size decreases (Cybis, Holtz and Faust, 2010). However, desktops as well as smartphones must have simple interfaces to aid the memory. For this, they require good structure and balance to relate elements and facilitate the subjective apprehension of information, without requiring too much effort.

The number of features should be limited to the most important ones, the layouts need to be cleaner and the buttons and touch areas should be larger. These measures meet the Golden Rules of Shneiderman and Plaisant (2005) and other usability principles (Hansen, 1971; Nielsen 1993; Molich and Nielsen, 1990; Jordan, 2001 e Bastien and Scapin, 1997). The authors argue that an interface cannot induce the user to commit serious mistakes.

Despite the ubiquity of smartphones, it is not always possible to have a good quality internet signal. This can happen due to an overloaded network or the mere absence of a signal. Wireless networks are far from being ubiquitous and when the speed is low, the page load rate is important as there is always a chance of failure. Speed is one of the reasons that contribute to users' poor performance on smartphones (Nielsen and Budiu, 2014). Mary Ellen Coe (2019), president of Google Customer Solutions, has found that "a one-second delay in mobile load times can impact the conversion rates by 20%." Additionally, designing mobile sites with performance in mind will also benefit the efficiency of the desktop design (Wroblewski, 2001).

11.4 MARKET AND USER CONTEXT

11.4.1 IN BRAZIL

According to The Nielsen Company report (The Nielsen Company, 2019), there are 4 billion people connected to the internet (53% of the world's population) and 1.66 billion are e-shoppers (Statista, 2018). Among the categories of durable goods, apparel ranks first in sales, 61% of consumers said they had already purchased at least one product from this category; it is followed by travel services, 59%, books/music/stationery, 59%, IT and mobile, 47% and event tickets, 45% (The Nielsen Company, 2019). Advances in technology, logistics and payment methods, combined with the rising access to mobile internet and consumer demand for convenience, have created a worldwide online shopping scenario that represented US$ 2.3 trillion (The Nielsen Company, 2019).

The Webshopper report showed that in 2018 (E-bit, 2017), e-commerce revenue reached R$ 53.3 billion* (123 million orders). In January 2018, m-commerce (mobile commerce) accounted for 42.8% of the total sales. Also, in 2018, 58.8 million consumers in Brazil made at least one online purchase and 10 million people bought online for the first time. The Perfumery and Cosmetics category was the leader in the number of orders, representing 17.8% of the total, followed by the Fashion and Accessories category, representing 16.5% of the total. The Southeast region registered the highest number of online orders, accounting for almost 60% of the total (E-Bit | Nielsen. 2019). In Brazil, 52.3% of online users are women and 47.3% are men, with a balanced distribution, and are on average 42.1 years old (E-Bit | Nielsen. 2019).

---

* Approximately € 11.1 billion.

In 2018, Health/Cosmetics and Perfumery passed Fashion and Accessories (a position that had been maintained since 2013) (E-Bit, 2017) and became the top-selling category in Brazil (E-Bit I Nielsen. 2019), representing 17.8% of all online sales. The Fashion and Accessories category falls from second to seventh place, when we analyze its financial volume, indicating that the monetary value invested by consumers is low. This suggests the consumer's unwillingness to make a significant financial investment, which may be related to the high-risk perception of buying clothing online. Consumers are willing to take more risks when purchasing low-price products (Kim and Benbasat, 2009).

11.4.2 In Europe

The 2017 Ecommerce Foundation report revealed that € 534 billion was spent on online shopping in 2017 (E-commerce Foundation, 2018). Europe's population was recorded at 840 million inhabitants in 2017, the gross domestic product, GDP, was € 17 trillion in 2017 (E-commerce Foundation, 2018). The internet penetration rate is 81%. The United Kingdom is Europe's largest e-commerce market, followed by Germany and France. In Europe, people between 16- and 24-years old buy most online, unlike Brazil, where the age group that buys the most is between 35 and 40 years old (E-bit I Nielsen, 2019). "Sportswear and clothing" is the most popular category representing 36% of the market (E-commerce Foundation, 2018).

11.4.3 In Portugal

In 2018, the population of Portugal was 10,283,822 people (Prodata, 2019); of these, 76% are internet users (Eurostat, 2018). According to the Correios de Portugal* (CTT) report (CTT, 2018), electronic commerce grew 12.5% (2 percentual points more than in 2016) in 2017, reaching an annual value of € 4,145 million, approximately 36% of the population buy online. The "Apparel and Footwear" category is the most popular in the country, over half of the buyers purchased products from this category. According to the CTT Report (CTT, 2019), four out of ten users stated that it is common or very frequent to abandon a purchase at the research stage, and the main reasons are (1) the products are expensive and (2) unclear or insufficient product information (70% and 40%, respectively).

Clothing retailers use a variety of tools, such as 3D tools and omnichannel strategies, to effectively address a problem that is harmful to not only the industry but also the UX: returns (Bozzi and Mont'alvão, 2019). The inability to physically interact with the product may result in return rates that reached 50% (Shopify, 2016). By offering a wide variety of channel options, retailers can compensate for the weaknesses innate to each one (Bell, Gallino, and Moreno, 2017). There are currently numerous solutions retailers are adopting to offer alternative channels to support the decision-making process of buying clothing online.

---

* The Portuguese post services.

11.5 THE PROJECT

Design acts as the mediator of people's relationships and their activities while interacting with the environment (Nardi and Kapetlinin, 2006). According to Davis (2008), the development of the design practice is becoming more complex and focused on the experience and behavior resulting from this interaction (Davis, 2008). This paradigm shift, from the object to the experience, demands a new body of knowledge as well as new methodologies to support the decision-making processes within the area. The author argues that the scope of research extends beyond the immediate interaction of users with artifacts to include the influence of design within more complex social, physical, economic and technological systems, and our research strategies must go beyond testing in ergonomics laboratories, and focus groups that separate people from environments where research-relevant behavior can be observed.

Humans have a multitude of ways to absorb and understand information, so designers need to use numerous tactics and methods to make that information meaningful (Baer, 2008). It is evident from this challenge that the traditional design knowledge base has its limits and that, for design practice to remain relevant in this rapidly changing environment, the field must generate new knowledge and methods (Davis, 2008), thus it is relevant to use design approaches to address UX projects as the one described in this study.

The following sections will describe the steps involved to develop the interactive size table, from benchmarking to developing the first version of a high-fidelity prototype.

11.5.1 BENCHMARK

As mentioned previously, almost half of what is bought online is returned, mainly due to the inability of consumers to determine their correct size from the information available on apparel e-commerce websites.

According to Shopify.com (Orendorff, 2019) the top reasons for returns are:

- Size too small: 30%
- Size too large: 22%
- Changed my mind: 12%
- Style: 8%
- Not as described: 5%
- Defective: 5%
- Other or not specified: 18%

More than half of the returns happen due to issues regarding size, which suggests that there might be a communication problem between the apparel companies and the users. As this factor is taken into account it was considered that assessing how size information is presented to consumers was highly relevant. The first step was to determine the websites that were going to be analyzed.

Table 11.3 provides the list of the 33 websites that were selected and analyzed. Besides the size information, the things such as if the website was either responsive or designed specifically for mobile devices, which products are sold, if there is an app and if it sells multi-brands or own products were noted.

The criteria used for the selection were:

- Websites suggested by specialists (Annexure 1);
- Top 10 most accessed women's clothing websites from Alexa.com ranks (Annexure 2); and
- Top 10 apparel websites from Alexa.com and Similarweb.com ranks (Annexures 3 and 4).

The size information page of every website was accessed, and this interaction was recorded for future reference. The analysis revealed that there are basically three types of sources from where size information can be retrieved:

- Size guides: interactive apps that require the users to input their size information and are usually embedded from external companies such as the Fit Finder used by Asos.com (Figure 11.3);
- Size tables: do not require users' interaction and serve as reference. Some of the tables that were analyzed did not fit the width of the smartphone's screen and required horizontal scrolling. Others were longer than the length of the screen and required scrolling vertically and sometimes it was necessary to scroll both horizontally and vertically;
- Measuring guides: diagrams or pictures combined with textual information that instructs the user how to measure main body parts, such as bust, waist and hips.

The different information sources answer three questions:

- What is my size?
- How do I convert my size from x to y (e.g., from FR to EUA)? and
- How do I measure myself?

In Table 11.4, there is a list of the positive and negative aspects found in the analyzed size information on the visited websites. This analysis was fundamental for the development of the first version of the iTMI.

The apps available to help users to determine their sizes often depend on the consumers' interest, the insertion of some of their body measurements into the system, the likelihood of finding an item of the appropriate size and shape, and the consumers' willingness to engage in this process prior to purchase (Gill, 2015). Both applications and online systems for virtual fit analysis require some details of body shape and assume that the user has enough understanding to classify their own body (Gill, 2015). Mason et al. (2008) found in their study that it is uncommon for people to have a good understanding of their own body measurements.

TABLE 11.3.
List of Analyzed Websites

Company	Website	Country	Type	Products		App	Obs.
H&M	hm.com	Sweden	Responsive	Women l Men l Divided l Kids l H&M Home		√	
Lulus	lulus.com	USA	Responsive	Women		x	
Revolve	revolve.com	USA	Mobile	Women l Men		√	Multi-brands
Freepeople	freepeople.com	USA	Responsive	Women		√	
Modcloth	modcloth.com	USA	App	Women		√	The website was offline
Ann Taylor	anntaylor.com	USA	Desktop	Women		x	
Nelly	nelly.com	Sweden	Responsive	Women		x	
Tory Burch	toryburch.com	USA	Responsive	Women		√	
Monki	monki.com	Sweden	Responsive	Women		x	H&M Group
Talbots	talbots.com	USA	Responsive	Women		x	Multi-brands
Wildberries	wildberries.ru	Russia	Responsive	Women l Men l Kids l Home l Pets		x	Multi-brands
Zara	zara.com	Spain	Mobile	Women l Men l Kids		√	
Trendyol	trendyol.com	Turkey	Mobile	Women l Men		√	
Macy's	macys.com	USA	Responsive	Women l Men l Kids l Beauty l Accessories l Home		√	Multi-brands
Next	next.co.uk	UK	Responsive	Women l Men l Kids l Home l Beauty		√	
Zozo	zozo.jp	Japan	Responsive	Women l Men		√	
Uniqlo	uniqlo.com	Japan	Responsive	Women l Men l Kids l Baby		x	
Myntra	myntra.com	India	Responsive	Women l Men l Kids l Home l Beauty		√	
Gap	Gap.com	USA	Responsive	Women l Men l Kids l Baby		√	
Zappos	Zappos.com	USA	Mobile	Women l Men l Kids		x	Multi-brands
Forever 21	Forever21.com	USA	Responsive	Women l Men		x	
American Eagle Outfitters	Ae.com	USA	Responsive	Women l Men		√	

(Continued)

TABLE 11.3. (CONTINUED)
List of Analyzed Websites

Company	Website	Country	Type	Products	App	Obs.
Victoria's Secret	Victoriassecret.com	USA	Responsive	Women	x	
Mango	Mango.com	Spain	Responsive	Women | Men | Kids	√	
J Crew	Jcrew.com	USA	Responsive	Women | Men | Kids	x	
6pm	6pm.com	USA	Mobile	Women | Men | Kids	√	Zappos Group Multi-brands
Acne Studios	acnestudios.com	Sweden	Responsive	Women | Men | Jeans | Face Collection	x	
Asos	asos.com	UK	Responsive	Women | Men | Jeans | Face Collection	√	Multi-brands
Études Studio	etudes-studio.com	France	Responsive	Women | Men	x	
Luisa Via Roma	luisaviaroma.com	Italy	Responsive	Men | Women | Boys | Girls | Home | Beauty Men | Beauty Women	√	
Net à porter	net-a-porter.com	UK	Responsive	Women	√	Multi-brands
Top Shop	m.eu.topshop.com	UK	Mobile	Women | Men	√	
Yoox	mobile.yoox.com	Italy	Mobile	Women | Men | Kids | Design + Art	√	

FIGURE 11.3 asos.com size information page.

TABLE 11.4.
Positive and Negative Aspects of the Size Information on the Analyzed Webpages

	Positive aspects
in. and cm option	Enables users to see measurements in centimeters or inches
Tabs	Using tabs to include different size tables is a good way to save screen space and avoid horizontal scrolling
Chats with "fit and size specialists"	Some websites had an option that enabled the user to chat with customer support assistants to help them understand the fit and size of the item
	Negative aspects
Use of "scroll"	Especially the horizontal scroll forces user to scroll back and forth to be able to get the desired information
Do not fit in the screen's width	Tables that do not fit the screen's width force users to have to scroll or swipe to be able to get the desired information
Loss of reference	Horizontal scroll may cause some information to disappear and the user loses reference
Size tables do not correspond to the selected item	When a user selects an item, a dress, for example, and wants to refer to the size table, the table for dresses should be the first option presented
Are not mobile friendly	Some tables are not designed to be seen on a smartphone
Difficult to read	Very small font size and some tables do not allow zooming in
Absence of size table or app	Some websites did not have a size table or guide
Unclear "how to measure" instructions	Instructions to help user to measure their body are often unclear and inconsistent

Online Selling and Buying of Women's Clothes

Some tables on the selected websites required horizontal and/or vertical scroll, thus demanding the user to use their short-term memory, which is not good at storing large amounts of information (Nielsen and Budiu, 2014). This load on the users' short-term memory may cause confusion and lead the users to select the wrong size or even prevent them from determining the most appropriate size and not being able to purchase the desired item due to lack of information.

11.5.2 IDEATION

"Ideation is only one step in the full UX design process; once ideas are generated; separate analysis has to follow to decide which ideas (or parts of ideas) to pursue. The more ideas the better: a broad pool to choose from increases the likelihood that one of the ideas will be the seed for a great design solution" (Harley, 2017).

Therefore, following the websites' analysis, a series of sketches (Figure 11.4) and a low fidelity prototype* were developed prior to the construction of the wireframes (Figure 11.5). The final step was constructing a high-fidelity prototype to carry out tests with users in the future. The sketches guided how the flow of the product would work (Figure 11.6).

Note that three types of tables were created: size, product measurements and conversion (Figures 11.7–11.9). The use of separate tabs in the size table (Figure 11.7) enabled the table to fit the width of the smartphone screen and also with options "pickers" for the user to select which size system they would like to visualize (FR, EU, IT, etc.) (Figure 11.7). The conversion table was then replaced by a system to enable the user to convert from one size to another, from IT (Italy) to FR (France), for instance (Figure 11.9).

It is expected that presenting clearer sizing and measurement information will enable users to make more conscious purchases; therefore, reduced return rates and product disposal – due to consumer dissatisfaction – mitigate the impact on the reverse logistics chain. This project is not intended to persuade the user, rather, to give them tools to be able to make more knowledgeable decisions.

The next step to this project is to test the table models with users and specialists, by means of an experimental research approach, and make adjustments according

FIGURE 11.4 Sketches.

---

* Access to the low fidelity prototype: https://marvelapp.com/1f4j3c6g/screen/56595678

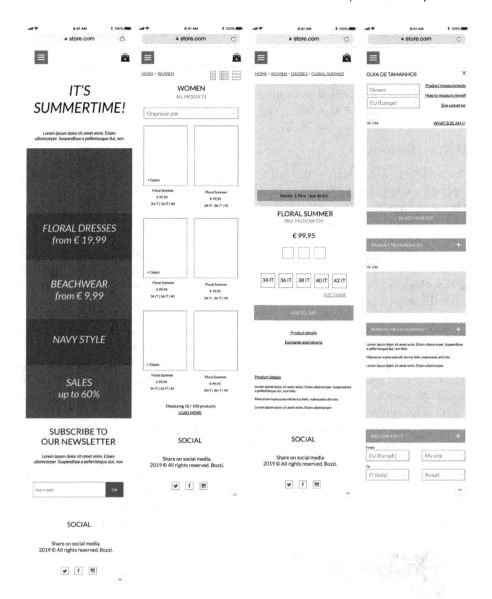

FIGURE 11.5 Wireframes.

to their responses. These tests will be carried out on mobile devices, preferably smartphones, and also on desktops to check how the tables work on different screen sizes and if there is any impact on how well the information is understood and on the overall UX. It is only then possible to determine if the suggested table layouts, the conversion system and the measuring guide are clear and contribute to users' decision-making process.

Online Selling and Buying of Women's Clothes

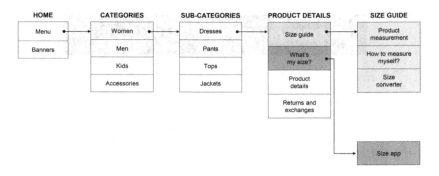

FIGURE 11.6 Product flow.

FIGURE 11.7 Table with tabs.

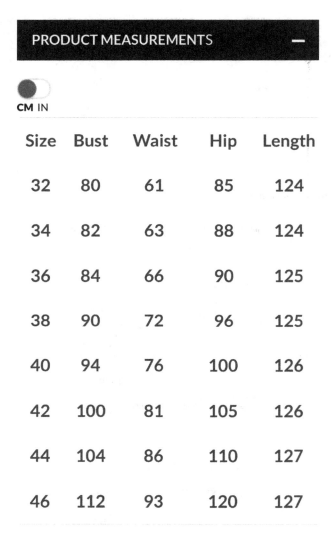

FIGURE 11.8 Product measurements.

FIGURE 11.9 Conversion system.

11.6 CONCLUSION

Design approaches are an extremely valuable investigative tool for surveying problems faced by users trying to buy clothes online. This paper contributes to the area because design research helps the development of the discipline, contributing to the consolidation of the design methodology and science (Cross, 2000).

In an increasingly connected world and the continuing emergence of new websites, Design can contribute significantly to the creation of more intuitive and user-friendly digital projects. There are few studies about it and most of them derive from marketing and fashion. These studies are mainly aimed at the commercial side and not the welfare of the user. Human-centric projects may increase a customer's lifetime value, as they do not focus on immediate results but the holistic experience of the consumer with the brand.

The goal is of this project is to enable consumers to make more knowledgeable decisions that might or might not lead to a purchase and towards a more sustainable view on consumption. These actions contribute to the Sustainable Development goals, as they attempt to reduce waste originated by unusable returned items; cut greenhouse-effect gas emissions as the reverse logistic chains are not overloaded; and show companies that transparency can lead to more loyal consumers, which in the long-term are more profitable. Therefore, the design methods must be used to organize and present information in a more orderly manner, validating with users and specialists throughout the process by means of experimental research to reach the expected objectives.

REFERENCES

Aldrich, W. 2007. History of sizing systems and ready-to-wear garments. In: Ashdown, S. P. (ed.), *Sizing in clothing: Developing effective sizing systems for ready-to-wear clothing*. Cambridge: Woodhead Publishing Ltd.

Alexa.com. 2019. Top 50 apparel websites (women's), accessed April 9, 2019, https://www.alexa.com/topsites/category/Top/Shopping/Clothing/Women's.

Alexa.com. 2019a. Top 50 apparel websites, accessed April 9, 2019, https://www.alexa.com/topsites/category/Top/Shopping/Clothing.

Araujo, A. C., Matsuoka E. M., Ung, J. E., Hilsdorf W. C., and Sampaio, M. 2013. Logística reversa no comércio eletrônico: um estudo de caso. *Gestão & Produção, São Carlos* 20, no. 2: 303–320.

Ashdown, S. P. 1998. An investigation of the structure of sizing systems: A comparison of three multidimensional optimised sizing systems generated from anthropometric data with the ASTM standard D5585-94. *International Journal of Clothing Science and Technology* 10, no. 5: 324–341. doi: 10.1108/09556229810239324.

Baer, K. 2008. *Information design workbook: Graphic approaches, solutions and inspiration + 30 cases studies*. Vacarra, J. (contribuiting writer). Beverly, MA: Rockport Publishers.

Banjo, S. 2013. Rampant returns plague e-retailers. *Wall Street Journal*, accessed November 10, 2018, https://www.wsj.com/articles/rampant-returns-plague-eretailers-1387752786.

Barclaycard. 2018. Return to Sender: Retailers face a 'Phantom Economy' of £7bn each year as shopper returns continue to rise, accessed April 1, 2019. https://www.home.barclayc

ard/media-centre/press-releases/Retailers-face-a-Phantom-Economy-of-7-billion-each-year.html.

Bastien, J. M. C. and Scapin, D. L. 1997. Ergonomic criteria for the evaluation of human-computer interfaces. [*Relatório técnico*]. *Behaviour & Information Technology* 16, no. 4/5: 220–231.

Bell, D. R., Gallino, S., and Moreno, A. 2016. Offline showrooms in omnichannel retail: Demand and operational benefits. *Management Science* 64, no. 4: 1477–1973.

Beuren, A., Gausmann, E., and Diedrich, H. 2015. Proposta de melhoria no processo de devolução ao fornecedor da empresa Alpha Ltda. *Revista Destaque Acadêmicos - CGO/UNIVATES* no. 1: pp: 138–149.

Bozzi, C. 2019. Mont'alvão. E-commerce de moda: uma reflexão sobre cenários futuros e atuais. In: *Anais 13º Congresso Pesquisa e Desenvolvimento em Design 2018, Joinvile*. São Paulo: Blucher, pp. 5113–5126. doi: 10.5151/ped2018-7.1_ACO_14

Brown, P. and Rice, J. 2001. *Ready-to-wear apparel analysis*. 3rd ed. Hoboken NJ: Prentice Hall.

Código de defesa do consumidor [CDC]. 2017. *Lei No. 8.078/90*, Art. 49, 11th September 1990. In: *Código de defesa do consumidor e normas correlatas*. 2nd ed. Brasília: Senado Federal, Coordenação de Edições Correlatas.

Coe, M. E. 2019. Milliseconds earn millions: Why mobile speed can slow or grow your business, accessed September 23, 2019, https://www.thinkwithgoogle.com/intl/en-gb/advertising-channels/mobile/milliseconds-earn-millions-why-mobile-speed-can-slow-or-grow-your-business/.

Correios de Portugal. 2018. e-Commerce report CTT 2018, accessed January 7, 2019, https://www.ctt.pt.

Correios de Portugal. 2019. e-Commerce report CTT 2019, accessed January 7, 2019, https://www.ctt.pt.

Criteo. 2018. *Global commerce review EMEA Q2 2018*, accessed October 15, 2019, https://criteo.com/wp-content/uploads/2018/09/18_GCR_Q2Report_EMEA.pdf.

Cross, N. 2000. Designerly ways of knowing: Design discipline versus design science. In: *Design+research symposium*. Politecnico di Milano, Italy.

CSCMP (Council of Supply Chain Management Professionals). 2013. *Supply Chain Management: Terms and Glossary*, [available at: https://cscmp.org/CSCMP/Educate/SCM_Definitions_and_Glossary_of_Terms.aspx access January, 7, 2019]

Cybis, W., Holtz, A., and Faust, R. 2010. *Ergonomia e usabilidade*. São Paulo: Novatec.

Davis, M. 2008. Why do we need doctoral study in design? *International Journal of Design* 2, no. 3: 71–79.

E-Bit. 2017. *Relatório Webshoppers* 36ª Edição, accessed September 30, 2018, http://www.ebit.com.br/webshoppers.

E-Bit | Nielsen. 2019. *Relatório Webshoppers* 39ª Edição, accessed September, 2018, http://www.ebit.com.br/webshoppers.

E-commerce Foundation. 2018. *Ecommerce report Portugal 2017*. Amsterdam: E-commerce Foundation.

Europe. n.d. Guarantees and returns. Your Europe: European Union, accessed May 16, 2019, https://europa.eu/youreurope/citizens/consumers/shopping/guarantees-returns/index_en.htm.

Eurostat. 2018. Internet use by individuals - % of individuals aged 16 to 74, accessed November 21, 2019, https://ec.europa.eu/eurostat/tgm/table.do?tab=table&plugin=1&language=en&pcode=tin00028.

Faust, M.-E. 2014. Apparel size designation and labelling. In: Gupta, D. and Zakaria, N. (eds.), *Anthropometry, apparel sizing and design*. Cambridge: Woodhead Publishing Limited in association with The Textile Institute.

Flavián, C., Guerra, R., and Orús, C. 2009. The effect of product presentation mode on the perceived content and continent quality of web sites. *Online Information Review* 33, no. 6: 1103–1128.

Gielser, M. and Veresiu, E. 2014. Creating the responsible consumer: Moralistic governance regimes and consumer subjectivity. *Journal of Consumer Research* 41: pp: 840–857.

Gill, S. 2015. A review of research and innovation in garment sizing, prototyping and fitting. *Textile Progress* 47, no. 1 (2015): 1–85. doi: 10.1080/00405167.2015.1023512.

Hansen, W. J. 1971. User engineering principles for interactive systems. *Fall Joint Computer Conference* 39: 523–532.

Harley, A. 2017. Ideation for everyday design challenges. NNGroup. Com, accessed November 26, 2019, https://www.nngroup.com/articles/ux-ideation/.

Hope, K. 2018. The people who return most of what they buy. *BBC News*, accessed March 6, 2019, https://www.bbc.com/news/business-46279638.

International Organization for Standardization [ISO]. 1991. *Standard sizing systems for clothes, (Technical Report ISO/TR 10652:1991)*, International Organization for Standardization, Geneva.

Jordan, P. W. 2001. *Introduction to usability*. London: Taylor and Francis.

Katawetawaraks, C. and Wang, C. L. 2011. Online shopper behavior: influences of online shopping decision. *Asian Journal of Business Research* 1, no. 2 (2011): 66–74.

Kennedy, K. 2009. What size am I? Decoding women's clothing standards. *Fashion Theory* 13, no. 4: 511–530.

Kim, D. and Benbasat, I. 2009. Trust-assuring arguments in B2C e-commerce: Impact of content, source, and price on trust. *Journal of Management Information Systems* 26, no. 3: 175–206.

Kim, M. and Lennon, S. 2008. The effects of visual and verbal information on attitudes and purchase intentions in internet shopping. *Psychology and Marketing* 25, no. 2: 146–178.

Kotler, P. 2000. *Administração de marketing*. 10th ed. São Paulo: Prentice Hall.

Kotler, P. and Keller, K. L. 2016. *Marketing management*. 15th Global ed. Boston, MA: Pearson.

Lal, R. and Sarvary, M. 1999. When and how is the Internet likely to decrease price competition? *Marketing Science* 18, no. 4: 485–503.

Lander, S. Push vs. pull supply chain strategy. Seidel, M. (Rev.). 2019. Chron.com. Last updated March 05, 2019, accessed December 19, 2019, https://smallbusiness.chron.com/push-vs-pull-supply-chain-strategy-77452.html

Leite, P. R. 2009. *Logística Reversa: meio ambiente e competitividade*. São Paulo: Prentice Hall, 2009.

Mason, A. M., De Klerk, H. M., Sommervile, J., and Ashdown, S. P. 2008. Consumers' knowledge on sizing and fit issues: A solution to successful apparel selection in developing countries. *International Journal of Consumer Studies* 32: 276–284.

McCollough, S. S. 1999. *Mastering commerce logistics*. Cambridge: Forrester Research.

Molich, R. and Nielsen, J. 1990. Improving a human-computer dialogue. *Communications of the ACM* 33, no. 3: 338–348.

Moreira, C. S. C., Casotti L. M., and Campos, R. D. 2018. Socialização do consumidor na vida adulta: desafios e caminhos para a pesquisa. *Cad. EBAPE.BR* 16, no. 1: 119–134.

Nardi, B. and Kapetlinin, V. 2006. *Acting with technology: Activity theory and interaction design*. Cambridge, MA: MIT Press.

Nielsen, J. 1993. *Usability engineering*. San Francisco, CA: Morgan Kaufman.

Nielsen, J. and Budiu, R. 2014. *Usabilidade móvel*. Rio de Janeiro: Elsevier.

Novaes, A. G. 2007. *Logística e gerenciamento da cadeia de distribuição*. Rio de Janeiro: Campus.

Orendorff, A. 2019. The plague of ecommerce return rates and how to maintain profitability. shopify.com, accessed November 21, 2019, https://www.shopify.com/enterprise/ecommerce-returns.

Petrova, A. 2007. Creating sizing systems. In: Ashdown, S. P. (ed.), *Sizing in clothing: Developing effective sizing systems for ready-to-wear clothing*. Cambridge: Woodhead Publishing Ltd.

Pordata. 2019. População Europa, accessed July 19, 2019, https:www.pordata.pt.

Ricker, T. 2017. I've discovered the secret to worry-free online clothes shopping. *The Verge*. https://www.theverge.com/2017/5/3/15527516/free-home-pickup-stress-free-returns

Rogers, D. S. and Tibben-Lembke, R. 2001. An examination of reverse logistics practices. *Journal of Business Logistics* 22, no. 2: pp. 129–148.

Shneiderman, B. and Plaisant, C. 2005. *Designing the user interface*, 4th ed. Boston, MA: Pearson.

Shopify. 2016. *Fashion & apparel, cosmetic, jewelry and luxury industry report*. Ottawa: Shopify.

Similarweb. 2019. Top websites per category, accessed April 9, 2019, https://www.similarweb.com/top-websites/category/shopping/clothing.

Solomon, M. R. 2011. *Comportamento do Consumidor: Comprando, possuindo, sendo*. 9th ed. Porto Alegre: Bookman.

Statista. 2018. Number of digital buyers worldwide from 2014 to 2021 (in billions), accessed December 30, 2018, https://www.statista.com/statistics/251666/number-of-digital-buyers-worldwide/

The European Parliament and the Council of the European Union. 2011. Directive 2011/83/EU of the European Parliament and of the Council of 25 October 2011 on consumer rights, amending Council Directive 93/13/EEC and Directive 1999/44/*EC of the European Parliament and of the Council and repealing Council Directive 85/577/EEC and Directive 97/7/EC of the European Parliament and of the Council Text with EEA relevance*, accessed October 4, 2019, https://eur-lex.europa.eu/legal-content/EN/TXT/?uri=CELEX:32011L0083&qid=1403274218893.

The Nielsen Company. 2019. *Connected commerce*. New York: The Nielsen Company.

Tufte, E. [1990] 1998. *Envisioning information*. 6th ed. Cheshire: Graphic Press.

United Nations [UN]. 1987. United Nations General Assembly, accessed August 10, 2019, https://sustainabledevelopment.un.org/content/documents/5839GSDR%202015_SD_concept_definiton_rev.pdf.

United Nations. n.d. Sustainable development goals, accessed October 30, 2019, https://sustainabledevelopment.un.org/sdgs.

Wood, D. J. 1991. Corporate social performance revisited. *Academy of Management Review* 16, no. 4: 691–718.

Wroblewski, L. 2001. *Mobile first*. New York: A Book Apart.

Yoo, J. and Kim, M. 2014. The effects of online product presentation on consumer responses: A mental imagery perspective, *Journal of Business Research* 67, no. 11: pp. 2464–2472. doi:10.1016/j.jbusres.2014.03.006

Zhang, F. 2018. Rolling out mobile-first indexing. *Google Webmaster Central Blog*, accessed November 12, 2019, https://webmasters.googleblog.com/2018/03/rolling-out-mobile-first-indexing.html.

Zwaga, H. J. G., Boersema, T., and Hoonhout, H. C. M. (eds.). 2004. *Visual information for everyday use: Design and research perspectives*. 1st ed. Philadelphia, PA: Taylor & Francis e-Library.

ANNEXURES

Annexure 1. Websites Suggested by Specialists

Brand	Website	Country	Type of website	Products	App	Obs.
Acne Studios	acnestudios.com	Sweden	Responsive	Women I Men I Jeans I Face Collection	x	
Asos	asos.com	UK	Responsive	Women I Men I Jeans I Face Collection	√	Multi-brands
Études Studio	etudes-studio.com	France	Responsive	Women I Men	x	
Luisa Via Roma	luisaviaroma.com	Italy	Responsive	Men I Women I Boys I Girls I Home I Beauty Men I Beauty Women	√	
Net à porter	net-a-porter.com	UK	Responsive	Women	√	
Top Shop	m.eu.topshop.com	UK	Mobile	Women I Men	√	
Yoox	mobile.yoox.com	Italy	Mobile	Women I Men I Kids I Design + Art	√	

Annexure 2. Top Ten Female Apparel Websites. Source: Alexa (2019)

Rank	Website	Daily time on site[1]	Daily pageviews per visitor[2]	% of traffic from search[3]	Total sites linking in[4]
1	Hm.com	8:45	7.65	23.60	18,839
2	Lulus.com	6:42	5.39	28.60	1,862
3	Revolve.com	5:52	5.72	24.50	1,222
4	Freepeople.com	5:47	5.14	28.50	4,330
5	Modcloth.com	5:48	5.12	24.90	5,023
6	Anntaylor.com	5:03	5.21	26.80	4,367
7	Nelly.com	6:46	6.18	26.20	1,701
8	Toryburch.com	4:23	4.50	35.00	2,110
9	Monki.com	5:18	6.76	23.50	876
10	Talbots.com	7:46	7.82	23.30	823

Annexure 3. Top Ten Apparel Websites. Source: Alexa (2019a)

Rank	Website	Daily time on site[1]	Daily pageviews per visitor[2]	% of traffic from search[3]	Total sites linking in[4]
1	Hm.com	8:45	7.65	24	18,839
2	Gap.com	7:26	4.09	26	10,556
3	Zappos.com	3:36	3.81	38	7,766
4	Forever21.com	7:59	6.62	31	8,652

5	Yoox.com	8:05	9.70	8	2,139
6	Ae.com	7:31	6.79	35	2,827
7	Victoriassecret.com	7:45	6.90	26	4,831
8	Mango.com	7:05	6.11	22	4,899
9	Jcrew.com	4:27	3.64	23	5,525
10	6pm.com	6:49	5.42	14	2,573

[1] Daily time on site: estimated daily time on site (mm:ss) per visitor to the site. Updated daily based on the trailing 3 months.
[2] Daily pageviews per visitor: estimated daily unique pageviews per visitor on the site. Updated daily based on the trailing 3 months.
[3] Percentage of traffic from search: the % of all referrals that came from search engines over the trailing month. Updated daily.
[4] Total sites linking in: the total number of sites that Alexa found that link to this site.

Annexure 4. Top Ten Websites *de vestuário*. Source: Similarweb (2019)

Rank	Website	Change[1]	Average visit duration[2]	Pages/visit[3]	Bounce rate (%)[4]
1	wildberries.ru	=	00:12:58	17.86	23.35
2	hm.com	=	00:07:29	12.98	25.37
3	asos.com	1	00:07:32	11.66	32.47
4	zara.com	−1	00:06:47	13.18	24,57
5	trendyol.com	=	00:07:37	9.63	34.47
6	macys.com	=	00:05:45	6.45	44,58
7	next.co.uk	1	00:07:07	38.78	33.41
8	zozo.jp	−1	00:07:53	12.07	30.09
9	uniqlo.com	=	00:03:54	6.61	48.66
10	myntra.com	=	00:06:04	9.16	46.92

[1] Change: change in ranking from the previous month.
[2] Average visit duration: average time spent by users on the website per visit.
[3] Pages/visit: average website pages viewed per visit.
[4] Bounce rate: percentage of visitors that view only one page on the website before exiting.

Section 5

Case studies in Usability and User Experience

12 User Testing in an Agile Startup Environment
A Real-World Case Study

Carlos Diaz-de-Leon, Sarah Ventura-Basto, Juliana Avila-Vargas, Zuli Galindo-Estupiñan and Carlos Aceves-Gonzalez

CONTENTS

12.1 Introduction 222
    12.1.1 The Startup Way 222
    12.1.2 The Lean Startup and Human-Centered Design 222
    12.1.3 Agile Software Engineering, User-Experience and Human-Centered Design 223
12.2 Case Study: Fanbot Places 223
    12.2.1 The Product 224
    12.2.2 The UX Team at Fanbot 224
12.3 User Testing Iterations 225
    12.3.1 Iteration 1: The Pilot 225
        12.3.1.1 Goal 225
        12.3.1.2 Highlights of the Test 225
        12.3.1.3 Learnings to Improve the Product 226
    12.3.2 Iteration 2 226
        12.3.2.1 Goal 226
        12.3.2.2 Highlights from the Test 226
        12.3.2.3 Learnings to Improve the Product 227
        12.3.2.4 Changes to the Product 227
        12.3.2.5 Pending to Assess 228
    12.3.3 Iteration 3 228
        12.3.3.1 Goals 228
        12.3.3.2 Highlights from the Test 228
        12.3.3.3 Learnings to Improve the Product 230
        12.3.3.4 Changes to the Product 231
        12.3.3.5 Pending to Assess 231
    12.3.4 Iteration 4 231
        12.3.4.1 Goal 231
        12.3.4.2 Highlights from the Test 231

DOI: 10.1201/9780429343513-17

      12.3.4.3 Learnings to Improve the Product 233
      12.3.4.4 How These Results Modify the Product 233
      12.3.4.5 Pending to Assess 234
  12.3.5 Iteration 5 234
      12.3.5.1 Goals 234
      12.3.5.2 Highlights from the Test 235
      12.3.5.3 Learnings to Improve the Product 236
      12.3.5.4 How These Results Modify the Product 236
      12.3.5.5 Pending to Assess 236
  12.3.6 Iteration 6 236
      12.3.6.1 Goals 236
      12.3.6.2 Highlights from the Test 236
      12.3.6.3 Learnings to Improve the Product 237
      12.3.6.4 How This Results Modify the Product 237
      12.3.6.5 Pending to Assess 237
12.4 Final Recommendations 237
12.5 Conclusion 239
Acknowledgment 239
References 240

12.1 INTRODUCTION

12.1.1 THE STARTUP WAY

Eric Ries (Ries 2011) defines a startup as an organization designed to create a new product or service under the conditions of extreme uncertainty (Ries 2011). A startup is based on an idea or a solution to a problem which is identified as a potential business with a positive impact on users. The idea is that the solution is presented as a minimum viable product (MVP) or with Little Design Up Front (LDUF) (Startupcommons 2019).

The goal of a lean startup is to work through iterative feedback with users during the development of the product/service (Maurya 2012), which allows testing the core assumptions of the business (Muller and Thoring 2012).

12.1.2 THE LEAN STARTUP AND HUMAN-CENTERED DESIGN

The concept of Lean Startup is nourished by the "customer development" method (Blank 2006, Muller and Thoring 2012) which suggests that a startup should develop a product properly through a process in which it understands its users. The above approach or process ties with the concept of Human-Centered Design (HCD). An approach to the development of interactive systems is one that seeks to make them usable and useful when focusing on users, their needs, requirements, and when applying human factors/ergonomics, knowledge and usability techniques (ISO 2019).

Interactive systems designed under an HCD approach improve usability and user-experience. These two concepts are critical to this case study. Usability is defined as a characteristic of systems, products and services that allows them to be easy to use and learn, with the minimum number of errors (effectiveness), a certain degree of satisfaction and using only the resources necessary to achieve an objective (efficiency) (Nielsen 1994) in a specific context of use (ISO 2018).

The user-experience encompasses the perceptions and responses that result from the use or anticipated use of a product; this includes emotions, beliefs, preferences, perceptions, comfort, behaviors and achievements that occur before, during and after the interaction with the product or service system (ISO 2018).

As the processes of a Lean Startup, the HCD process is iterative and consists of repeating a sequence of steps until the desired result is obtained. The iterations should be used progressively to eliminate uncertainty during development, especially where there is human–computer interaction. Due to the complexity of interactions, it is not possible to cover every detail of the system at the beginning of its development (ISO 2019). It is worth noticing that the HCD can only intervene in the design aspects of the user-experience (UX) (ISO 2018).

12.1.3 Agile Software Engineering, User-Experience and Human-Centered Design

Agile software engineering (ASE) has as a priority user satisfaction, which seeks through short iterations where software improvements are worked on (Jurca et al. 2014). This mode of work is similar to the Lean Startup approach and HCD. The ASE itself does not cover the improvement of software usability, nor that of the user-experience. Therefore, there are efforts to integrate usability and UX practices into the ASE because the HCD processes improve both the experience and the usability of the systems (Jurca et al. 2014).

An agile process has short iterations and seeks to work on small elements of the product. On the opposite, the HCD spends efforts and resources on research and analysis before starting the product development, which has been seen as necessary (da Silva et al. 2012). However, it should be noted that the ASE and the HCD have common goals: user satisfaction. Also, both are human-centered and cyclic (da Silva et al. 2013).

It is assumed that human-centered design can improve ASE with a systematic approach to examine user needs. Agile processes provide more and more frequent iterations of tests with users, obtaining feedback that translates into changes in the product that responds to the findings or problems found in the evaluations (Jurca et al. 2014).

These three themes help in understanding the relevance of considering users under the development of a startup and while using an ASE methodology. The case study allows us to understand particular aspects of the value of user testing within the iteration process of the interface components and the user interaction.

12.2 CASE STUDY: FANBOT PLACES

Fanbot is defined as a startup company where software is offered as a service (Software as a Service or SaaS). Its main product focuses on customer loyalty programs for restaurants and other businesses. The characteristic of SaaS is to offer a license to use software through a subscription dynamic, and it is responsible for providing maintenance and all services related to the operation of the product (Ma 2007).

The value proposition describes Fanbot Places as a platform that increases businesses' *frequent client base in an automated way through the clients they already have;* this chapter is based upon this product. For Fanbot Places users, it is defined as *a community where users can choose what they can instantly win when visiting their favorite places.*

12.2.1 THE PRODUCT

Fanbot Places is an interactive system that is embedded in the Facebook Messenger® platform. Interactive systems are characterized by the combination of hardware and/or software and/or services and/or people with whom users interact to achieve a specific objective (ISO 2018). Its user interface (UI) has a sticker (placed in a table) with instructions to access the Chatbot through the scanning of a Facebook Messenger Code (FMCode) (Figure 12.1 shows a Fanbot Facebook Messenger Code).

A Chatbot is part of the interface where the interactions occur in a chat-type scenario, a human–computer dialogue with a natural language (Jia 2004). The user sends text messages, and the interaction takes place in a conversational way (Goyal et al. 2018). The Fanbot Places Chatbot tells the user through text messages, what are the steps to win a prize through playing a game and once the user selects the prize, he can access the game. At the end of the game, the Chabot shows the results and invites users to return to the same place (i.e., a restaurant) in the future.

12.2.2 THE UX TEAM AT FANBOT

The UX and usability concepts are part of the daily work of a UX designer. It is not the objective of this chapter to define the position or the different roles of the UX; it will

FIGURE 12.1. Fanbot Messenger Code (FMCode)

describe the role that the UX team played in Fanbot. The user-experience design team participated in the integration of an HCD approach to Fanbot and had a total of three experts, one of the team members served as a coordinator. The UX team did face-to-face work with the Fanbot team at least three days a week. The rest of the work was performed remotely, and the communication was done by messaging applications or email.

At the end of each iteration, a presentation was made with all other Fanbot members. The results of the iteration, the changes suggested to Fanbot Places and the feasibility of these were shown. Based on the results and feedback of Fanbot and the UX team, the next iteration and modifications to Fanbot Places were planned.

This chapter aims to illustrate the process of adapting user testing in a small agile startup. It describes the different user test design and their adjustments, the process of implementing them, how the results were translated into actionable steps, and how it changed the product design.

12.3 USER TESTING ITERATIONS

During nine months, the UX team accompanied the Fanbot team along with six user testing iterations. One of the aims of this intervention was to improve and adapt the protocol for user testing in each iteration. Through this process, the results allowed us to build a better version of Fanbot Places centered on users' needs.

In this paragraph, the description of the structure of each iteration is presented. Each one has five sections except the first one which has only three of the sections. First, the *goal* of each iteration is presented. Second, the *highlights of the test* describe the general characteristics of the participants, the instruments used or their modifications and other aspects of the test design. Third, *learnings to improve the product* highlights the most relevant results of each test that impact Fanbot Places or the test design. Fourth, *changes to the product* present an overview of the changes made according to business decisions or changes that emerged from the user testing to improve Fanbot Places experience. The last section, *pending to assess* mentions the elements that were not addressed in the current iteration, or business decisions that had to be considered as a starting point for the next stage.

12.3.1 ITERATION 1: THE PILOT

12.3.1.1 Goal

The goal was to assess the content of the questions on the protocol and the time that the participants were spending doing each test. It was important to determine the time that the UX team will need for scheduling and executing the tests efficiently.

12.3.1.2 Highlights of the Test

The UX team modified the sticker instructions based on an expert review. A new version of the sticker was designed and used for the pilot test.

Two participants collaborated in this pilot test. They were asked to read the sticker instructions and play the game Fanbot Places. Then, each participant was asked to enunciate the general steps they followed to play. This information served to fit the user's cognitive model with the Hierarchical Task Analysis (HTA) as previously made by the UX team.

After the test, they answered three questions for a qualitative assessment of the user-experience, related to their perception of Fanbot Places. The questions were: Do you have some trouble or difficulty playing? Are the sticker instructions clear? And, would you like to play again?

12.3.1.3 Learnings to Improve the Product

Regarding the questions, the most evident issue was the need to change the approach. The first question should have focused on difficulties during the interaction with Fanbot Places and the reasons for it. The question related to the reuse of the game or the Fanbot Places was eliminated. On the other hand, the UX team decided to add four assertions: (1) the sticker instructions are clear, (2) the chat messages are clear, (3) the speed with which the chat messages appear is suitable, and (4) the prizes were easy to identify. These affirmations were planned to be answered considering a seven-point scale, where one represented totally agree and seven was totally disagree. A final question was added to close the next test, asking for additional comments or ideas for improving Fanbot Places.

The time taken for each test lasted almost 20 minutes. This result allowed to add the four assertions and helped to schedule the participants for the next test and allowed them to run the tests more efficiently. This time was considered as a preset for the next iterations by the UX team.

The last achievement was the validation of HTA; this helped to analyze the errors (efficacy) in future tests and it contributed to relate errors that occurred during the user testing with specific elements of the user interface of Fanbot Places.

12.3.2 ITERATION 2

12.3.2.1 Goal

A lab test scenario was designed to assess the user-experience of new users of Fanbot Places and to identify the difficulties (efficacy) that arise from the starting point of the interaction (reading the sticker instructions) until the end of the game and the interaction with the Chatbot.

12.3.2.2 Highlights from the Test

In this iteration, 11 participants between the ages of 19 and 35 were recruited; this age range was determined because it represented the segment with the highest interaction with Fanbot Places according to Facebook® analytics. The only requirement was that the participants had not previously used Fanbot Places.

All participants used an Android OS phone provided by the UX team. All interactions were recorded and observed in real-time from a computer. The same sticker with textual instructions and the FMCode was used by all participants (Figure 12.2).

Similar to the previous iteration, participants were asked to describe the steps they took to play, in order to reinforce the UX team's HTA model of Fanbot Places.

User Testing in an Agile Startup

FIGURE 12.2. Iteration 2 sticker design

Since the pilot test showed that the time was sufficient, the modified questionnaire proposed in section *learnings to improve the product* of Iteration 1 (pilot) was applied.

12.3.2.3 Learnings to Improve the Product

As in the previous iteration, the scanning action was the most challenging task for most of the participants and was reflected in the number of errors recorded. Printed instructions were not clear for the participants, especially one that explained how to scan the FMCode. Also, the participants could not identify the FMCode, although it was printed on the sticker. This may be because they were new users and did not know how to do a code scan with the Facebook Messenger ® camera.

In contrast, participants mentioned that the chatbot and game instructions (explained by the chatbot) were not clear enough and that the messages were displayed too quickly. They also pointed out that the images of the prizes were difficult to comprehend, probably because they did not have visual consistency (some were photos, and others were drawings).

12.3.2.4 Changes to the Product

It is important to note that the scanning task could not be modified to make it easier for the users, because the scanning action is embedded in the way Facebook Messenger ® operates. For this reason, the changes to the product were focused on improving the clarity of the sticker instructions, since it is an element that could be modified, and it was the first element of Fanbot Places interface that users interact with and could facilitate the interaction with new users. Therefore, the sticker design, instructions (content and writing) and distribution of the elements (texts and the FMCode) were modified by the UX team.

The texts of the chatbot messages and the game instructions in the chatbot were modified, the speed with which the messages appear was reduced in order to emulate the time it takes for a real person to write them.

The images of the prizes were changed, the user interface (UI) designer made different illustrations so that they were shown with a uniform visual language.

12.3.2.5 Pending to Assess

Considering the results and modifications, in the next iteration (Iteration 3) the UX team focused the test design on assessing the new sticker design, the chatbot messages and the new illustrations of the awards.

One of the problems identified in the development of the test is that some users used the iOS operating system daily and they were asked to perform it on an Android device. This may have somehow affected the test results. This is a call for action to provide the user with any of the two popular operating systems (Android or iOS) when a lab scenario-type test is performed.

12.3.3 ITERATION 3

12.3.3.1 Goals

Based on what was found in the previous test, two instruction options were designed for the sticker, the first one had textual instructions and indicated the user to scan through the camera feature in Facebook Messenger® (Type A), Figure 12.3 shows Type A sticker. The second one had textual instructions with icons and indicated the user to scan through a contacts function button in Facebook Messenger ® (Type B), Figure 12.4 shows Type B sticker. The objective was to determine which of the two types of instructions were clearer and which process facilitated scanning the FMCode.

In addition, within the user interaction with Fanbot Places a Facebook "check-in" was added for users to exchange it for a second chance to play the game; this feature was proposed in an ideation workshop between the previous iteration and this one. The objective of this change was to evaluate whether interrupting the first opportunity to play at the second mistake to form the figure in the game was perceived positively.

This test also evaluated the chatbot instructions (clarity and speed) and the comprehension of the new illustrations of the prize.

12.3.3.2 Highlights from the Test

A total of 14 people participated who were recruited virtually and personally. Of the 14 tests that were carried out, half of them used the Type A instruction and the other half the Type B instruction.

The post-test questionnaire included the same items as the previous iteration and only two questions were added: *In the first game, making a mistake twice offers you*

User Testing in an Agile Startup

FIGURE 12.3. Type A sticker design

FIGURE 12.4. Type B sticker design

another chance, would you have liked to finish the game? And, *had you scanned codes with your phone? What kind?*

12.3.3.3 Learnings to Improve the Product

The textual instructions (Type A) reflected more errors in the scanning task as opposed to instructions with icons (Type B). Once the two instructions were shown to the participants, participants commented that the textual instructions were more precise, since there were fewer steps to reach the same goal. It may be that the number of errors in scanning was lower in the instruction with icons because the scanning route was different. When using this route, the camera viewer already had the form that suggested the code scan (Figure 12.5).

One of the objectives of the test was to know if users felt motivated to publish a "check-in" on Facebook in exchange for another opportunity to play the game, this was offered when they make two mistakes and the game was stopped immediately, leaving the first game unfinished. Given this, most participants would have liked to finish the game even if they assumed they would lose, but their argument for continuing their playing was that *it's a very short game to get cut before finishing it*. Some

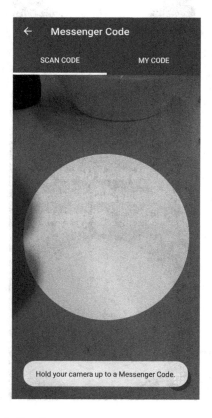

FIGURE 12.5. Scan code function

User Testing in an Agile Startup   231

participants were left wanting to continue playing or made them think that the game was fixed and generated distrust.

The modified chat messages were perceived as clear and distinct. However, the speed at which they were displayed was still perceived as quick even though they provided a positive rating considering the score of this feature.

The new illustrations of the prizes were also validated as more than half of the participants correctly remembered the type of prizes when asked to do so.

12.3.3.4 Changes to the Product

It was recognized that according to users' comments, the option of using icons on the printed instructions was discarded. In the end, the Type A instructions route was selected to go forward with; this is because the "Scan code" function, a step from type B instructions, was disabled from the contacts section, as well as the Facebook Messenger application. At the time of writing this chapter, the use of codes remains available despite Facebook's statement about the removal of this feature.

The decision was made to reduce the speed with which the chatbot messages appear, as the participants perceived that they were showing up very quickly.

Regarding the check-in, the user was allowed to finish the first game and then offered him a second chance to win his prize by redeeming for it a Facebook "check-in" in the place where Fanbot Places was playing.

12.3.3.5 Pending to Assess

Based on a design sprint after this iteration, the "star accumulation" dynamic was added to motivate users to return to the places they already visited in order to accumulate stars (one correct action in the game equals a star) and then redeem them for a prize. Therefore, it was necessary to include the evaluation of this new feature in the next iteration.

12.3.4 ITERATION 4

12.3.4.1 Goal

Once again, the goal was to identify opportunities resulting from the changes made in the user graphic interface or to the instructions. Although this iteration focused mainly on assessing how users perceive the status bar (Figure 12.6) and if they understand the dynamics of star accumulation (Figure 12.7). For the first time, both new users and experienced users of Fanbot Places were evaluated.

12.3.4.2 Highlights from the Test

A total of seven people participated in this iteration, three of them were new users and the rest had previously interacted with Fanbot Places. After reviewing the Fanbot database, participants with the highest number of interactions were invited to participate.

To achieve the objective of the test, the post scenario questionnaire was modified as follows: only the question regarding the general difficulties of the interaction was

FIGURE 12.6. First design iteration for the status bar.

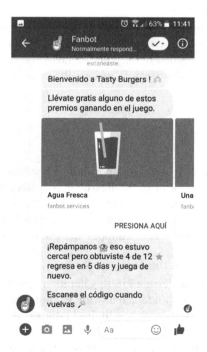

FIGURE 12.7. Star accumulation dynamic explained by the Chatbot

User Testing in an Agile Startup 233

maintained, and instead of asking about comments to improve the experience, participants were asked how can Fanbot improve Fanbot Places to satisfy their needs. The following seven questions were added: Did you see the status bar that was at the bottom of the screen during the game? What kind of information did the status bar give you? Do you remember what the score was accumulated in your game? How did you win the stars? What was the main benefit you have received from Fanbot Places? What would you change in order to improve the status bar? and Have you recommended Fanbot Places to someone? In addition to the above questions, three statements were included that were answered and assessed with the same Likert scale of the previous iterations: (1) The status bar indicates the missing points to earn the prize; (2) I like to be rewarded when I visit a place; and (3) it motivates me to visit a place when I am few points away to earn a reward (prize).

12.3.4.3 Learnings to Improve the Product

Participants who had never interacted with Fanbot Places experienced difficulties during the test, especially when scanning the FMCode. In contrast, experienced users had no problem scanning the FMCode since they had done it previously. This could suggest that the difficulties faced in the first interaction might be overcome after a couple of interactions.

The fact of not being able to modify the Facebook Messenger interface to eliminate the difficulties of scanning tasks suggested the responsibility for providing proper communication in all other elements of the Fanbot Places interface, the sticker and the chatbot.

Users did not relate a correct action in the game with the win of a star; this may be because no visual or auditory feedback helps to generate this association: *correct action – good consequence*. Additionally, it was identified that users focused their attention on the game, and because of this, they might not have seen additional elements of the interface such as the status bar. No participant identified the function of the status bar or the relationship to the dynamics of star accumulation because they did not know how they obtained the stars.

12.3.4.4 How These Results Modify the Product

Based on the reported lack of understanding of the dynamics of star accumulation and the status bar, additional messages were added to the chatbot before starting the game. The intention was to represent clearly and explicitly the star accumulation dynamics. The post-game messages were modified accordingly. Additionally, an animation was added to provide feedback and facilitate the association between successes and winning stars (correct action – good consequence).

Regarding the status bar, it was decided to modify the design within the user interface (Figure 12.8) to evaluate if it was easier to perceive it. Also, the number of stars to be collected to win a prize was modified. This change was done based on the question, how many times should a user visit a place with Fanbot Places to win a prize? As a result of this, the data scientist made calculations to reduce the number of stars required to earn a prize. The idea was to achieve a balance between the user opportunity to win a prize and the number of prizes that could be awarded per place.

FIGURE 12.8. New design for the game interface and status bar.

In this way, the Fanbot Places system keeps customers excited to get the prize they have been after and business owners do not give more prizes than the estimated ones.

12.3.4.5 Pending to Assess

The following iterations focused on the evaluation of the design of the new graphic interface of the game; this included the new design of the accumulation bar and feedback animations that sought to reinforce the action of winning a star. Also, the chatbot texts were assessed, especially those that referred to the status bar and the dynamics of star accumulation.

12.3.5 Iteration 5

12.3.5.1 Goals

In this iteration, instead of performing the test in a controlled environment, it was carried out in the different restaurants that had paid for Fanbot Places subscription intending to simulate more real user interaction with Fanbot Places.

The user-experience was assessed based on the results of the previous iterations. This time, it was aimed to assess the level of understanding of the new design of the accumulation bar and the new Chatbot texts.

In this iteration, the sticker instructions were not assessed because the version of Iteration 4 was defined as the standard with a positive perception by users. It was

User Testing in an Agile Startup

decided by the Fanbot team that this version will not change unless the scanning process in Facebook Messenger is modified.

12.3.5.2 Highlights from the Test

Ten places were selected for the testing; the requirement for restaurant selection was that they must have placed the instruction sticker with the latest authorized design by Fanbot. The UX team scheduled testing times based on the number of visitors each restaurant received. Fanbot's internal analytics data and Google Maps data (Web Desktop Version) were crossed to identify the day and time with most Fanbot Places interactions and general visits.

In total, 33 participants were recruited for this iteration, 13 men with an average age of 27 years and 20 women with an average age of 24 years. The UX team invited people who were closer to the stickers or people who were waiting at the counter for their order to be delivered. If the person agreed to take part, the team members proceeded further to apply the test.

This test used the stickers that were already placed at the restaurants (Figure 12.9). The phones used for the tests were owned by the participants with the only requirement that they should have installed the Facebook Messenger application. In some cases, the UX team provided a mobile phone and recorded it on the database.

Researchers, using their mobile phones, created Google Forms to record the results of each test and the participants' responses. The same questionnaire of Iteration 4 was used but the three statements were replaced by the following two statements: the instructions on this screen are clear (referring to chatbot messages), and the difficulty level of the game is adequate.

FIGURE 12.9. Sticker design for iteration 5 and 6.

12.3.5.3 Learnings to Improve the Product

Only one-third of the participants detected the status bar and only six among them identified the status bar function. Participants suggested that the bar should be highlighted more and that instructions regarding the status bar and the game dynamics should be improved.

Half of the participants were able to remember their accumulated stars and 23 of them were aware of the information that they won the stars when performing a correct action during the game.

Among the suggestions recorded to improve the game, users suggested increasing the variety of games to have more opportunities to play in a single visit. It was recommended that the FMCode to be easier to scan and the participants be able to play the Fanbot Places game without the need for a sticker. This last comment unleashed a new Fanbot Places functionality in which the user had the opportunity to play from anywhere with the goal of completing stars to get one of the prizes.

12.3.5.4 How These Results Modify the Product

From this iteration, the UX team decided to improve the design of the graphic interface of the status bar. Also, the Chatbot texts that refer to the star accumulation dynamic at the end of the interaction should explain more clearly how users won the stars, how many they have won and how can accumulate more.

12.3.5.5 Pending to Assess

After this iteration, the pending is the evaluation of the redesign of the accumulation bar and the chatbot texts at the end of the interaction that refers to the dynamics of the accumulation of stars.

12.3.6 Iteration 6

12.3.6.1 Goals

The main goal of this iteration was to assess the chatbot texts that explained the dynamics of the accumulation of stars at the end of each interaction. This evaluation included the assessment of all texts (messages) of the chatbot.

It also evaluated the users' perception of the game difficulty level; if participants were able to perceive the status bar, and how they named or identified the different elements of the Fanbot Places interface. The intention was to include those names in later versions of all texts of the interface.

12.3.6.2 Highlights from the Test

The test was carried out in a lab-type scenario again. A total of 11 participants without any previous interaction with Fanbot Places were recruited (two men and nine women). The average age of the participants was 38.80 years.

For this iteration, the questionnaire retained only the items related to the difficulty of the interaction, the accumulated score and how the stars were obtained. Also, the two statements of Iteration 4 were maintained. In the same questionnaire,

User Testing in an Agile Startup 237

participants were asked to name the different elements of the user interface (sticker, FMC, prices, the name of the game and the star bar).

12.3.6.3 Learnings to Improve the Product

The first instructions of the chatbot conversation were rated as clear and simple by eight participants, while one did not read them and two commented that they were not clear; Seven participants pointed out that the chatbot instructions were clear. Six participants rated as clear the messages that explained the dynamics of star accumulation; three commented that they had to read it twice to understand them, and just one commented that messages were not clear. Three participants commented that they had not seen the bar. However, this time seven respondents commented that they were able to remember their final score, which can be explained by the fact that the final number of accumulated stars was indicated at the end of the chatbot conversation.

The level of difficulty of the game was perceived as challenging by four participants, and two of them commented that it seemed to them that the game was tricked. One participant said that he had to guess how it worked while playing. The task of scanning the code still gives the perception of difficulty. Finally, the instructions about the game were perceived as not clear.

After the user interaction, the different elements of the UI were presented individually. Users responses were analyzed, and as a result the names of the elements were modified. This was considered for re-writing the text messages of the chatbot.

12.3.6.4 How This Results Modify the Product

Based on the test results, chatbot copies and the names given to different elements of the Fanbot Places interface were modified. The UX team proposed to create a GIF or an animation to explain the game's dynamic to make it clearer to users. This animation should be displayed before starting the game as a set of instructions on how the game works.

12.3.6.5 Pending to Assess

After this iteration, the intervention of the UX team concluded. However, the evaluation of the new chatbot messages and the creation of the instructional animation of the game remained pending.

12.4 FINAL RECOMMENDATIONS

In this section, some recommendations are presented to help the UX practitioners to improve the implementation of user testing in ASE or Startup environment:

Starting the user testing process with a pilot test allows identifying the different elements that integrate the product and helps to define an effective and efficient way to focus the after-test questionnaire to the users. Also, it defines the test timetable and the adjustments to the protocol.

In each iteration, the findings of the previous test, the improvements to the test protocol and the changes in the business model were considered to determine the adjustments that would be made in the next test design. Thus, each iteration was a learning opportunity, not just to improve the product; it was also a way to improve what the UX practitioners apply and how they implement their techniques and skills. For the Startup or agile environment, each iteration is an opportunity for embedding a human-centered design culture.

These iterations made it possible to identify those elements in the user interface that do not have the possibility of being modified – such is the case of the Facebook Messenger interface – which suggested the need and responsibility of the startup for improving other interface elements like the sticker and the Chatbot messages. That situation steered the UX team efforts, both in interface design and user testing, focusing on improving such elements to help the user in achieving the objectives of the interaction.

As a lesson learned, the UX practitioner should consider providing the participants equipment with the characteristics of the operating system that they regularly use to avoid unfamiliarity errors or biases, unless the goal of the test is to assess the adaptation to new characteristics or a different product. For example, if a user uses an iOS device, a highly similar device must be provided for testing.

Applying a post-test questionnaire as an interview allows UX designer to deepen the participants' responses and determine specific reasons why individual decisions are made by the user during the interaction or knowing the reasons behind lack of satisfaction.

In the in-between stages of each iteration, different ideation workshops were held at Fanbot, such as design sprints, gamification workshops and empathy maps. The results of these workshops were new features that were implemented in Fanbot Places and were tested with users in the next iteration. Product innovation did not necessarily come from user participation but the user testing allowed validating and adjusting the design to user characteristics, needs and expectations.

The results of each iteration helped to build different hypotheses and assumptions regarding the user-experience. The iterative process helps to test if they were correctly carried out; and if they were not, changes should be made. Assumptions and hypotheses can come not only from the user testing but also from different places such as within the business or startup, but this process (user testing) is an efficient way to validate this.

The post-testing meetings to present and discuss the results of the user testing with the full team or principal stakeholders at Fanbot allowed to improve more efficiently the features of Fanbot Places and focus on actionable steps that would impact Fanbot Places design and the user-experience. Structured meetings allowed the whole team to be on the same page, working toward the same goals. It is suggested that meetings and face-to-face time with the UX team should always have a defined structure and an opportunity for idea discussion during these times.

The usability was not assessed as a whole in all iterations since all its dimensions (efficiency, effectiveness and satisfaction) were not taken into account. In most of the iterations, just effectiveness was assessed, but this does not affect the results the UX team was trying to achieve. The UX team must know what commitments can be obtained at the time of planning each user testing. This is an excellent practice to efficiently focus the user test and obtain information that will help to justify decision-making in the design of products, services or systems.

12.5 CONCLUSION

In this chapter, different iterations of user testing were shown in which changes in the different aspects of Fanbot Places were assessed (the user interface and the user-experience), as well as the new functionalities that emerged in the process. User testing made it possible to determine which elements help or block users from having a satisfactory experience.

One of the achievements of the intervention in Fanbot was to permeate the UX processes and HCD scope to other elements of Fanbot services besides Fanbot Places. One of the reasons why it was achieved was that the intervention started with a small and concrete goal (the improvement of the sticker instructions). Hence the intervention was scaled up to a snowball effect which allowed the UX team to gain confidence, voice and vote within Fanbot.

The iterative process not only allowed to improve the elements of Fanbot Places design, but it also allowed the UX team to improve the quality of the test design and make it more efficient. All the improvements of the tests were based on the previous iteration. At the end of the intervention, the ability to adapt, change and plan those tests was performed at the same pace as the other teams within Fanbot.

Also, during the iteration process, it was possible to share face to face the results of each iteration with all Fanbot members and thus obtain objective and actionable feedback. In this way, the UX team was to be conceived as a fundamental element for decision-making concerning the product, with a clear focus on benefits for the business and mainly for the users.

It should be noted that not only user testing can demonstrate the importance of users in this project, but also other activities carried out by the UX team, which were briefly mentioned throughout the chapter and it allowed different modifications to Fanbot Places. Some of UX team activities during the intervention at Fanbot were: user research, analytical review, planning and execution of user tests (usability and user-experience), hierarchical task analysis, user journeys, creation of proto-personas, manage ideation and design sprint workshops, generation of gamification canvas, value proposition canvas, empathy canvas, and development of service blueprints. Each activity allowed to permeate all Fanbot members with the concepts of UX and HCD, also helped the UX team to adapt traditional user testing process to the lean and agile methodologies. Each step on this intervention showed how the UX team proceeded and made decisions in a defined time frame in order to achieve the expected results for improving the service.

ACKNOWLEDGMENT

We thank the entire Fanbot team for allowing us to experience the startup way and also for letting us implement our user-centered vision in their products and for permitting us to freely share these experiences, learnings and data.

REFERENCES

Blank, S. 2006. *The Four Steps to the Epiphany*. K & S Ranch.

Da Silva, T., Silveira, M., Maurer, F., and Hellmann, T. 2012. User experience design and agile development: From theory to practice. *Journal of Software Engineering and Applications* 5, 743–751.

Da Silva, T., Silveira, M., Melo, C. O., and Parzianello, L. 2013. Understanding the UX designer's role within agile teams. In *International Conference of Design, User Experience, and Usability*, ed. Marcus A., 599–609. Springer.

Goyal, P., Pandey, S., and Jain, K. 2018. Developing a chatbot. In *Deep Learning for Natural Language Processing*, 169–229. Apress.

ISO. 2019. ISO 9241-210:2019. *Ergonomics of Human-System Interaction. Part 210: Human-Centred Design for Interactive Systems*.

ISO. 2018. ISO 9241-11:2018. *Ergonomics of Human-System Interaction. Part 11: Usability: Definitions and Concepts*.

Jia, J. 2004. The study of the application of a web-based chatbot system on the teaching of foreign languages. In *Society for Information Technology & Teacher Education International Conference*, eds. R. Ferdig, et al., 1201–1207. Association for the Advancement of Computing in Education (AACE).

Jurca, G., Hellmann, T., and Maurer, F. 2014. Integrating agile and user-centered design. In *Agile Conference*, ed. IEEE, 24–32. IEEE, Kissimmee, FL, USA.

Ma, D. 2007. The business model of "software-as-a-service". In *IEEE International Conference on Services Computing*, ed. IEEE, IEEE Salt Lake City, Utah, USA, 701–702.

Maurya, A. 2012. *Running Lean: Iterate from Plan A to a Plan that Works*. O'Reilly Media.

Müller, R. M., and Thoring, K. 2012. Design thinking vs. lean startup: A comparison of two user-driven innovation strategies. In *Leading through Design*, eds. E. Bohemia, J. Liedtka and A. Rieple, 91–106. DMI: Design Management Institute.

Nielsen, J. 1994. *Usability Engineering*. Elsevier.

Ries, E. 2011. *The Lean Startup: How Today's Entrepreneurs Use Continuous Innovation to Create Radically Successful Businesses*. Crown Books.

Startupcommons. 2019. What is a startup. https://www.startupcommons.org/what-is-a-startup.html

13 User-Experience and Usability Review of a Smartphone Application
Case Study of an HSE Management Mobile Tool

*Marcello Silva e Santos, Sebastian Graubner,
Linda Lemegne, and Bernardo Bastos da Fonseca*

CONTENTS

13.1 Introduction 241
13.2 Theoretical Framework 242
13.3 Development of a Mobile App for HSE Management 245
13.4 The PrüfExpress System Case Study: Planning, Design, Operation,
       User-Experience and Usability Review 248
       13.4.1 State-of-the-Art Mobile Application for Testing and
               Documenting Work Equipment 248
       13.4.2 LegalDocumentation 249
       13.4.3 Testing and Documentation Process 250
       13.4.4 Concept, Technology and Programming Languages 250
       13.4.5 The Architecture of the System 251
13.5 Final Considerations 256
References 257

13.1 INTRODUCTION

This book chapter starts with a comprehensive description of related texts covering the user-experience (UX) and usability of various types of mobile applications. Then it addresses issues related to recent developments in the design and evaluation process, proposing some quite appropriate and opportunist discussions as Industry 4.0 advances in the transformation of human and, needless to say, social fabric as well. At last, it presents a case study of the implementation of a mobile app called PrüfExpress, developed by Graubner Industrie-Beratung GmbH. This mobile application is part of a Health Safety and Environmental (HSE) Management System

DOI: 10.1201/9780429343513-18

outlined by one of the largest automotive manufacturers in the world, with a pilot project in Gaggenau, Germany.

It is important to emphasize that the theoretical background describes not only particularities in the study of usability and user-experience subjects. The advent of the digital – now more mobile – society of the past few decades shortened the conceptual gap that used to exist between the product design and its use. A mobile device is, as some describe, an extension of a modern man. And as men and women are forced to evolve along their "extensions," the creative process of mobile apps is somewhat "tainted" by user needs. In fact, a great portion of this chapter deals with the issues related to the implementation of the mobile system, approaching even indirectly the Usability and UXD implications within the process.

As it will be read further, the development of an industrial mobile application always starts with "organizational user" requirements or needs. This user will most likely never try the app once it is implemented within a particular work system. But it will definitely receive feedbacks that should encourage or not the pursuing of similar new strategies of bringing mobile functionalities to management systems. In the case of the PrüfExpress system, the operational results and its integration within the production process – by employing Radio Frequency Identification(RFID) sensors, for example – and also as an Enterprise Resource Planning (ERP) resource, are definite proofs that mobile applications are not only a trend tool but rather a muscular accessory for the manufacturing and management systems.

13.2 THEORETICAL FRAMEWORK

This text intends to reach multiple aspects of the system's design from its preconception to its use phase and further evaluation. So, a general overview of user-experience and Usability in the Design development is necessary to introduce the specific issue of mobile application for industrial environments, which in fact emerges as another important topic to be addressed in this theoretical framework.

The ISO 9241-11 standard defines usability as the "extent to which a product can be used by specified uses to achieve specified goals with effectiveness, efficiency and satisfaction in a specified context of use." A more pragmatic definition of usability would be the property of an object to adequately serve its purpose by all means, with qualities ranging from technical efficiency to its conformity and configuration or as being easy to use when looked at as a function of the context in which the product is used. Or as Nielsen (1994) puts it, "a property of the system: it is the quality of use in context." He suggests five attributes of usability:

- Efficiency: the quality of resources appropriated in a product in terms of the relation to the accuracy and completeness by which users achieve goals;
- Satisfaction: the quality of positive attitudes towards the *comfortable* use of the product, emphasizing individual susceptibility as for the highlighted condition;
- Learnability: the quality in which a product should be easy to learn so that the users can rapidly start getting work done with it;

- Memorability: the quality in which characteristics of the product are easy enough to remember so that the casual user is able to return to it after some time without having to learn everything all over again;
- Errors: the quality of having a low error rate, so that users can easily recover from them and that gross errors must not occur.

As heuristics to generic usability standards is employed along with Apple's, Google's and Nielson's usability rules for mobile apps, Table 13.1 sums up some guidelines that can be followed by developers in any OS or platform.

In the case of the PrüfExpress 2 mobile application, whose details will be further addressed in a specific session, it was based on a previous version, called PrüfExpress 1, a personal digital assistant (PDA) application for Windows Mobile. PrüfExpress 1 evolved over more than 10 years. It was developed and perfected in close communication with the users. The development team was permanently on-site and met very frequently with the user team, at least monthly, sometimes weekly or even daily once technical issues would arise. In some events, the chief developer would interact with a leading user, responsible for the process, and by this mutual learning approach they managed to continuously improve the application. Because it derived from an extensively tested and validated previous version, PrüfExpress 2 did not go through a formal usability verification process. Instead, it was developed in close exchange with power users, by observing how they got along with the programmer's ideas and then rectified if necessary. It is fair to say that it also followed, intuitively, Google Usability Guidelines.

User-experience, on the other hand, covers the whole product or service acceptance. It encompasses all aspects of an object, from its pragmatic, practical features to the hedonic aspects of a given product. The pragmatic or instrumental refers to the utilitarian aspects, such as usefulness and ease of use. The hedonic or

TABLE 13.1.
Sample of Mobile Apps Usability Heuristics

Usability rule	Nielsen rule (1993)	Google usability guidelines	Apple usability guidelines
User control	User control and freedom	Decide for me, but leave the final decision to me	User control
Consistency	Consistency and standards	Apps use consistent design, typography and contents	Consistency
		Apps use familiar patterns, same iOS, UI standards (navigation bars, buttons, etc.)	
		Looks alike; performs alike	
Information and visual hierarchy	–	Information hierarchy and structure	–

Source: Swaid; Suid, 2017, adapted by Authors.

non-instrumental to the emotional and experiential aspects of product use (Partala; SAARI, 2015). The fact is that research related to user-experience has gained considerable interest from both scholars and practitioners. Generally, the research field of user-experience is seen to include all factors which affect the user's interaction and experience of a system or product (Yazid; Jantan, 2017). Based on a comprehensive bibliographical review (de Paula; Menezes; Araujo, 2014; Ibrahim; Ahmad, Shafie, 2015) and Nielsen's Usability criteria (Nielsen, 1993 & 1994) the UXD elements that should be taken into account for mobile application development is summarized – not necessarily in that particular order – in Table 13.2.

As Yazid and Jantan (2017) point out, it is no surprise that UXD has become so important in the development of mobile applications, thus having a huge influence on product success or failure. The combination of utilitarian and non-utilitarian properties in a mobile app application brings a new paradigm in UXD, since the app cannot be evaluated without a context. Likewise, a context depends on time and scope issues, which brings into play the product and the system in which the app becomes fully functional. In other words, the device that runs the app and its "drivers" become part of the assessment context. Hence, this creates a totally different scenario for assessing how a user feels and how he or she perceives the kind of relationship with a mobile app. In fact, it is difficult to define a successful mobile application because it is tied directly to user acceptance, which may vary immensely.

Mobile apps are basically software applications designed to run on smartphones, tablets and other mobile devices. In a recent paper, as Swaid and Suid (2019) cited info on the growth of the mobile apps business, they went ahead and outlined some projections. Back in the year 2016, it was expected that by 2020, mobile apps would generate around 189 billion US dollars in revenues via app stores and in-app advertising. But the explosion of mobile apps that came in just about every industry and the multitude of possibilities of their use was about to amaze the most skeptical tech-savvy geek on the planet. As of 2018, the updated prediction shows almost half-trillion dollars of revenues for 2020 and just a bit shy of US$ 1 trillion by 2023 (Statista, 2019).

TABLE 13.2.
UXD Elements for Mobile Apps Design

Number	UXD elements
1	Ease of use
2	Learnability
3	User interface
4	User satisfaction
5	Security
6	Behavioral intent
7	Environment

Source: Swaid; Suid, 2017.

UX and Review of a Smartphone Application

Due to facts like these and bound to the numerous advantages of mobile applications, designing a good mobile application has become a primary issue for companies, software programmers and designers, which gave birth to an interdisciplinary development effort to maximize the effectiveness of mobile apps. For industrial applications, for example, there are additional requirements or constraints, such as the need for integration with Integrated Management Systems, ERP models and so forth. Plus, sometimes it may involve change management issues, particularly in situations in which mobile apps eliminate human tasks or even human efforts. This is fast becoming a major problem as artificial intelligence (AI), automation and the whole IoT evolution come to play.

13.3 DEVELOPMENT OF A MOBILE APP FOR HSE MANAGEMENT

In order to contextualize the discussion presented herein, an HSE (Health Safety and Environmental) Management support system, the PrüfExpress Mobile App, was used as a case study. The goal is to illustrate how a mobile app can deliver an effective UXD approach that is required to achieve both user acceptance and organizational expectation when it incorporates the end user input and provides adequate usability. This section deals with issues involved in the process of ideation of such HSE Management control support system. In other words, how did all different elements in the system design came about, like motivation, planning and the development itself of a Mobile App, specifically designed for optimizing HSE Management?

Since 2006, the PrüfExpress software is being used by Daimler Corporation to manage HSE equipment and routines, especially for documenting the testing of work equipment. Currently, the system is only designed for Daimler and is used in six plants. In general, the plants differ in the types (or groups) of plants that are managed with PrüfExpress. While the Gaggenau(Germany) plant documents the testing of load handling attachments (see the left picture in Figure 13.1) or shelving units with PrüfExpress, the Ulm plant checks cranes (see the right picture in Figure 13.1) and documents them with PrüfExpress.

Each system type has its own characteristics and properties that must be observed and documented during testing. These properties are also known as master data. In the case of cranes, for example, the residual usability of the tested system must be

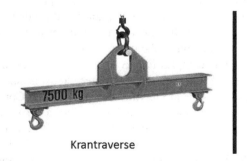
Krantraverse

Pick-Up Kran

FIGURE 13.1 Source: Graubner Industrie GmbH.

calculated after each test –the residual usability communicates how long the tested system can still be used. In addition, there are situations in which two plants are dealing with the same asset group. This is where the difference between these plants lies in the way the data is entered (the name of the installation, the required characteristics). The data recorded at Gaggenau or the designations of Gaggenau do not necessarily correspond to those of Ulm or Rastatt. In other words, the general conditions for the use of PrüfExpress vary from factory to factory.

Previously, each work was treated as a separate case. This means that each plant has its own system (from the database to the client application) and this is maintained independently of the others. The problem in the current situation is the inefficient maintenance and expansion of these systems. However, it often happens that new features must be implemented for the six works. In such cases, this function was written six times for the six systems. With only one system, one could have saved the time for the implementation in five other plants. A universal system for all these works and possibly for external companies must be developed within the framework of this work, since the demand for mobile applications on the shop floor increases at a fast pace, never seen before with other technologies.

In fact, if the advancements in mobile technology may surprise many, the fact is that most people do not realize the speed of the transformations the world is going through, as everyone becomes a mobile citizen of planet Earth. As eerie as it may sound, mobile phones only started to be used as personal computing devices in 2007. Thus, the mobile apps industry coincides with Apple's introduction of the iPhone, although its phenomenal growth is due to the entry of several competitors into the marketplace, notably Motorola, Samsung and LG, to cite some. This competition has given rise to an entirely new product that has become known as smartphones (Rakestraw;Eunni;Kasuganti, 2013).

Smartphones brought far greater functionality and usefulness than normal mobile phones exactly due to their ability to run mobile apps. According to Mobilewalla.com, a website dedicated to cataloging and rating apps, the one-millionth app was made available to users in December 2011 (Newark-French, 2011). Amazingly, it was not long ago that today's Smartphones and their associated mobile software applications or "apps" became a ubiquitous part of all human being's daily life. Most mobile apps are basically compact versions of regular computer software, designed for facilitating peoples' lives as they are on the run. However, some apps are much more useful, doubling as work tools for operational performance and productivity. It is possible to cite numerous examples of such applications, from physician's aids that can help to diagnose a disease by analyzing a photo taken with a smartphone to managerial tools that can be embedded in complex ERP Systems of large industries.

As one of the largest statistical mobile data repositories states, "most articles on mobile marketing and mobile usage statistics are terribly outdated" (Blue Corona, 2020). Thus, taking into consideration that mobile technology adoption is speeding at an ever-increasing pace, any data older than a couple of years might be considered only informative, not a definite trend. Take for instance the data on Spending on mobile apps worldwide 2009–2015, published by Statista Research Department, in 2018, on their website. It shows a stairway shape graph of evolution from 4 to 35 billion dollars spending on the sector. However, it fails to consider the geometric progression as the timeframe advances and other important variables, such as the number of free apps available and special usage apps, such as the ones designed for the industry. They "escape" the reach of the commercial data statistics.

With the new decade approaching, it is wise and quite timing to analyze the past, assess the present and embrace the future of data. Most people involved with high-tech research agree that those who do not aggressively embrace artificial intelligence (AI) and machine learning to optimize the use of data will be left behind and lose performance and profitability. Data collection is more than widespread; it shapes the way that the best companies in the world do business nowadays. When compared to non-data-driven companies, data-driven organizations are 23 times more likely to acquire customers, 6 times as likely to retain customers and 19 times more likely to be profitable (Bokman et al., 2014).

There are two main types of mobile applications: native and mobile Web. Native applications integrate directly with the mobile device's operating system and can interact with its hardware much like the software on a personal computer. Mobile Web applications are apps that run directly from an online interface such as a website. Some applications are hybrids that combine the interface and coding components of a web-based interface with the functionality derived from native applications. Institutional applications would be a subcategory that has only recently started to spread out throughout businesses. Rakestraw, Eunni and Kasuganti (2012) present a Graph showing a breakdown of the various categories of applications used within a span of 30 days (Figure 13.2). There is no reference to an organizational or institutional category. This leads to a double conclusion: (a) the novelty of using endogenous apps as management control tools in the shop floor and (b) the more recent growth in the development of industrial apps has probably affected this graph trends.

Mobile computing devices are smart consumer products that are usually used by a heterogeneous group of users. This introduces three main reasons that signify the importance of an integrated approach to usability heuristics for mobile apps. First, mobile devices have inherited limitations due to the nature of mobile devices themselves such as the small screen size of the device, display resolution, limited input mechanisms, connectivity-based issues, security, and limited performance capabilities. Other constraints to consider when discussing mobile devices are the huge variability among the different brands and variability within one brand. For example, when designing apps for iPhones, the user interface is standardized. However, when designing for Android or Blackberry phones, there are different screen sizes, and interaction models to consider (Swaid; Suid, 2018).

Despite the fact organizational users do not adhere to the same rules as regular consumers, one should consider the possibility that work systems' apps must follow some basic usability principles. Florence and Liotta (2007) pointed out that users or consumers take a significant role in the success/failure of any software. Therefore, as for the PrüfExpress System App, used to contextualize this chapter is presented in the following section, there will be more emphasis given to the user-experience and usability issues reported during its implementation in the shop floor, rather than in its development process.

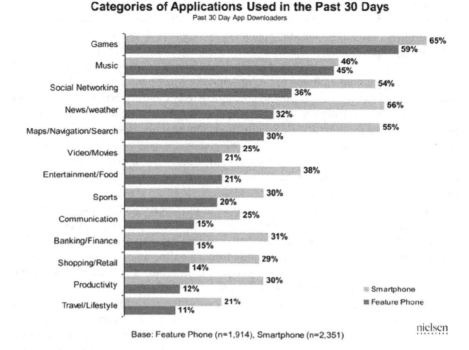

FIGURE 13.2 Source: Rakestraw, Eunni & Kasuganti (2012).

13.4 THE PRÜFEXPRESS SYSTEM CASE STUDY: PLANNING, DESIGN, OPERATION, USER-EXPERIENCE AND USABILITY REVIEW

13.4.1 STATE-OF-THE-ART MOBILE APPLICATION FOR TESTING AND DOCUMENTING WORK EQUIPMENT

A perfectly functioning tool is the most important prerequisite for high-quality services in the industry. Therefore, the periodic checking of work equipment for functionality and accuracy is indispensable. This means that the inspection of work equipment in the company has not become a new topic, yet it has become routine for a long time. Before the spread of mobile devices, especially smartphones, many companies, including Daimler, had documented the test on paper. In the previous inspection, an inspector would receive an Inspection Report, outlining in detail each system to be inspected. The test report contained information about the plant, the client, the inspector, the test questions and a part for the feedback. The documentation of the test is called in this case, the completion of the test protocol.

In contrast to that, with the PrüfExpress System App the examiner receives a mobile device for the electronic documentation of the exam, with the help of which he can document the exam. This alternative is now very popular and represents the electronic documentation of the testing of work equipment. This form offers a noticeable advantage over the paper form solution as it can be carried out much faster. In addition, it frees up qualified personnel to work on more important things related to the production process.

The electronic documentation of the testing of work equipment is used nowadays in many companies, with numerous applications that convey similar innovations and changes on the way paperwork is being dealt in the shop floor today. PrüfExpress is not only developed for the control of work equipment but also oriented towards maintenance. According to Germany's Ordinance on Industrial Safety and Health, by maintenance one must consider the totality of all measures for maintaining or returning an installation to a safe condition (BJVDE, 2015). This means ensuring the functional condition of the equipment or restoring it in the event of a failure. From this point of view, maintenance is an issue that can be subdivided into inspection, maintenance and repair. In view of this definition, the following question arises: What happens if a company does not want to take care of the complete maintenance (the maintenance or repair, for example, is usually done by external companies)?Would the company still be responsible for buying all the software or there are already applications that only take care of the inspection (and control) of the work equipment?

Apart from PrüfExpress, there are few applications that deal with documenting the testing of work equipment, for example, the "Darwin CheckMaster" from the Darwin company and the "Dekra application" from the Dekra company. The Darwin application differs from PrüfExpress in functionalities such as voice control and digital signature. Since this application only runs on Windows Mobile 5.0, PrüfExpress has the advantage that it can run as a cross-platform application on all operating systems. Due to time constraints and the need to stay on course to the main theme, a detailed comparison of these applications is not included in this text.

13.4.2 LEGAL DOCUMENTATION

The testing of work equipment is carried out in accordance with the Industrial Safety Regulation. Each test must be documented. The resulting document must be kept in a safe place as it will be used as recognized evidence in court in the event of an accident or personal injury(Arbeitssicherheit, 2017). Since the resulting documents can be used in court, they must be audit-proof. The term revision security refers to electronic archiving systems that point to the following characteristics after pounds: (a) the contents of the document are stored unchanged and forgery-proof, (b) the contents of the document are recallable through a search and (c) all actions in the archive are logged for reasons of the comprehensibility (Pfund, 1995).

In the event of an accident, the competent court system will evaluate in detail whether the work equipment has been regularly inspected, when the last inspection was carried out, what was inspected (the activity of the inspector) and who carried out the inspection (identity of the inspector) (Neumann, 2015).

13.4.3 Testing and Documentation Process

The testing of work equipment and its documentation is a two-stage process that must be carried out by a trained inspector. However, the inspector must know which parts of the work equipment are to be inspected and how the inspected parts are to be documented. The process consists of a passive and an active component. In the first phase, the examiner is an active component that does all the work. For example, he can carry out a visual inspection. The application, on the other hand, plays a passive role because it provides the test questions that the examiner can use as a guide to perform the test.

The second phase is the documentation itself of the test carried out. In contrast to the first phase, the application plays an active role in the process. The examiner describes what he has done (responding to the question, entering the time feedback and describing his activity) and the application takes care of storing the registered information in a database. In this chapter, the term performance of the audit is used to refer to the audit and documentation process.

13.4.4 Concept, Technology and Programming Languages

The system to be developed should be as flexible and modular as possible in order to be open for extensions and maintenance. This requires a clearly structured and documented architecture. In the following, the architecture of the new application is presented considering the technical requirements and trying to adhere to Usability and User-Experience issues. The system focuses on the following two technologies: RFID and barcode. These are contactless identification methods and are used for automatic data transfer, which saves the user the trouble of entering information manually. This technology makes it easy to read and store data from the database.

While the identification with RFID is done via electromagnetic waves, the data is scanned in the barcode. It is not possible to use Apple's NFC mode in this work for some reasons, for instance, an NFC interface would have to be implemented for Android and Windows. Since the new application must be platform-independent, the idea of using NFC for all platforms was no longer pursued. Instead, a different strategy was used to continue to allow automatic data transfers between the application and the object.

The most recent alternative is to use the Bluetooth function from all mobile devices. An external Bluetooth reader is used in addition to the smartphone. This reader has the advantage that it can read all RFID tags and can be connected to iOS as well as Android and Windows Smartphones via Bluetooth. After the pairing between the smartphone and the reader, the reader is recognized as a "keyboard input" (Bluetooth keyboard) device on the smartphone. Another reason for using this alternative is the reduction of development costs. There is no need to develop a new interface between the smartphone and the application, only an entry field in the application with a focus on where the data is entered.

UX and Review of a Smartphone Application 251

Naturally, the use of this alternative generates additional costs for the company, as Bluetooth readers must be acquired. However, during the discussions prior to the system's implementation, it was decided by Daimler that the application had to be platform-independent. The Xamarin tool with Visual Studio was used to implement the requirements. For the structure of this application, the Xamarin Forms is used using Portable Class Library because it not only offers more flexibility than the others, but it is also easy to test and extend.

An important aspect of the application is the ability to process and store data with or without an Internet connection. A SQLite database is used for the local storage and processing of the data. An external service (Zumero) is then used to implement automatic synchronization between the local database and the server-side database. This text also does not deal with the structure of Zumero. It is only necessary to understand that Zumero offers a web service and a management tool to synchronize data between databases. It also takes care of resolving conflicts that may arise during synchronization. For conflict resolution, rules are defined during the configuration of the managers. Zumero thus offers bi-directional synchronization. The use of this framework offers the advantage that the developer saves a lot of time during development.

SQLite is used for the local database. This defines a limited SQL program library designed for embedded use. It fulfills most of the standardized SQL voice commands and does not require an SQL server or SQL software. Likewise, no process is started in the background that is to be used for database administration. Another advantage of SQLite is that there is only one database file where all database entries are stored. Xamarin is a component of Visual Studio and since Visual Studio is associated with the C# programming language, the C# programming languages are as specified.

In summary, the identification of work equipment is not read by the NFC chip of the smartphones, but by the RFID tag using a Bluetooth reader and passed on to the application for processing. The barcode technology is used as before, since the application only needs to access the camera API of the respective devices. The task of synchronizing the data between the devices and the server is done by an external component –the Zumero.

13.4.5 The Architecture of the System

The application was meant to be implemented as a distributed architecture, more precisely as a client-server architecture. It consists of a mobile application for collecting the data and a desktop application for managing the collected data. The testing of work equipment is carried out by means of a mobile application. This work does not deal with the development of the desktop application, but due to the distributed architecture of the new system, the desktop application must be built with little effort. This implementation is not done within the scope of this work. This section presents a synthesis of the implementation results for the PrüfExpress App. At the beginning of the operation, the inspector must start the application on a smartphone. Once the app initializes itself, Figure 13.3 is displayed.

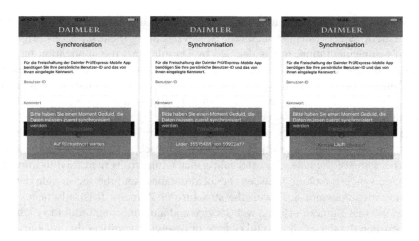

FIGURE 13.3   Source: Graubner Industrie GmbH.

Then, the user must wait until the data has been completely synchronized before logging in. During the first synchronization process, the user must expect a waiting time of approx. 5 seconds because the total database is transferred locally to the device. After the synchronization process has been successfully completed, the registration form is displayed. The user must register now before beginning to work. In the next image (Figure 13.4), the user is asked to enter his username and password in order to log in. With a successful login, he or she can then access the application. Once the logon information is set, all attachment groups are displayed, and proper authorization of access is required.

FIGURE 13.4   Source: Graubner Industrie GmbH.

UX and Review of a Smartphone Application   253

Then, a list of all installation groups is shown to the logged user (Figure 13.5). In this case, the user has only one installation group: the "LASTAUFNAHMEMITTEL" (load handling attachment) and is abbreviated as "A." Here the user has a business role (KMadmin) for about 466 cost centers. This means that he can not only create the documentation for the tested work equipment but also manage the data. If he clicks on "LASTAUFNAHMEMITTEL," he is forwarded either to the administration view (left picture in Figure 13.6).

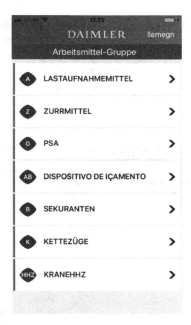

FIGURE 13.5   Source: Graubner Industrie GmbH.

FIGURE 13.6   Source: Graubner Industrie GmbH.

The administration layer (left picture of Figure 13.6) offers the user four possibilities to further interact with the system. He can call the "Execute Job" function in order to be able to carry out the inspection of a piece of equipment (center image – scanning layer – in Figure 13.6). Or he can call up the administration function for the work equipment, the order or the inspection plan. Figure 13.6 shows the management, scanning, and attachment detail view. A user is directed to the plant detail view after searching the plant for his transponder, barcode or inventory number.

The button "ORDER DETAILS" on the left side of Figure 13.7, on the plant detail page, is used to generate the test order for this work equipment. If the user clicks on it, then the right image of Figure 13.7 is displayed (in the PruefPlan Select page), where he must select the test plan (template) of the order to be generated in order to be able to continue working. The user selects the test plan and presses "Next." The order is generated in the background.

FIGURE 13.7 Source: Graubner Industrie GmbH.

As a result, the user sees the image in Figure 13.8 and can navigate through "tabs" between the pages to complete the order and save it at the end.

UX and Review of a Smartphone Application 255

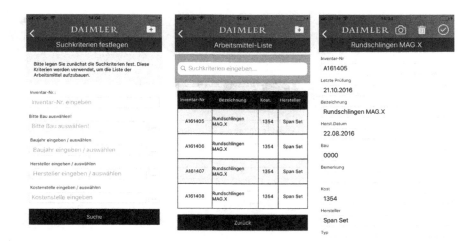

FIGURE 13.8 Source: Graubner Industrie GmbH.

On the left side of Figure 13.7, the test questions are displayed on the test frequency (cycle). The test questions each have a checkbox that allows the user to answer the question. By default, the checkboxes of the check questions have the "On" status. The "On" or "Off" status here corresponds to the "Yes" or "No" answer. The image in the middle of this figure represents the confirmation page. The user can then select predefined feedback from a picker. The selected confirmation is then written to the editor text. The picture on the right sketches the last step of the audit documentation. If necessary, the user is asked for time feedback. If there is this information then he clicks on the order save button to save the information of the check, feedback and save page.

Figures 13.6–13.8 represent the administrative view of the application and they were designed using Daimler's usability and user-experience requirements for upper management software systems. While Figure 13.6 shows the work equipment view, Figure 13.8 depicts the order view. Left images of both Figures 13.6 and 13.8 allow the user to filter the work equipment or the orders. Clicking on an element in the list displays the details of that element. It is important that the details of an order cannot be processed.

The right picture in Figure 13.7 shows the details of selected work equipment in this case a "C-hook." Since the logged-in user has write access, he can edit the details of the work equipment and save the changes in the database with the Save button.

The last diagram (Figure 13.9) illustrates all the inspection plans and the details of an inspection plan.

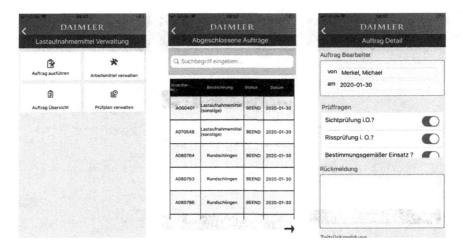

FIGURE 13.9 Source: Graubner Industrie GmbH.

13.5 FINAL CONSIDERATIONS

The primary motive of this research was to entice a reflection. How important are current Usability and User-experience (UXD) design guidelines and parameters when dealing with the planning and design of mobile applications? Interface issues, for example, are crucial for commercial (leisure) usage but are somewhat useless on the shop floor, where other elements of the mobile system's architecture are more relevant. Yes, the methodological aspect of software engineering is still important, but maybe not always a key factor in successfully developing mobile applications.

The study case presented to contextualize this chapter shows that innovation and technology are not transforming and reshaping the social world as we know, but it is also transforming technological innovation itself. Maybe the software development life cycle is losing its importance as a development strategy tool, in the case of mobile apps. Some of the main observations found in this research are mobile app is not a tiny/simple piece of software anymore; mobile apps are continually growing in number, usage possibilities and complexity; instead of modern PM techniques, old process-oriented approaches and techniques are required to handle successful mobile application development in the shop floor.

Usability and User-Experience (UXD) are the key factors for achieving quality in the development of mobile apps, especially for its users, all of them. However, research on mobile apps usability is still fragmented and inconsistent. In this study, it was presented a usability heuristics model. It is based on an integrative approach using Nielsen 10-rules heuristics, and the design guideline put by Google and human-factors design of Apple for their mobile apps developers. This guidance approach has been proved successful in the development of commercial, ordinary apps.

Then, a pragmatic overview of mobile applications for industrial applications was outlined. The subjective assessment resulting from that shows that researchers and software engineers are yet to guide themselves based on any guideline. The particularities and peculiarities of mobile applications are just too broad to allow for a single methodology or development approach. When asked if Nielsen's rules were used in the planning and development of the PrüfExpress Mobile App, the general project manager in charge of the system responded that they did not follow any specific usability guidance, other than regular PM processes carried out by Daimler. However, he added, as I became aware of those guidelines, I must say that all those "rules" were somewhat followed by the work team.

Thus, when organizational idiosyncrasies are added to the formula, it becomes noticeable that a system's overview is needed. The case presented to contextualize this chapter showed the power of mobility as a management tool, which reinforces the need to properly address usability and user-experience issues, especially in cases where the applications are embedded with ERP-like informational systems. The description of the development and implementation process of the PrüfExpress Mobile App is solid proof that process-oriented approaches are required in order to improve design quality. Moreover, it must be emphasized that design quality cannot be dissociated from Usability and User-Experience characteristics.

REFERENCES

Arbeitssicherheit, Arbeitssicherheit und Gesundheitsschutz (Occupational Safety and Health Protection Regulations) by Wolters Kluwer Deutschland (2017), Available at: https://www.arbeitssicherheit.de/de/html/library/document/7331539%2C20, Retrieved on 12 Dec. 2019.

BJVDE - BundesamtfürJustiz, Deutschland, (Federal Bureau of Justice of Germany) Bundesministerium der Justiz und fürVerbraucherschutz, juris GmbH (2015), Available at: https://www.gesetze-im-internet.de/betrsichv_2015/BJNR004910015.html, Retrieved on 5 Dec. 2019.

Blue Corona, 75+ mobile marketing statistics for 2020 and beyond (2019), Available at: https://www.bluecorona.com/blog/mobile-marketing-statistics/, Retrieved on 15 Jan. 2020.

Bokman, A.; Fiedler, L.; Perrey, J.; Pikersgill, A., Five facts: How customer analytics boosts corporate performance (2014), Article in McKinsey website, Available at: https://www.mckinsey.com/business-functions/marketing-and-sales/our-insights/five-facts-how-customer-analytics-boosts-corporate-performance, Retrieved on 15 Dec. 2019.

de Paula, D.F.; Menezes, B.H.; Araujo, C.C., Building a quality mobile application: A user-centered study focusing on design thinking, user experience and usability, in *International Conference of Design UXD. Usability*8518 (2014): 313–322.

Florence, A.; Liotta, A., Addressing user expectations in mobile content delivery, *Mobile Information Systems* 3, no. 3 (2007), 153–164.

Ibrahim, N.; Ahmad,W.F.; Shafie, A., User experience study on Folktales mobile application for children's education, in *Next Generation Mobile Applications, Services and Technologies*, 2015, pp. 197–200.

Neumann, T., Die Drei Todsünden des Elektronitechnikers vor Gericht, Article presenting court cases of work accidents (2015), Available at: http://www.dds-cad.net/fileadmin/redaktion/PDF, Retrieved on Dec. 11, 2019.

Newark-French, C., Mobile apps put the web in their rear-view mirror, *Flurry Blog* (2011), Available at: http://blog.flurry.com/bid/63907/Mobile-Apps-Put-the-Web-in-Their-Rear-view-Mirror, Retrieved on 14 Dec. 2019.

Nielsen, J., Finding usability problems through heuristics evaluation. In: *Proceedings of the SIGCHI Conference on Human Factors in Computing Systems*, June 1992 Pages 373–, pp. 373–380. ACM Press, New York, 1992.

Nielsen, J., Heuristic evaluation. In: Nielsen, J.; Mack, R.L. (eds.), *Usability Inspection Methods*. New York: Wiley, 1994.

Nielsen, J.,Severity Rankings for Usability Problems (1994), Available on http://www.useit.com/papers/heuristic/severityrating.html, Retrieved on 28 Dec 2019.

Nielsen, J., *Usability Engineering*, San Diego: Morgan Kaufmann, 1994.

Partala, T.; Saari, T., Understanding the most influential user experiences in successful and unsuccessful technology adoptions, *Computer Human Behavior* 53 (2015),381–395.

Pfund, A., Document classification in Germany, Andreas P. Informations manager (1995), Available at: http://andreaspfund.de/archivierung/elektronische_archivierung/revisionssicherheit.php, Retrieved on 8 Dec. 2019.

Rakestraw, T.L.; Eunni, R.V.; Kasuganti, R.R., The mobile apps industry: A case study, In *Journal of Business Cases and Applications*, Youngstown State University, 2013.

Silva e Santos, M.S.; Vidal, M.C.R.; Moreira, S.B.; Almeida, M.M., *The RFad Method - A NON Invasive Fatigue Measurement Method Based on Ergonomics evaluation Techniques*. SAE Technical Paper Series, v. 1, p. 2011-36-0012, 2011.

Statista, Total global mobile app revenues 2014–2023, *Statista Data Platform* (2019), Available at: https://www.statista.com/statistics/269025/worldwide-mobile-app-revenue-forecast/, Retrieved on 2 Dec. 2019.

Swaid, S.I.; Suid, T.Z., Usability heuristics for M-commerce apps. In: Ahram, T.; Falcão, C. (eds.), *Advances in Usability, User Experience and Assistive Technology. AHFE 2018. Advances in Intelligent Systems and Computing*, vol 794. Springer, 2019, Available at: https://doi.org/10.1007/978-3-319-94947-5_8. Retrieved on 8 Jan. 2020.

World Mobile Applications Market - Advanced Technologies, Global forecast report (2010–2015), *Elaborated by Markets & Markets* (2015), Available at: http://www.marketsandmarkets.com/Market-Reports/mobile-applications-228.html, Retrieved on 12 Dec. 2019.

Yazid, M.A.; Jantan, A.H., User Experience Design (UXD) of mobile application: An implementation of a case study, *Journal of Telecommunication, Electronic and Computer Engineering* 9, no. (3) (2017), 197–200, Available at: https://jtec.utem.edu.my/jtec/article/view/2902. Retrieved on 8 Jan. 2020.

14 Encounters and Difficulties when Gathering User Experience Data

Arminda Guerra Lopes

CONTENTS

14.1 Introduction 259
14.2 Conceptual Context 260
    14.2.1 Digital Inclusion 260
    14.2.2 Human Work in Human–Computer Interaction 262
    14.2.3 User-Experience 263
14.3 Case Studies 265
    14.3.1 Description 265
    14.3.2 Methods for Data Collection and Analysis 266
14.4 Results and Discussion 270
14.5 Conclusions 276
Acknowledgments 276
References 276

14.1 INTRODUCTION

Technology plays an important role in the full participation of persons with disabilities and older people in a digital society. Technology continues to upset the world of work. The tendency over the last decade has been for increasing Internet use among the older population.

Available research indicates that the use of technologies has the potential to a positive motivation towards the use of digital tools (Picton and Clark 2015). However, technology is not easy to use and, at this moment, still many people worldwide are disabled by inaccessible technology, or do not have access to assistive technology (AT)-based solutions that could help them to participate in an equal position in modern society (Hoogerwelf 2016).

The research presented in this paper describes experiences from people with special needs: the old people and those with disabilities. These people could be adequately supported by alternative interaction modalities to deal with the digital environment in order to simplify their life routines. The appropriate technology permits them to remain independent, socially connected and with a better quality of life in their homes and communities.

In this paper, five case studies are described in order to settle the constraints and challenges that either authors or users encountered along the development of the artifacts. Users intervened during and after the design process phases, as well as, in the artifacts evaluation tests.

The approach followed to analyze data was the concept of work, in human–computer interaction (HCI) field, which focuses on tasks, activities and user interactions. According to Preece, the concept of task has many meanings: goals, tasks and actions (Preece et al. 2019). The main importance is the goal focus on the thing to be achieved. Tasks are the series of activities or actions required to achieve the goal and the action is a simple task, which requires no problem-solving.

14.2 CONCEPTUAL CONTEXT

14.2.1 Digital Inclusion

People aren't born with digital skills. The development of digital skills requires time to explore, together with help to learn. People may lack the access, support, technical ability, confidence or motivation to develop their basic digital skills.

Digital inclusion has been expressed specifically to address issues of opportunity, access, knowledge and skills. Abah states that it articulates the policy, and practical efforts to look beyond issues of access to computers and the Internet are towards a more robust understanding of the skills, content and services needed to support individuals, families and communities in their abilities to truly adopt computers and Internet (Abah 2019).

The Internet poses challenges and opportunities for individuals and communities alike. These challenges and opportunities have not been uniformly disseminated. Digital technology has opened new domains of exclusion and privilege for some, leaving some populations isolated from the vast digital land. They enable people to access information and services that were previously unavailable to them, opening new opportunities for income generation, personal development and engagement with community and political decision-making.

Presently, in Portugal, there are more than 2 million Portuguese who have never used the Internet and around 4 million who make an even basic use. In the last report of the European Commission – Digital Economy and Society Index – the level of digitization is compared among the 28 countries of the European Union. Portugal is in 16th place, below the European average, mainly due to the low digital competencies of the Portuguese:

> The risk of digital exclusion for certain population groups is particularly high in Portugal. The biggest challenge in Portugal lies in improving the digital skills levels of its citizens, especially among the elderly and people with low levels of education or with low incomes.
>
> **(EDPR 2017)**

In this context, several governmental initiatives to promote digital inclusion have been created. There are some programs that will help to change the present portrait: the MUDA – Movement for Active Digital Use (2019), the INCoDE.2030 (Heitor 2020) and the National Initiative for Digital Competences (INcoDE 2030). Conversely, the Portuguese Government signed a Memorandum of Understanding (MoU) with Cisco that over the next two years Portugal and Cisco will cooperate to capture opportunities presented by the digital economy, as well as, social inclusion and quality of life. The main goal of this memorandum is the contribution to the country's digitalization acceleration (Hinchliffe 2018).

MUDA is a Portuguese educational initiative promoted by a group of companies from the most relevant sectors of the economy, universities and the Portuguese Government that jointly undertake the commitment to encourage Portuguese participation in the digital space and help to take advantage of the associated benefits to the use of digital services made available by companies and the State, contributing to the reality of a more advanced, inclusive and participatory country (MUDA 2019). This initiative will support Portuguese people who want to acquire basic digital skills.

INCoDE.2030 is a new Portuguese program to improve the digital literacy and skills of the population. The initiative is divided into five areas, which include inclusion, education, qualifications, specializations and research. It is focused on providing different groups of the population with access to technology infrastructure and knowledge that are vital for their personal growth and work opportunities (INCoDE 2030).

The National Initiative for Digital Competences e-2030 – Portugal CoDigital 2030, version 2017, intends to insert Portugal into the European group of countries with digital competencies. The main social challenges are "digital literacy and inclusion for the full exercise of citizenship," encouraging "specialization in digital technologies and applications" to qualify human resources and make the economy more competitive, and "produce new scientific knowledge" through international cooperation (INCoDE 2030).

Despite the availability of inclusion programs and the opportunities offered by digital technology experimentation, there are still many problematic situations for the inclusion of older people and those with disabilities.

The tendency over the last decade has been for increasing Internet use among the older population. However, it is known that the majority of them do not use it. The main reasons are lower income, older age, living alone, mobility challenges and problems with memory or ability to concentrate.

One way to contribute to digital inclusion is through assistive technologies. Assistive technology (AT) is a term indicating any product or technology-based service that enables people of all ages with activity limitations in their daily life, education, work or leisure. AT can be specifically designed for persons with disabilities, such as an electronic wheelchair, or not, for example, a tablet or smartphone. The term thus does not relate to a specific category of products but their enabling function. However, there are still barriers to the use of these technologies by people with special needs (Borg et al. 2011). These barriers are organized into three categories: use and interface (application design requires effort and specific skills), organization and support (inexistence of appropriate support and need to learn new digital skills) and financial costs (technologies are not cheap and frequently not available for free).

14.2.2 Human Work in Human–Computer Interaction

HCI contributes to usability efficiency and a good work environment for skilled professionals in IT-supported work. For Kim, it studies primarily to what extent the users are able to interact with the computers. HCI consists of three components: the user, the computer and how these two work together (Kim 2018).

The term "work" is today one of the shared words by people all over the world. People talk of going to work, coming from work and looking for work or a job. People leave their families, their homes and even their children, with strangers, so as to go to work. It is also usual to come across people who express that they do not have the time to think, be with their friends or be with their children because of work. Work, therefore, has become a culture of a kind. Work is, obviously, important in human life: it is an act of communion for the common good, not only an act to help satisfy the common, material need.

The concept that we form from the term work is that of an activity which requires of the agent that there be some effort in self-application to something with the view of realizing something or of obtaining something. Work is a process with an end. Since man is a social being, work is also a service to others.

Work in the context of human–computer interaction (HCI) is the study of the way in which computer technology influences human work and activities (Dix 2009).

Storrs describes it as:

> people ... act to achieve their purposes. This means that they are constantly engaged in goal-driven, sometimes planned behavior. For some of this behavior, the plan, or the goal structure is fixed to a greater or lesser degree and this kind of behavior is what we normally call tasks. Some of this task-related behavior will involve interactions and some of these interactions may involve computers.
>
> **(Storrs 1994)**

So, the human action that contributes to a useful objective, aiming at the system, is a task (Diaper 2006). The analysis of tasks and workflows is a longstanding tradition in human–computer interaction (HCI). Task analysis defines the performance of users. HCI activities are proposed based on the concept of tasks. Tasks are characterized as the means by which work is performed (Diaper and Sanger 2006). One of the goals of research on interfaces aims at improving interface utility, accessibility, performance, safety and usability. For these purposes, utility is defined with reference to the task to be performed. A useful system contains the necessary functions for the completion of tasks users are asked to perform. Norman's evaluation/execution model is a useful way of understanding the nature of user interaction with interactive systems (the user work) (Norman 2002).

Difficulties in Gathering UX Data 263

The work concept presented in this section supports the user-experience (UX) approach followed in this research (see Section 14.4).

14.2.3 USER-EXPERIENCE

This section presents definitions of user-experience (UX), examples of different approaches followed in literature and the elements of UX. The goal is to contextualize our study. Figure 14.1 shows an example of the concepts that one can link to the UX domain.

In terms of approach, user-experience (UX) concept is handled from several points of view. Law considers three perspectives to take UX: as a phenomenon (circumstances and consequences of UX explanation), as a field of study (investigating and developing UX design and assessment methods) and as a practice (envisioning UX as part of a design practice), (Law et al. (2009). The study described in this paper follows the practice approach.

This extensive field encompasses all the aspects of how people use an interactive product: the way it feels in their hands, how well they understand how it works, how they feel about it while they are using it, how well it serves their purposes and how well it fits into the entire context in which they are using it (Alben 1996).

According to Nicolas (2011), there are four elements of UX: user, interaction, artifact and context. The actor – the user – is identified with different names: user (Hassenzahl 2010) or as a person or people (McCarthy and Wright 2004; Hektner, Schmidt, and Csikszentmihalyi 2007; Hekkert and Schifferstein 2008). Interaction has been seen in different perspectives: as part of the dynamic element (Hassenzahl 2010) or interaction with the social context (Forlizzi and Battarbee 2000). The artifact has different synonyms: products, technologies, interactive systems, technological artifacts, forms or objects. In this paper, the terminology used for these outputs

FIGURE 14.1. Concepts in UX domain.

is artifacts to describe all the applications, systems, and developed platforms. Concerning the context, it can be approached as context or environment.

Morville declares the existence of seven factors to describe user-experience: useful, usable, findable, credible, desirable, accessible and valuable. An artifact is useful when it has a purpose; it is usable when it enables users to effectively and efficiently permits to achieve their end objective with the artifact; it is findable when it is easy to find; it is credible by permitting the user to trust in the artifact that was provided; it is desirable by the design characteristics conveyed through branding, image, identity, aesthetics and emotional design; an artifact provides an experience that can be accessed by all users and the artifact delivers also value. Morville (2004) dealt this subject in more detail and through different approaches.

Mekler presented a framework of meaning in interaction, based on a synthesis of psychological meaning research. The framework outlines five distinct senses of the experience of meaning: connectedness, purpose, coherence, resonance and significance (Mekler et al. 2019).

Another focus is on the consequences, i.e., a consequence of a user's internal state (predispositions, expectations, needs, motivation, mood, etc.), the characteristics of the designed system (e.g., complexity, purpose, usability, functionality) and the context (or the environment) within which the interaction occurs (e.g., organizational/social setting, meaningfulness of the activity, voluntariness of use) (Hassenzahl & Tractinsky 2006).

Figure 14.2 synthesizes the focus from different authors about UX.

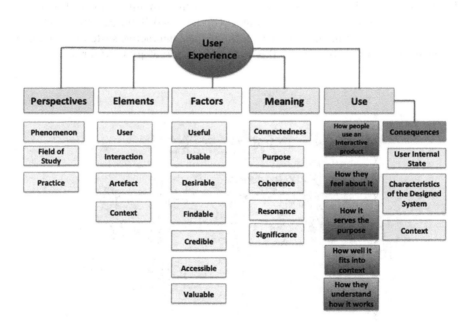

FIGURE 14.2. Different authors' approaches about UX.

14.3 CASE STUDIES

This section presents the description of five case studies developed in authors' collaboration. The authors chose digital artifacts that were developed for people with special needs.

14.3.1 Description

Case 1: Deaf Learn – A digital inclusion application, for windows phones (Lopes 2014). The project consisted of video format gestures, voice application and Web service SignalR structure's improvement. The process of communication occurred through a chat and it permitted to translate text to voice, voice to text and also text to sign language, in real time. The application facilitated the communication between people with hearing difficulties, as well as, blind people. The developers were two students (one of them was deaf). They also had the role of users, which facilitated the consideration of their own needs.

Case 2: Mobile interface for blind users – This interface made use of near field communication technology (Ramos & Lopes 2012). The goal was to help visually impaired and blind people, to have more independency, during shopping activities. When they were at home it could help to make a grocery list by approaching the mobile device, such as a smartphone or a tablet, to buy the product he/she wanted. The product was automatically added to the list. In a supermarket, the user only needed to approach a product with his mobile device. Spoken information was returned with the product's name, price and period of validity.

The authors studied people in their "naturally occurring settings by means of methods which capture their social meanings and ordinary activities, involving the researcher participating directly in the setting" (Brewer 2000).

Case 3: Gestural Language Multimedia Application – This multimedia tool permitted auditory handicapped children to learn and improve gestural language communication. The application's use started with an infantile story narrated to children in order to get into the child's imaginary world. The intention was to contextualize the gesture's meaning and understanding. This tool contributes to fulfilling some existing gaps, in didactic terms, concerning the available software in the market, and it assists children, their teachers and parents to communicate with gestural language (Lopes, A. et al. 2002). The motivation for the development of this application was the insufficient resources available in the market, the complexity to use them and their high prices.

Case 4: Gestural Recognition Interface for Intelligent Wheelchair Users – This artifact consisted of the development of a new human–machine interface – an interface for a wheelchair controlled by the recognition of human hands' static gestures. It allows the occupant of an intelligent wheelchair to communicate with certain objects in order to facilitate their daily life. The methodology drew on the use of computational processes and low-cost hardware. The development of the artifact involved dealing with computer vision issues in a comprehensive way. It was based on the steps of video image capture, image segmentation, feature extraction, pattern recognition and classification. In terms of its relevance and impact, the artifact promotes a more natural and intuitive mode of interaction for disabled individuals, which is expected to improve their quality of life (Proença & Lopes 2013).

Case 5: A Multiuser Interaction Platform for Elderly Community – The platform intended to contribute to the strength of human capabilities, based on the sharing of best practices, innovation, cooperation and exchanges. It helps to combat isolation by reducing or substantially delaying situations of depression and dementia including Alzheimer's, for example. The service comprises a database of activities, sources of knowledge and methods, for people's quality of learning improvement. It gives access to a social network and permits the use of cognitive stimulation facilities, as well as, a monitoring system at home. The platform was considered by psychologists as a support tool to help the elderly community to exercise their mental and physical activities. This is a work in progress project.

The four elements of user-experience considered by Nicolas (user, interaction, artifact, context) were considered in the description of the five case studies in the analysis (Nicolas 2011) (Table 14.1).

14.3.2 METHODS FOR DATA COLLECTION AND ANALYSIS

The methods used in the presented case studies were those mainly used in HCI. The HCI methods were not subordinated to utilitarian assumptions but started from a framework of methods capable of understanding a broader context as in the investigations that translate well the different possibilities of combining methodologies from various authors (Gaver, Boucher, Pennigton & Walker 2004; Mendes et al. 2012; Hoven, Eggen, & Mols 2014). The works from these authors focused on empathy and engagement of users' participative action using a playful approach to study design, adopting and adapting creative methods to develop their productions through inspiring responses for their projects.

The use of various methodologies allowed us to better understand the phenomena involved in the use of interactive systems in context (Raldall and Rouncefield 2013). This process of combining methodologies since the design process phase involved a participatory practice. The results were the sharing of individual knowledge and experiences, which came to be reflected in the project as continuous dialogues through various expectations (Sharp, Rogers & Preece 2007; Muller & Dayton 1997).

The questionnaire as a data collection instrument was used covering its qualitative and quantitative characteristics, in a less participatory relationship such as those that involved conversations and interviews (recorded), which were shown to be more intensified regarding the participation between researcher and interactors. However, this departure was essential to provoke other situations that evoked unexpected information.

Participatory Observation – The term participative has a communicative and reciprocal interaction, while the use of "participant" brings the idea of a researcher "who takes part in something," just the opposite of what was intended. The intention was to consider the wealth of communication in different relationships and languages during the moments of interaction.

TABLE 14.1.
Four Elements of UX

User (People, person, interactors)	Interaction (Relationship between the user and the artifact and action accomplished by a user on an artifact)	Artifact (Product, object, item and system)	Context (Place and time)
Deaf, blind and developers	Ask the artifact to find contacts Ask by names Communication between people with disabilities Communication between people without disabilities	Mobile phone Mobile artifact Communication: Voice to text Text to voice Text to sign language Chat	Real-time application People have different disabilities They want to communicate with people with the same disabilities They want to communicate with those with other disabilities They want to have a common language to communicate Access to the application is important for social and cultural inclusion
Blind	At home: Making a grocery list Approaching a mobile device to the product to be bought At supermarket: Approaching with the mobile device to a product Making gestures to the touch screen device Listening to speech information returned about product's name, price, validity	Interface with near field communication technology. Shopping activity and grocery list	Blind people use mobile phone They need to make a grocery list They go to supermarkets They need help from family, friends or supermarket employees
Auditory handicapped children Parents Teachers	Listening to an infantile story narrative Finding gestures meaning Attention to the gestures understanding Communication by voice Communication by gestures	Computer Parrot story Gestures learning	Computer story Learning by playing At home, at nurse schools

(Continued)

TABLE 14.1. (CONTINUED)
Four Elements of UX

User (People, person, interactors)	Interaction (Relationship between the user and the artifact and action accomplished by a user on an artifact)	Artifact (Product, object, item and system)	Context (Place and time)
People with mobility problems	Communication by gestures Image captation Human hand segmentation Feature extraction Gestures classification	Intelligent wheelchair Laptop with a built-in webcam Robotic arm	Static gesture recognition performed in real time Hand located in front of the webcam Low-cost hardware
Elderly community Caregivers Doctors Families	Playing with videos and tutorials Making alerts Chat Checking agenda Playing online games Pressing emergency button	Sensors Statistics Help and support Notifications	Elderly leaving alone Caregivers need to visit them (sometimes, far away from cities) Monitorization, entertainment

Interviews were another qualitative method used to complement the data gathered through questionnaires and observations.

Case 1 – Before the application design, the authors did 12 interviews with blind people and 8 with deaf participants. The goal was to obtain the user's opinion about the application, namely, about its functionalities. After the prototype development, usability and user tests were carried out with the hearing/blind-impaired people to test it.

Case 2 – The methods used were participant observations, interviews and discourse analysis of natural language, as well as personal documents' analysis. The focus was a study group of impaired blind people. A total of 17 blind people were invited for the study, 11 male and 6 female (Ramos & Lopes 2012). The blind participants were defined as people who typically used a screen reader to access a computer. Blind participants were invited via local blind organizations and face-to-face.

Case 3 – An association for deaf people was contacted and staff members were interviewed. The data from interviews was analyzed and was taken into consideration for the development of this application. A deaf young boy and 2 girls participated as characters in the video recorded for the application. Participants took a proactive role during the project development. The application's validation was also made with the participants: developers and users.

Case 4 – The methods used to collect data were observation and questionnaires. People who used a wheelchair were observed in their settings. They gave their opinions and expressed their wishes about the proposed artifact.

Case 5 – Concerning data gathered, this case involved several participants with different backgrounds and professional performances.

Interviews were a research method of data collection and analysis. Interviews were conducted with caregivers (12 participants) and with elderly people (22 participants – 8 male and 14 female participants). The questions were mainly open type and the issues discussed with caregivers were about the number of elderly they had in charge to take care of; the difficulties and challenges they had with them; types of tools and equipment that were available for use; personal suggestions to include in an interactive application to help the collaboration and communication processes among the intervenient. The content of the end users' questions was about: whether they lived alone or with family; the ways they spent time; the number of visitors per day; having access to technological tools – laptop and mobile phone; and Internet access. Interviews were face-to-face and with interviewees' authorization. They were video recorded. Subsequently, a transcript of the interview was made. After that, memo writing was elaborated to highlight the main interviewee's concerns. The data analysis results were considered on the application development following a user-centered design strategy. User-centered design is a design approach where the user has a close involvement in order to meet users' expectations and requirements. The design process sketches the phases of the design and development and it explicitly presents users, tasks and environments (Norman 2002, Benyon 2014, Lowdermilk 2013 and Batenburg et al. 2010).

Table 14.2 presents the methods of data collection and analysis of each case study.

TABLE 14.2.
Data Collection Methods

Case	Methods	Participants
Deaf learn	12 interviews	Blind
	8 interviews	Deaf
Mobile interface for blind users	Participant observation	Blind
	17 interviews	
	Discourse analysis	
Gestural language multimedia application	9 interviews	Deaf
Gestural recognition interface for intelligent wheelchair users	3 observations	Disable people – mobility
	8 questionnaires	
A multiuser interaction platform for the elderly community	Interviews	Caregivers
	12 caregivers	Elderly people
	22 elderly	

For Case 5, there were other types of data collection. Data was collected using three different methods: the technicians and the users inserted data, manually, in the system; data collected through sensors spread at home (without any human interaction); whenever the user interacted directly with the platform, for example when requesting a call with an occupational therapist or when playing "chess" with other interactors.

14.4 RESULTS AND DISCUSSION

The author encounters several challenges and barriers, especially when the user was observed executing the tasks to operate with the artifacts. Figure 14.3 is the author's designed framework, which was followed to obtain and analyze data along the artifacts development process.

For each case, the user expectations, the work or task to be executed, the user-experiences, as well as, the perceptions and reactions were analyzed (Table 14.3). The perceptions and reactions of user-experience were observed during artifacts use.

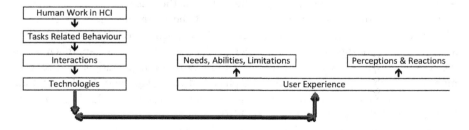

FIGURE 14.3. Framework for data collection and analysis.

Difficulties in Gathering UX Data

TABLE 14.3.
Data Analysis

	User expectations	Work (tasks)	User-experiences (needs, abilities, limitations)	Perceptions and reactions (Positive emotional responses)	(Negative emotional responses)
Case 1	*Communication* between auditory handicapped people; communication with others. *Translation*: text to voice; voice to text; gestural language	*Sending messages*: 2 buttons – for voice, for text. To send voice message: choose application, choose contact person, choose the type of message (voice, text). In case of voice: record and convert into text	Meaningfulness of the activity. Use of mobile phone. Chat with others. Press buttons or talk	Functionality	Mood
Case 2	Be independent; Quality of life improvement. Voice software (information about product's name, price, validity)	*Approach a Mobile to a product*; Press letters; listen to words. *Draw gestures* associated with actions. Each gesture corresponds to the first letter of the action (C – Buy; L – List; quantity – draw a number)	Autonomous shopping. Gestures associated with letters. Difficulties using the mobile touch screen	Purpose Usability	Performance difficulties

(Continued)

TABLE 14.3. (CONTINUED)
Data Analysis

	User expectations	Work (tasks)	User-experiences (needs, abilities, limitations)	Perceptions and reactions (Positive emotional responses)	Perceptions and reactions (Negative emotional responses)
Case 3	Learning gestural *language*; communication; *entertainment*	Press objects; hear words; *observe gestures*; play educational games; listen to a story	Aesthetic experience. Feelings and emotions. Rewards. Advice	Time	Goals misunderstanding
Case 4	Permits the *individual action*; freedom and autonomy	*Make six gestures*; pause; stretch mechanical arm with speed 1; stretch mechanical arm with speed 2; open claw; close claw; move the arm to the starting positions	Autonomy. Gestures communication. Static gestures identification	Effort	Ambiguity
				Personal past experiences	

(Continued)

TABLE 14.3. (CONTINUED)
Data Analysis

	User expectations	Work (tasks)	User-experiences (needs, abilities, limitations)	Perceptions and reactions (Positive emotional responses)	Perceptions and reactions (Negative emotional responses)
Case 5	Talk frequently with *others; be monitored;* entertainment	Move inside the house. *Interactions through Technologies*	Elderly want to stay at their homes. Memory exercises. Chat. Instant occupational prescriptions. Surveillance	Engagement	Bad personal past experiences

People expressed different attitudes. The entire perception of the artifact's use, as well as, their relationship with it was based on their past experiences. Those people who had experiences, for example, using a mobile phone, touch screen, voice recognition systems, had better experiences than those who were not familiar with these technologies. This situation occurred in each case, i.e., past experiences revealed to be important for the understanding of new experiences.

User clicked or pressed buttons, completely obvious to what exactly was written on them. However, in some cases (Cases 1 and 2) the experience of previous use caused the opposite effect since these new interfaces were more complex. The context of use affected the user's perception. The goals about what they expected to hear, see or feel had an influence on their perceptions.

TABLE 14.4.
Challenges and Barriers that Users Face Using an Artifact

Challenges	Barriers		
	Competences	Personal motivation	Security
Technology can enable a person to reach outcomes at work	Lack of appropriate personal support	Technology perceived as not useful and not responding to actual needs	Risks related to safe Internet use
Technology can increase people's level of self-confidence and independence in daily life	Uncertainty about how to use artifacts	Low self-confidence	Risks related to personal data treatment
Technology can help to build peer networks	Difficulties for people assessing their needs	Fear that technology use might reduce human care and contact	Fear of losing control over technology
Technology can encourage increased access to the community			
New paradigms of interaction between humans and technology	Lack of user-centered and user-experience approaches	Alternative strategies to avoid the use of technology	Privacy implications
Technology might make disabled persons less dependent on others	Technology without accessibility	Difficulties using digital devices	
Technology can help to develop basic skills, including making choices		Lack of hardware and Internet access	
	An experience is personal		
Technology can support the safety, independence, social connectedness			
...	...	...	...

Difficulties in Gathering UX Data 275

The main features of perception analyzed were ambiguity, consistency and understanding of goals. It was found that the maximum number of users perceived and interpreted the interface in the same way. The interface's design was organized in a consistent way. The information was organized based on the principles of similarity and proximity (objects were organized and placed in groups). Elements of interaction were in specific places and their functions were working clearly. Finally, the artifacts' design followed the user-centered design approach; the user was involved during the whole phase of the process. This contributed to getting them understanding the goals, previously defined. So, the information the user needed to achieve the goal was visible and accessible, as expected. Concerning perceptions and reactions, the author analyzed the following parameters: time duration of each task, user effort to execute the task, ambiguity, user's personal past experiences and engagement.

The challenges and barriers that people (users) face when using an artifact (product, object, system) on the scope of the described cases are summarized in Table 14.4. The artifact is the output of human work, which can have different functions: aesthetics, technical or social. The five described cases were developed using technology either as a tool or as a medium of interaction among users. So, the challenges and barriers are focused on the use of technology and its positive and negative aspects.

The user competencies, personal motivation and health conditions have interfered with the interaction with the artifacts and, consequently, on the user's positive and negative experiences.

When using an artifact, the user faces challenges and barriers, which are important feedbacks for the development team. The goal is to take that information and consider it in the artifact's development. During the process of data gathering, in work domains, developers also find some challenges and barriers. Examples are presented in Table 14.5. Tasks and subtasks were evaluated to understand the user facilities/difficulties using the artifact. As described in Section 14.3, the development of each artifact involved different types of participants: deaf people, blind people,

TABLE 14.5.
Challenges and Barriers when Gathering User Data in Work Domains

Challenges	Barriers
Teamwork (user, researcher, caregivers)	Lack of collaboration among institutions
Digital inclusion of community members	Uncertainty on appropriate personal support
Processes simplification	Difficulties of users assessing their needs and matching the needs with the artifact
Knowledge exchange among the stakeholders	Lack of awareness on the opportunities offered by the artifact for activities
Innovative approaches encouraging more independence	Communication problems due to participants variability
Multiple media of communication	Disabled people had constraints executing new tasks
Psychologist support to combat isolation	Elderly people vulnerable due to social isolation

caregivers, common users and other stakeholders. They had different kinds of interventions on the artifact design process and progress: some of them were involved before the idea elaboration, others since the briefing, and others during the prototyping phase. The different users' backgrounds, desires and experiences brought, on the one hand, some challenges and, on the other hand, some barriers.

14.5 CONCLUSIONS

This paper describes five case studies where five artifacts were developed for people with different disabilities. Those artifacts' development contributed to innovative solutions for the usage of these people and their digital inclusion. The users and stakeholders were involved in different processes' phases and they contributed enormously to have artifacts developed based on user-centered design and user-experience design approaches.

Authors found some difficulties but also challenges during the artifacts development, especially on the phases of getting data from the work domains (tasks' execution) and from the stage when the user was experiencing the artifact.

The challenges and barriers were handled either by the users or by the development team. For the formers, through mixed methods research approaches, author settle positive and negative aspects found during users' technology usage. The challenges were general evidence, clichés, which one can consider as an asset when using the available technology by society, in general and individually (Table 14.4). The barriers were analyzed considering the users' competences, their personal motivation to use the artifact and the security that they affected. From the development teams, the challenges were focusing on the work, the simplification of processes, the knowledge exchange and the creative process/innovation on the artifacts' design. Conversely, the barriers reflected the problems of collaboration, communication and user characteristics considering their variability and vulnerability.

ACKNOWLEDGMENTS

This research was funded by LARSyS (Projeto - UIDB/50009/2020).

REFERENCES

Abah, Joshua. (2019). Theoretical and conceptual framework for digital inclusion among mathematics education students in Nigeria. In: *Global Perspectives on Educational Issues*, 1 (1), M. J. Adejoh; A. D. E. Obinne; A. B. Wombo, pp. 79–111, 2019, Global Perspectives on Educational Issues, 978-978-967-290-7.

Alben, Lauralee. (1996). Quality of experience: Defining the criteria for effective interaction design. *Interactions*,3(3), 11–15,.

Batenburg, R., and Koopman, G. (2010). The conditional benefits of early user involvement at employee self-service applications in four Dutch ministries. *International Journal of Business Information Systems*, 5(2), 162–174.

Benyon, David. (2014). *Designing Interactive Systems: A Comprehensive Guide to HCI, UX and Interaction Design*. 3/E Pearson, Napier University, Edinburgh.

Borg, Johan, Larssow, Stig, and Östergren, Per-Olof. (2011). The right to assistive technology. *Disability & Society*, 26(2), 151–167.
Diaper, Dan, and Sanger, Colston. (2006). Tasks for and tasks in human–computer interaction. *Interacting with Computers*, 18(1), 117–138.
Dix, A. (2009). Human computer interaction. In: Liu, L. and Ozsu, M. T. (eds) *Enciclopedia of Database Systems*. Springer, Boston, MA.
EDPR. (2017). *Europe's Digital Progress Report (EDPR) 2017 Country Profile Portugal*. https://ec.europa.eu/digital-single-market/en/desi
Forlizzi, J., and Ford, S. (2000). The building blocks of experience: An early framework for interaction designers. In: *Proceedings of Designing Interactive Systems (DIS 2000)*, New York.
Gaver, W., Boucher, A., Pennington, S., and Walker, B. (2004). Cultural probes and the value of uncertainty. *Interactions - Funology*, 11(5), 55–56. Goldsmiths Research Online. doi: 10.1145/1015530.1015555
Hassenzahl, Marc. (2010). *Experience Design: Technology for All the Right Reasons*. Morgan & Claypool Publishers, Penn State University. doi: 10.2200/S00261ED1V01Y2 01003HCI008
Heitor. (2020). Manuel higher education, research and innovation in Portugal, perspectives for 2020. https://www.portugal.gov.pt
Hekkert, P., and Schifferstein, H. N. J. (2008). Introducing product experience. In: H. N. J. Schiffer-Stein and P. Hekkert (eds) *Product Experience* (pp. 1–8). Amsterdam, Elsevier
Hektner, J. M., Schmidt, J. A., and Csikszentmihalyi, M. (2007). *Experience Sampling Method: Measuring the Quality of Everyday Life*. Sage Publications, Inc. Thousand Oaks, London, New Delhi
Hinchliffe. (2018). https://portugalstartups.com/2018/03/portugal-cisco-digitization/
Hoogerwelf, Evert. (2016). *Digital Inclusion – A White Paper*. Entelis – European Network for Technology Enhanced Learning in an Inclusive Society. European Commission under the Lifelong learning programme.
Hoven, E. van den, Eggen, B., and Mols, I. (2014). Making memories: A cultural probe study into the remembering of everyday life. In: *NordiCHI14 Proceedings of the 8th Nordic Conference on Human-Computer Interaction: Fun, Fast, Foundational*. https://dl.acm.org/citation.cfm? doi: 2639189.2639209
INCODE. (2030). https://www.incode2030.gov.pt/sites/default/files/redeoblid_literacia_digital.pdf
Kim, A. (2018). https://medium.com/@annjkim/what-is-human-computer-interaction-hci-3020e5c29e5b
Law, Lai-Chong, Roto, Virpi, Hassenzahl, Marc, Vermeeren, Arnold, and Kort, Joke. (2009). Understanding, scoping and defining user experience: A survey approach. In: *Proc. CHI '09* (719–728), ACM, New York. doi: 10.1145/1518701.1518813.
Lopes, A., and Valente, J. (2002). Gestural language multimedia application. IASTED Conference on Internet and Multimedia Systems and Applications (*IMSA) Hawaii*, USA, 248–251.
Lopes, E., Aleluia, B., Santos, H., and Guerra, A. (2015). Deaf learn – An application for digital inclusion. In: *Livro de Resumos do VIII Semime (Exclusão Digital)*, (eds) Carlos Ferreira e José Alvez Diniz Edition. Faculdade de Motricidade Humana. FMH Editions, Cruz Quebrada, Lisboa. Portugal.
Lowdermilk, T. (2013). *User-Centered Design*. 1st ed., O'Reilly, Beijing, Sebastopol, CA.
McCarthy, J., and Wright, P. C. (2004). *Technology as Experience*, The MIT Press Cambridge, Massachusetts; London; England.

Mekler, Elisa D., and Hornbæk, K. (2019). A framework for the experience of meaning in human-computer interaction. In: *CHI Conference on Human Factors in Computing Systems Proceedings (CHI 2019)*, May 4–9, 2019, Glasgow, Scotland, UK. ACM, New York, NY, 15 pages. doi: 10.1145/3290605.3300455

Mendes, M., Ângelo, P., Nisi, V., and Correia, N. (2012). Digital arts, HCI and environmental awareness: Evaluating play with fire. In *NordiCHI 2012, Nordic Conference on Human-Computer Interaction*. Copenhagen, Denmark, pp. 408–417. doi: 10.1145/2399016.2399079

Morville. (2004). Available at https://intertwingled.org/user-experience-honeycomb/

MUDA. (2019). Available at https://www.muda.pt 27-10-2019

Muller, M. J., Haslwanter, J. H., and Dayton, T. (1997). Chapter11 - Participatory practices in the software life cycle. In: Helander, M. G., Landauer, T. K., and Prabhu, P. V. (eds) *Handbook of Human-Computer Interaction*, (2nd ed., pp. 255–297), Taylor & Fancis Group, Boca Raton; London; New York.

Nicolás, Ortiz Juan, and Aurisicchio, Marco. (2011). A scenario of user experience. In: *ICED 11–18th International Conference on Engineering Design - Impacting Society Through Engineering Design*. 7, 182–193.

Norman, D. A. (2002). *The Design of Everyday Things*. Basic Books, New York.

Picton, I., and Clark, C. (2015). The impact of ebooks on the reading motivation and reading skills of children and young people: A study of schools using RM books. *Final Report*. London: National Literacy Trust.

Preece, Jennifer Jackson, Rogers, Yvonne, and Sharp, Helen. (2019). *Interaction Design - Beyond Human-Computer Interaction*. John Wiley & Sons Inc., United States of America

Proença, R. Guerra A. & Campos, P. (2013). Gestural recognition interface for intelligent wheelchair users. *International Journal of Sociotechnology and Knowledge Development*, 5(2), 63–81. IGI Publishing Hershey, PA.

Ramos, J., and Lopes, A. (2012). Mobile interface for blind users. In: *4th International Conference on Software Development for Enhancing Accessibility and Fighting Infoexclusion (DSAI 2012)*, Douro, Portugal.

Randall, D., and Rouncefield, M. (2013). Ethnography. In: Soegaard, Mads and Dam, Rikke Friis (eds) *The Encyclopedia of Human-Computer Interaction* (2nd ed.). The Interaction Design Foundation, Aarhus, Denmark.

Sharp, H., Rogers, Y., and Preece, J. (2007). *Interaction Design: Beyond Human-Computer Interaction*. John Wiley & Sons, New York.

Storrs, G. (1994). A conceptualization of multi-party interaction. *Interacting with Computers*, 6(2), 173–189.

Brewer, John. (2000). *Ethnography (Understanding Social Research)*. Open University Press, Buckingham; Philadelphia, PA.

15 Parametric Design Method for Personalized Bras

Yuanqing Tian and Roger Ball

CONTENTS

15.1 Introduction 280
    15.1.1 Background of Bra Design 280
    15.1.2 Problems of the Traditional Methods 280
    15.1.3 Motivation and Goal 280
15.2 Design Exploration 281
    15.2.1 Anthropometrical Study 281
        15.2.1.1 Breast Arc Length Approach 281
        15.2.1.2 The Folding Line Approach..282
    15.2.2 Bra Design 282
    15.2.3 Parametric Design Method...282
15.3 Design and Development 283
    15.3.1 The Self-Measuring Approach 283
        15.3.1.1 Landmarks 283
        15.3.1.2 Arc Lengths 283
    15.3.2 Parametric Algorithm 284
        15.3.2.1 Original References and Base Bra Model.........................284
        15.3.2.2 Mechanism of Rhino/Grasshopper 285
        15.3.2.3 Parameters and Algorithm..285
        15.3.2.4 Generating Personalized Bra Models 287
15.4 User Test and Evaluations...287
    15.4.1 Purpose of the User Test...288
    15.4.2 Procedures 288
    15.4.3 Evaluation Results 288
    15.4.4 Discussion 288
15.5 Future Study 289
    15.5.1 Optimization of the Parametric Algorithm 289
    15.5.2 Prototyping and Manufacturing 291
    15.5.3 Fit Tests 291
    15.5.4 Personalization Era for Women's Bras 291
Acknowledgment 291
References..292

DOI: 10.1201/9780429343513-20

15.1 INTRODUCTION

15.1.1 Background of Bra Design

Modern bras were invented in the nineteenth century, developed for hundreds of years (Crandall, 2018) and have been evolved into today's mass production situation where the predominant sizing and grading methods take the lead. The traditional methods of bra sizing and grading take standardized normal distribution of body measurements to develop band sizes and cup sizes. However, such methods vary between companies, standards, regions and populations with various factors involved (Hardaker & Fozzard, 1997). This leads to an important problem of body fit, where over 70% of women wear the wrong size bras (Pechter, 1998).

15.1.2 Problems of the Traditional Methods

The traditional bra sizing method requires two main measurements taken with tapes: the under-chest circumference defined by the horizontal circumference right below the busts around the torso and the upper chest circumference defined by the horizontal circumference through two nipple points. The under-chest circumference indicates the bra band sizes, such as 70 cm/30 inch, 75 cm/ 32 inch and 80 cm/34 inch bands. The difference between the upper and under-chest circumferences indicates the cup sizes in letters, such as A cup (2.6 cm/ 1 inch difference), B cup (5.1 cm/ 2 inch difference) and C cup (7.6 cm/3 inch difference) (Pechter, 1999). Using this measuring and sizing method, designers then consider the standardized population distribution of women's body shapes as a reference to proportionally scale up or shrink down the bra patterns to produce bras (Shin, 2015). However, this conventional method causes fitting issues due to its sizing limitations, the variety of morphological differences in women's body shapes and the different standards used by different companies and brands (Perling & Colizza, 2019).

15.1.3 Motivation and Goal

Although the current bra sizing and grading method has been developing to carry more and more elaborated sizes such as AA and DD cups, it still cannot address the problem of body shape morphology such as asymmetry breasts and cannot cover the maximal range where outliers exist such as women with gigantomastia (Castillo, 2015). Therefore, today's women's bra design is waiting for a new world of personalization with more scientific and technological approaches, where designers can provide better fitting and more comfortable bras for women with various body shapes.

The goal of this study is to develop a parametric bra design algorithm that can create a personalized bra design for individual woman using their self-taken body measurements. This personalized algorithm requires acquiring accurate body measurement data, manipulating and analyzing the data and morphing the CAD (Computer-Aided Design) bra model accordingly. With the algorithm, women wearers won't need to sacrifice their comfort by wearing standard sizes. Instead, the

Parametric Design for Personalized Bras 281

algorithmic system will generate the bra model in exact sizes to fit women wearers. To accomplish the goal, we specifically focus on three tasks:

- Construct a self-measuring approach that can collect adequate and accurate data on breast shape and size.
- Establish a parametric algorithm for personalized bras.
- Evaluate the algorithm by comparing the dimensions between the self-taken measurements and a 3D-printed bra prototype.

15.2 DESIGN EXPLORATION

The tasks of this project require a fundamental knowledge of the women's breast anthropometry, commercial bra design and parametric design methods. A well-balanced synthesis of these elements will help us acquire accurate univariate self-measurement data from the subjects and integrate their data into the 3D parametric design algorithm.

15.2.1 ANTHROPOMETRICAL STUDY

Many anthropometric experts have been studying body fit issues using data from 3D body scan technologies. We identify two methods that can serve as potential supports for the parametric design, namely, the breast arc length method (Oh and Chun, 2014) and the folding line method (Lee et al., 2004).

These two studies offer practical methods to locate the important landmarks and dimensions on women's breasts for measurements, the breast arc length approach (Oh & Chun, 2014)and the folding line approach (Lee et al, 2004).

15.2.1.1 Breast Arc Length Approach

Oh and Chun suggested that the reason for the traditional bra cup sizes causing poor fit is due to the incorrect cup sizes. This is because "the bra cup size does not accurately indicate actual breast volume" (Hong, 2002; Park and Lim, 2002). Oh and Chun propose a new technique of measuring the breast arc lengths by landmarks on 3D body scanned data.

Measuring the breast arc lengths requires the correct positioning of breast landmarks. Although the definitions of the breast landmarks vary among researchers, two points of landmarks are standardly well-defined, namely, the inner breast point and the outer breast point.

The inner breast point (Pi) and the outer breast point (Po) are placed on the upper chest circumference contour slicing through two nipple points (P_{BP}). The breast arc length also depends on finding the right landmarks. They measure four fundamental arc lengths, namely, the full arc length (Pi–Po), cross arc length (Pi–P_{BP}), outer arc length (Po–P_{BP}) and the upper arc length (Pu–P_{BP}). We take this procedure of landmarking and arc lengths measuring as references to apply to our study of defining new measurements as parameters. To maximally provide the easiest way for self-measurement, we mainly focused on two necessary arc lengths: the inner arc length and the outer arc length.

The inner breast arc length is the surface length from the inner breast point (Pi) to the nipple breast point (P_{BP}).

The outer breast arc length (Po–P_{BP}) is the surface length from the outer breast point (Po) to the nipple breast point (P_{BP}).

15.2.1.2 The Folding Line Approach

Considering the common issue of mastoptosis – sagged breasts, the folding line approach presented by Lee and his research team offers an impressive solution.

For bra cups, the volume is largely decided by the bottom arc since the upper arc can be varied due to different cup styles (full cup, ¾ cup and semi-cup, etc.). In Lee's study, they scan 37 women's nude breasts using 3D scan technology to get a reliable boundary of the breasts and thereby provide new shape parameters for breasts. The folding line approach is useful for finding a continuous and natural boundary for breasts so that the breast base can be measured more accurately. The folding line is the natural boundary between the bottom of the breast and the torso. "The position of the folding line on the skin surface does not change after releasing the force by hands due to the fact that the skin surface and the mammary gland forms a tight interconnection" (Lee et al., 2004). With the natural folding line easy to locate, we can then confirm the bottom breast point (P_{BBP}) and define the bottom breast arc length as the surface contour length from the nipple point (P_{BP}) to the bottom breast point (P_{BBP}).

Combining with the arc length method, we can then define the bottom arc length.

The bottom breast arc length (P_{BBP}–P_{BP}) was the surface length from the bottom breast point (P_{BBP}) on the folding line to the nipple breast point (P_{BP}).

15.2.2 BRA DESIGN

Kristina Shin, a PhD researcher, introduces common bra patternmaking methods widely used in the industry and offers her own innovative methods in her book *Pattern Making for Underwear Design.*

According to her book, we choose the basic T-shirt bra style without underwire to mainly concentrating on the size fitting.

We also select other detailed essential measurements for parametric design, including the strap length, band size and cup coverage.

- The strap length is adjusted by the surface length from the nipple point to the shoulder point, which refers to the end of the collarbone.
- The band size follows the traditional way of measuring the under-chest circumference.
- The cup coverage influences the bra style, defined by the surface length from the front center point (FCP) to the neck center point (NCP).

15.2.3 PARAMETRIC DESIGN METHOD

"Parametric design, or parametric modeling, is a CAD technique that uses parameters or variables (numbers, length, points or curves, etc.) to rule, clarify, and encode

Parametric Design for Personalized Bras 283

the relationship between design intent and design response" (Jabi, 2013). Baek and Lee, in their study of parametric human body shape (Baek and Lee, 2012), utilize the parametric design method to integrate the human body shape and size into multiple applications. Other researchers, such as Wang and Shatin, also explore the parametric design methods using mannequins to study human bodies (Wang and Shatin, 2005). These cutting-edge studies provide us valuable experience on how to manipulate the body data, define the parameters and construct the relationship between design inputs and outputs. In this study, our software platform is the Rhino CAD program with Grasshopper Plug-in, which allows us to visualize the logic behind the parametric algorithm on a flowchart-styled processor, adjust the parameter values and reform the 3D models in real-time.

15.3 DESIGN AND DEVELOPMENT

Based on the previous research, we propose a new self-measurement approach and construct the parametric algorithm. The self-measuring approach allows women users to easily measure themselves in maximal privacy. The parametric algorithm is the core system for obtaining the measurement data and generating the personalized bras accordingly.

15.3.1 THE SELF-MEASURING APPROACH

The new self-measuring approach includes seven dimensions around the breast area. To make the procedure clearer and easier, we name each landmark and dimension as follows. Figure 15.1 illustrates these landmarks and dimensions in the self-measurement approach.

15.3.1.1 Landmarks
- BP refers to the nipple points on both sides.
- IBP refers to inner bust points on both sides.
- OBP refers to outer bust points on both sides.
- BBP refers to bottom bust points on both sides.
- SP refers to shoulder points on both sides.
- FCP refers to the front center point, which is on the center between two nipple points on the sternum.
- NCP refers to the neck center point, which is on the center between two collarbones on the front neck.

15.3.1.2 Arc Lengths
- BP-IBP refers to the inner arc length on both sides.
- BP-OBP refers to the outer arc length on both sides.
- BP-BBP refers to the bottom arc length on both sides.
- BP-SP refers to the surface arc length from nipple points to shoulder points on both sides.
- FCP-NCP refers to the surface arc length from the front center point to the neck center point.

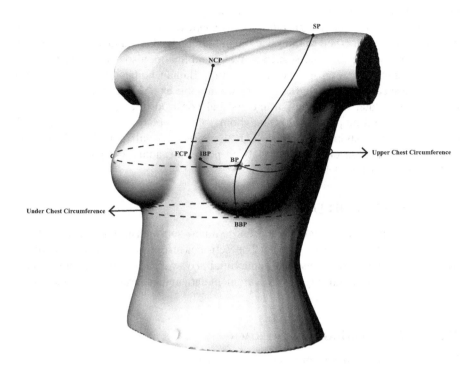

FIGURE 15.1 Defined landmarks and dimensions of the self-measuring approach.

- The upper chest circumference is the circumference through two nipple points around the chest.
- The under-chest circumference is the circumference through two bottom breast points (BBP) and the bottom folding line around the torso.

15.3.2 Parametric Algorithm

15.3.2.1 Original References and Base Bra Model

We construct the parametric personalized bra algorithm with Rhino and Grasshopper program. In this algorithm, we build an initial basic T-shirt-styled bra 3D model, which is derived from a 3D scanned mannequin form. This scanned mannequin form serves as an original reference for centering and positioning the bra model in the algorithm. More importantly, the contours extracted from the mannequin form accord with the needed dimensions in our self-measuring approach and composite the initial bra model structure. By simply adjusting the "parameters" (length and distance between landmarks) of the contours, we can morph the bra model accordingly. Figure 15.2 shows the extracted contours and the initial bra model derived from the scanned mannequin form.

Parametric Design for Personalized Bras

15.3.2.2 Mechanism of Rhino/Grasshopper

In the algorithm, the output is the geometrical 3D bra model displayed on the Rhino interface and the inputs are the parameters on the Grasshopper processor interface. The parameters are namely the measurement values that are self-measured from women users. By changing the values of each measurement, the contour lengths on the initial bra model will expand or shrink to the correct size and scale. Consequently, the output of the bra model morphs in coordinating with the real measurements from women users. Figure 15.3 shows the general layout of the parametric algorithm. On the left side is the generated bra contours displayed on Rhino screen, while on the right side is the parametric processor on the Grasshopper interface.

15.3.2.3 Parameters and Algorithm

With the consideration of the asymmetry breasts, we design the parameters bilaterally, which requires taking measurements on both breasts as well. Therefore, for inner arc length, outer arc length, bottom arc length and shoulder to nipple arc length, users need to measure both sides of their breasts. The bra model consists of four modules of parameters: the two bra cups on both sides, the central panel between the two cups, band panel and the detail size on the upper cup shape coverages. Each module has its core parameters to work functionally on each part of the bra model. The overall algorithm relies on the coordination and cooperation across all the modules to completely morph the bra model.

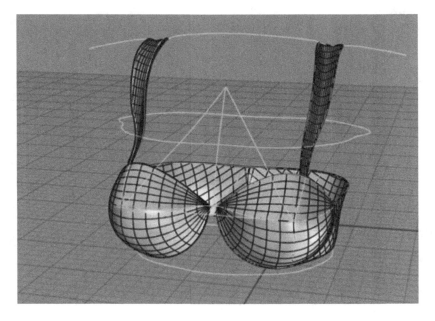

FIGURE 15.2 The contours extracted from the scanned mannequin as original reference.

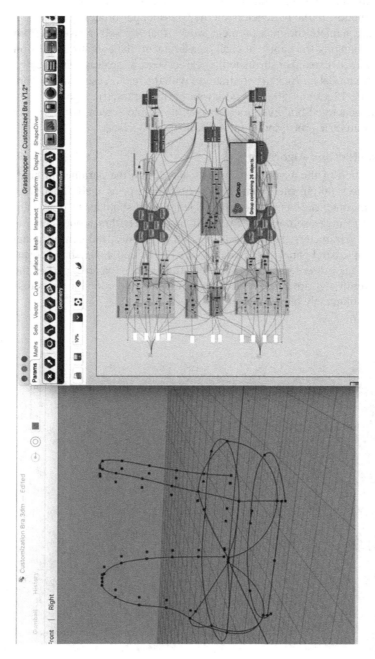

FIGURE 15.3 The Rhino and Grasshopper program layout.

15.3.2.3.1 Left and Right Bra Cups Algorithm

The left and right side modular algorithm control the left and right cup sizes and the shoulder straps lengths. The fundamental logic behind the algorithm is by inputting the measurement value of arc lengths, the base bra contours will be scaled by the ratio of the input value and the original value (the original measurements on the mannequin form). To connect the edge contours of the cup shape is also an important procedure which outcomes the edge cup contours of IBP-BBP, BBP-OBP, OBP-UBP (UBP is the connection point of the upper cup and the strap which is defined by the fixed ratio of the length of BP-SP).

15.3.2.3.2 The Center Panel Algorithm

The center panel controls the patterns between two cups. It basically posits the middle point of two cups, which can help center the model and generate the bra band contours.

15.3.2.3.3 Band Panel Algorithm

The band panel is the bra band which wraps around the body torso under the bra cups. In this modular algorithm of band panel, the upper chest circumference and the under-chest circumference take crucial parts in determining the band size. By mathematical operation with other measurements together, the band contour lengths will change to compose the new band contours for the bra model.

15.3.2.3.4 Cup Coverage Algorithm

By inputting the curve length of FCP to NCP, we define the upper cup arc position. By adjusting the number slider of the ratio, we adjust the arc radian properly to fit the cup shape.

15.3.2.4 Generating Personalized Bra Models

In total, 16 aggregated parametrically controlled contours form the basic bra structure. Patching these structural contours together into surfaces produces the personalized bra model digitally.

15.4 USER TEST AND EVALUATIONS

To evaluate the capacity and accuracy of the parametric algorithm, we conducted a user test by recruiting three female participants with different body sizes and making 3D-printed bra prototypes based on the parametric models. By examining the size range of the models and comparing the dimensions between real measurements and bra prototypes, we comprehensively assessed the algorithm's efficiency.

15.4.1 PURPOSE OF THE USER TEST

The bra shapes and sizes can be manipulated by adjusting the values of the parameters. The values of parameters are supposed to be derived from self-measurements from women participants.

We recruited three women participants and asked them to take self-measurements following a video instruction without others' help. We then collected the measurements, input them into the parametric algorithm, generated personalized bra models, 3D printed the bra prototypes and eventually measured the prototypes to justify the accuracy of the algorithm.

15.4.2 PROCEDURES

The three participants vary in body sizes including relatively small size, average size and relatively large size. They were assigned to measure themselves according to our measuring instructions. We then collected their measurement data, input them into the parametric algorithm, generated personalized bra models, 3D printed the bra prototypes and eventually measured the prototypes to justify the accuracy of the algorithm. All of the personally identifiable information of participants is NOT included in the study.

15.4.3 EVALUATION RESULTS

The algorithm correctly generated three different sized bra models. We used white SLA filament for 3D print prototypes. To reinforce the structure, we thicken the bra surface inwards by 3 millimeters to ensure the measurements taken on the outer surface are not biased. Figure 15.4 shows the 3D-printed bra prototypes.

We then measured the 3D-printed bra prototypes with masking tape on each designated dimension and compared the results to the participants' measurements. Table 15.1 illustrates the comparison results between three bra prototypes and three participants' measurements.

15.4.4 DISCUSSION

In general, the measurement results of 3D-printed bra prototypes turned out to be accurate. For all three sets of comparisons, most of the measurements on the cup arc length are exactly as same as the inputted measurement values, except for the difference within 1cm of the BP-IBP arc length on medium and big size bras. This might be caused by the manual operation using masking tape on the curvy surface.

However, for all three sets, the lengths of the upper and under-chest circumferences have a relatively higher value of errors. On small and big ones, the differences are confined within 1cm, while on medium one, the difference is 4 cm on both girths. A probable reason for this difference is the default set in the algorithm. When scaling up or down the circumference length, the curves keep the arc radian as defaulted setting and only expand or shrink the length. Thus, it causes the overlapping on the

Parametric Design for Personalized Bras

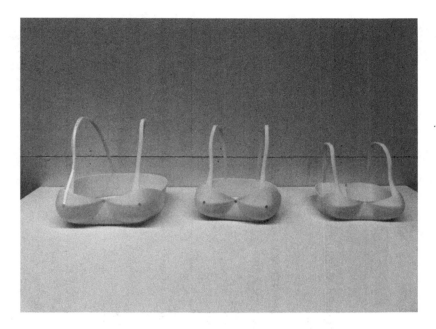

FIGURE 15.4 Three 3D printed bra prototypes in different sizes.

back-center part, which affects the length when patching the surfaces together. We consider this difference is within the limit of tolerance, since for real knitted bra manufactures, we need to leave the space for the back hook, which is about 5 cm long (Shin, 2015). The back hook is for users to adjust the tightness of the band.

Using the 3D print prototypes allows us to directly visualize the algorithm performance in terms of the lengths and dimensions. The algorithm still needs to be optimized and improved in many aspects in terms of accuracy and capacity.

15.5 FUTURE STUDY

So far, we have already accomplished the core tasks of establishing the first-stage parametric algorithm and evaluating usability. However, some trade-offs and compromises lead us to further explore the study.

15.5.1 Optimization of the Parametric Algorithm

The current parametric algorithm specifically aims at the size fit of personalized bras. It is based on the mannequin-scanned data as the reference and solves the potential problem of size fitting, particularly based on a length-driven method. In a further study, we expect to extend the algorithm with more parameters that can cover more extreme cases with higher flexibility. We also expect to include the 3D body scan technology to capture the body shape in order to maximize the algorithm capacity by considering the body morphology.

TABLE 15.1.
Comparison of Results between Bra Prototypes and Participants' Measurements in Three Different Sizes. All the Measurements Below Are in Centimeters (cm)

Size	Object	Body side	BP-SP	BP-OBP	BP-BBP	BP-IBP	FCP-NCP	Upper chest circumference	Under chest circumference
Small	Participant	Left	20.0	6.3	5.5	8.0	10.9	67.6	55.0
		Right	20.0	6.3	5.5	8.0			
	Prototype	Left	20.0	6.3	5.5	8.0	N/A	68.0	55.0
		Right	20.0	6.3	5.5	8.0			
Medium	Participant	Left	24.0	9.5	6.5	8.0	17.5	78.0	67.0
		Right	24.0	9.5	6.5	8.5			
	Prototype	Left	24.0	9.5	6.5	8.0	N/A	74.0	63.0
		Right	24.0	9.5	6.5	9.5			
Large	Participant	Left	25.8	10.5	8.2	9.0	9.0	88.0	74.3
		Right	25.8	10.8	8.2	9.3			
	Prototype	Left	25.8	10.5	8.2	9.5	N/A	89.0	50.7
		Right	25.8	10.8	8.2	9.3			

Parametric Design for Personalized Bras

In the current design, the bra is in a relatively basic and plain T-shirt bra style with only cup coverage that can be changed. Developing a fully personalized algorithm also requires users' preference for customizing styles, colors and details. Therefore, to further optimize the current stage of the algorithm, we propose to include the parameters that can adjust the bra styles and patterns to offer more available options for women users.

15.5.2 Prototyping and Manufacturing

We propose to utilize 3D knitting as a producing technique to offer high-quality bras. In this way, we can conduct a real trying-on experience for users. The current smart 3D knitting technique requires a 2D patterned design specification for manufacture, which indicates that we need to develop 2D bra patterns with dimensions from the digital 3D models. We will make efforts to get access to the 3D knitting manufacture to understand the mechanism of how 3D knitting works and further study the method of transforming between 3D models and 2D patterns.

15.5.3 Fit Tests

In the future study, we expect to bring real 3D knitted prototypes for user fit tests. With the optimized parametric algorithm, we will include and invite more participants with a larger size range for tests.

15.5.4 Personalization Era for Women's Bras

Our vision is not only limited to improve the personalized fit bra design but to provide a new model of service for women's bra personalization. This requires a comprehensive development on user-experience, service design and manufacture. With the technologies emerging into our daily life and changing the way people thinking, designing and manufacturing, personalization will be a new world ahead to conquer. We will be continuously working on this research topic to embrace the new realm where people can enjoy personalized wearable products with high quality, maximal fit and elaborate service.

ACKNOWLEDGMENT

I would like to express my deepest appreciation to all those who provided me the opportunity to complete this project.

I owe a special gratitude to my advisor, Professor Roger Ball, whose excellent guidance throughout my works and encouragement on my design decision helped me on coordinating my project especially in exploring the topic, communicating with stakeholders and writing this paper. I would also like to acknowledge with many thanks the crucial roles of my Master's committee members, Lisa Marks and Yaling Liu, who contributed their knowledge to help me with Grasshopper parametric methods and gave me high support on bra pattern design. Without their devotions, I would have not completed this project with such accomplishment.

Furthermore, many thanks go to those lovely schoolmates who assisted me in finishing the tasks of user test study; Xiaoshen Wang as my friend and classmate who had always been discussing the parametric methods with me; And my dear boyfriend Mr. Fan Geng who intensely encourages me all the time. Finally, I appreciate my parents who were always having my back and gave me the warmest support and wise advice.

REFERENCES

Baek, S., & Lee, K. 2012. Parametric Human Body Shape Modeling Framework for Human-Centered Product Design. *Computer-Aided Design* 44, 56–67.

Castillo, S. 2015. The Average Bra Size in America, Plus 4 Other Breast Size Facts You May not Have Known. *Medical Daily.* https://www.medicaldaily.com/.

Crandall, D. 2018. 100 Years of Brassiersres: The Historical Evolution of the Bra. *Insidehook.* https://www.insidehook.com/article/history/100-years-brassieres-inside-historical-evolution-bra.

Hardaker, C.H.M., & Fozzard, G.J.W. 1997. The Bra Design Process – A Study of Professional Practice. *International Journal of Clothing Science and Technology* 9(4), 311–325.

Hong, K.H. 2002. Engineering Design Process for the Sense-Friendly Comfort Brassiere Using Various Techniques of Human Technology. *The Journal of Korean Society of Living Environment System* 9(3), 226–237.

Jabi, W. 2013. Parametric Design for Architecture. *International Journal of Architectural Computing* 11, 4.

Lee, H., Hong, K., & Kim, E.A. 2004. Measurement Protocol of Women's Nude Breasts Using a 3D Scanning Technique. *Applied Ergonomics* 35, 353–359.

Oh, S., & Chun, J. 2014. New Breast Measurement Technique and Bra Sizing System on 3D Body Scan Data. *Ergonomics Society of Korea* 33(4), 299–311.

Park, Y.S., & Lim, Y.J. 2002. A Study on Establishment of Brassiere Size and Clothing Pressure for the Twenties-aged Women. *Journal of the Korean Society of Costume* 52(8), 15–27.

Pechter, A.E. 1999. Tape Measure for Bra Sizing. US Patent No. US6272761B1. Washington, DC: U.S. Patent and Trademark Office.

Pechter, E.A.M.D. 1998. A New Method for Determining Bra Size and Predicting Postaugmentation Breast Size. *Plastic & Reconstructive Surgery* 102(4), 1259–1265.

Perlinng, A., & Colizza, C. 2019. Are 8 Out of 10 Women Really Wearing the Wrong Bra Size? *The New York Times.* https://www.nytimes.com/2019/07/10/style/lingerie-are-8-out-of-10-women-really-wearing-the-wrong-bra-size-a-bra-myth-busted.html.

Shin, K. 2015. *Patternmaking for Underwear Design* (2nd edition).

Wang, C.L., & Shatin, N.T. 2005. Parameterization and Parametric Design of Mannequins. *Computer-Aided Design* 37(1), 83–98.

16 Dimensional Aspects of Usability of the Beds

Aleksandar Zunjic

CONTENTS

16.1 Introduction 294
16.2 Existing Dimensions of Beds 294
16.3 Usual Use of Certain Types of Beds... 295
      16.3.1 Small Single 297
      16.3.2 Extra Small Single 297
      16.3.3 Single/Twin 297
      16.3.4 Single/Twin Extra-long 297
      16.3.5 Large Single 297
      16.3.6 King Single 298
      16.3.7 Extra-long Twin 298
      16.3.8 Super Single 298
      16.3.9 Small Double 298
      16.3.10 Double/Full 298
      16.3.11 Queen 298
      16.3.12 King 299
      16.3.13 Super King 299
      16.3.14 California King 299
      16.3.15 Super King Size 299
16.4 Procedures for Determining Dimensions of the Beds................................. 299
      16.4.1 The First Procedure for Determination of the Dimensions
              of Beds 300
      16.4.2 The Second Procedure for the Determination of the
              Dimensions of Beds 303
      16.4.3 The Third Procedure for Determination of the Dimensions
              of Beds 305
16.5 Data Collection Procedure... 308
16.6 Results... 308
      16.6.1 Analysis of the Results 311
16.7 Conclusion 315
References 317

16.1 INTRODUCTION

The time that people spend in sleeping on a daily basis depends on a number of factors, such as age, professional obligations, physiological state of the organism, environmental conditions, as well as on certain other determinants. The World Health Organization (WHO Europe 2009) cites the data that adults spend on average 7.5 hours in bed, while the time that people spend effectively sleeping is somewhat less (according to the results of a study conducted by the Centre for Time Use Research). This indicates that people spend approximately 30% of their lives in interaction with the bed. With that in mind, it is of great importance to design usable beds in a way that will provide adequate comfort and health of users.

Although the bed undoubtedly plays a huge role in people's lives, surprisingly small number of ergonomic research has been done on the topic of usability of beds. In over 30 ergonomics books, the word bed is not even mentioned (which is not the case with the chair). The results of the extensive search of the literature in the field of ergonomics point out that so far only a few ergonomic studies have been done, which had a bed in the focus of attention. However, these papers did not take into consideration the size of the bed (dimensions) as one of the most important factors in influencing its usability. Also, it is difficult to find literature that deals with designing of beds if we exclude the primary aesthetic aspect. These circumstances have led to a situation where the bed often is not designed on the principles of usability. It is not seldom that making beds comes down to the application of the knowledge accumulated over the years in the field of handicraft production. This "good practice" has particular application in the field of mass-production related to the furniture industry.

The comfort and health of users not only depend on the properties of mattresses. An important characteristic that also affects users' comfort and their health is the dimensions of beds. It can be said that so far several surveys are conducted with subjects with regard to the determination of certain characteristics of beds and mattresses. However, there is no research in the field of usability that has been done so far, in order to determine the required dimensions of beds. This chapter aims to provide the procedure for determining the appropriate measures of beds that relate to the lying position of users, as well as the creation of usability recommendations for designing, based on the obtained results of research. This approach should ensure that the dimensions of the beds, from the aspect of sleeping, be consistent with the expectations and needs of users.

16.2 EXISTING DIMENSIONS OF BEDS

In order to have the ability to assess the adequacy of dimensions of beds from the aspect of usability, it is necessary to start from the available dimensions that are encountered in practice. In this connection, attention is focused on the dimensions of the available beds for the lying position of users. In the nineteenth century and earlier, there were significant variations in the dimensions of beds. However, in recent times, efforts are made in the direction of standardization of dimensions of beds. One of the successful examples of standardization of dimensions of beds and mattresses is the area of North America. Also, certain efforts have been made in other parts of the world, oriented toward the standardization of these types of furniture.

Depending on the type of beds, in Table 16.1, conventionally adopted dimensions of beds are given depending on the region of the world or country. These dimensions have frequent application. Some of them are used more frequently than others and can be considered conditionally standardized, such as those relating to the area of North America (Twin, Twin XL, etc.). However, it should be noted that currently there is no official standard with a binding application, which refers to the mandatory dimensions of the bed. For example, ISO standards do not prescribe the dimensions of beds, but only individual characteristics that relate to the safety of beds (for example ISO 9098-2: 1994 standard). Table 16.1 shows the names of the types of beds that are encountered in practice around the world, although there are certain types of beds that do not have to be represented in all regions or countries. The field is kept blank in the table (represented by –) when some type of bed is not represented in a particular region or country.

Table 16.1 shows the widths (in cm) and lengths (in cm) of beds that are used in all five continents. Included are also some countries where there are some deviations from the dimensions that are prevalent in some regions to which they belong but which have achieved a high level of standardization within its borders (for example the UK, which belongs to Europe). After analyzing and comparing data that are available from a variety of sources from the Internet, the data that are obtained primarily from the manufacturer and distributor of beds and mattresses are presented. It should be mentioned that there are minimal variations in dimensions that individual manufacturers and distributors represent for certain types of beds. These small differences may occur primarily as a result of conversion of measurements from one unit to another unit. In addition, within certain regions (continents), there are countries that offer the dimensions of beds, which, to some extent, deviate from the dimensions that are characteristic in the observed area. With the aim of focusing on the conventionally adopted sizes of beds with frequent practical application, these variations are not covered in Table 16.1 (they are generally the consequence of the offer of various dimensions that provide certain major manufacturers of the furniture, which are present on the markets of these countries).

16.3　USUAL USE OF CERTAIN TYPES OF BEDS

In order to be able to perceive the scope of usability of certain types of beds for certain categories of people or their combinations, it is necessary to start from the common use of these beds that is observed in practice. A basic description of certain types of beds, which includes the default use of these beds in practice, is discussed in the following sections. Recommendations of manufacturers and distributors of beds in this regard are also taken into account. In this way, for each type of bed, it will be noted whether its use is predicted for one or two people (if in the name of the bed, it is not already highlighted). In addition to the comparative analysis of the

TABLE 16.1.
Conventional Dimensions of Certain Types of Beds around the World

Type of a bed (cm × cm)	US/Canada	EU	UK	Australia	New Zealand	Japan	China	Latin America	South Africa
Small single	–	–	75 × 190	–	–	85 × 195	–	–	–
Extra small single	–	75 × 200	–	–	–	–	–	–	–
Single/Twin	99 × 191	90 × 200	90 × 190	99 × 191	92 × 188	97 × 195	99 × 190	90 × 200	91.5 × 188
Single extra-long	–	–	–	91 × 203	92 × 203	–	–	–	–
Large single	–	99 × 200	–	–	–	–	–	–	–
King single	–	–	–	106 × 203	107 × 203	–	–	–	–
Extra-long Twin	99 × 203	–	–	–	–	–	–	–	–
Super single	–	–	107 × 190	–	–	110 × 195	–	–	107 × 188
Small double	–	–	120 × 190	–	–	120 × 195	120 × 190	–	–
Double/Full	137 × 190	140 × 200	137 × 190	137 × 190	135 × 188	140 × 195	–	140 × 200	137 × 188
Queen	152 × 203	160 × 200	–	152 × 203	150 × 203	152 × 195	150 × 200	160 × 200	152 × 188
Super King	–	–	183 × 198	–	183 × 203	–	–	–	–
King	193 × 203	180 × 200	152 × 198	183 × 203	165 × 203	180 × 195	180 × 200	180 × 200	183 × 188
California King	183 × 213	–	–	–	–	–	–	–	–
Super King Size	–	–	–	203 × 203	–	194 × 205	–	–	–

dimensions of a bed, for each bed type will be noted whether its use is intended for children or adults.

16.3.1 Small Single

A small single-sized bed is primarily intended for toddlers and younger children. In practice, these beds are additionally used to accommodate adults who are private visitors (guests). Although there are other types of beds with the same length, their width is the smallest compared with other types of beds.

16.3.2 Extra Small Single

This type of bed has the same purpose as the Small single bed type. Although its name (extra) suggests that its dimensions are smaller than of the Small single bed type, its actual dimensions do not confirm that. The European version of this type of bed has a length slightly greater than the UK or Japanese version of the Small single bed (5–10 cm). The width of the Extra small single bed, as well as the Small single type of bed, belongs to the least category.

16.3.3 Single/Twin

Although the name Twin does not indicate that this is a bed for one person, this is yet a term used in the practice in North America to indicate a bed that is designed for one person. Depending on the region of the world, the width of these beds is in the range between 90 cm and 100 cm. It is especially recommended for small children who are still growing but within 165 cm body height. It is believed that the width of the bed is sufficient for almost all the width of users, particularly children.

16.3.4 Single/Twin Extra-long

This type of bed in North America has the same width as the Single/Twin bed type. In Australia and New Zealand, Single/Twin extra-long beds can be few centimeters narrower than in North America. However, in all cases, the length of this bed (203 cm) is bigger than the above-mentioned type of bed. Manufacturers and distributors of beds consider that this size is sufficient for the accommodation of adults up to 183 cm of body height.

16.3.5 Large Single

This type of bed that is designed for the European market is approximately similar in dimensions to that of the Twin extra-long bed type, which is present in the territory of North America (the second is 3 cm longer). The Large single bed represents 9 cm wider variation of the Single type of bed (for the EU market). It is intended for people of 180 cm in height and offers a slightly higher comfort in width compared to the Single variant for the EU area.

16.3.6 King Single

It is primarily offered in the area of Australia and New Zealand. The length of the King single bed corresponds to the length of the Single/Twin extra-long bed. However, this type of bed offers a somewhat greater width than Single/Twin extra-long. It is believed that this size of the bed is enough for the accommodation of adults or teenagers up to 183 cm of body height.

16.3.7 Extra-long Twin

This type of bed is exclusively designed for the area of North America. The main characteristic is that it has the same length as that of the Single extra-long, as well as the King single bed type. However, its width is slightly smaller than the previously mentioned two types of beds. The width of this bed is identical to the width of the Twin type of bed. Usually, it can be used to accommodate adults or teenagers up to 183 cm of body height, which do not have higher needs in terms of the width of the bed.

16.3.8 Super Single

This type of bed has a greater width than all the previously mentioned types of beds intended for the accommodation of one person. The width of this type of bed reaches 110 cm (in Japan). However, the Super single bed does not have the maximum length and its length is in the range of the Single/Twin type of the bed. Due to its dimensions, it is intended primarily for growing children up to 165 cm of body height, who require a higher level of comfort in the lateral plane.

16.3.9 Small Double

Small double bed type is primarily intended for two persons with lower body height, for example, two children. It is designed for people who do not mind sleeping in a very small distance. This type is used primarily in small bedrooms. However, since the width of this bed amounts to 120 cm, it can be used for individual use, when a user needs a higher width of the bed.

16.3.10 Double/Full

The length of this bed is similar to the length of the Single/Twin bed. It is intended for two people. Due to this, its width is in the range of 135–140 cm. Although this type of bed is also recommended for adults because of the greater width compared to the Small double, this type of bed is not suitable for use by a male person that is higher than average. It may be an appropriate choice when it is decided that two children will sleep together.

16.3.11 Queen

The Queen bed type has a slightly greater width than the Double/Full bed type. As such, it is suitable for the accommodation of two teenagers or two adults. However, the available width of this bed does not offer as much comfort as can be provided by the types of beds that follow. The length of this type of bed varies depending on the region of the world.

16.3.12 KING

The King bed is a type of bed that is intended for two people. However, there are certain variations in the length and width of this type of bed depending on the regions of the world. For this reason, it is very difficult to specify a general recommendation for use. However, this type of bed possesses considerable length, so that it can be used by people of about 180 cm and below. King and Queen types of beds are the commonly used types. According to Rosenblatt (2006), approximately 78% of couples choose one of these two types of beds. In addition, with the exception of the region of the UK and New Zealand, the King bed type provides a relatively acceptable width for the accommodation of two adults.

16.3.13 SUPER KING

This type of bed is used in the UK and New Zealand. The dimensions of this type of bed in general correspond to the dimensions of the King type of bed in other parts of the world. Given that, all that has been said for the previous type of bed also applies to the Super King bed type.

16.3.14 CALIFORNIA KING

The California King bed type is used in the North American region and is recommended for two people. This bed is 10 cm narrower than the King bed in North America. However, California King has a length of 213 cm, so that it is longer than all of the previously mentioned types of beds. It is recommended for people with maximum height.

16.3.15 SUPER KING SIZE

It is intended for two people. According to the length, the Super King Size bed type is closest to the King bed type (variations are almost minor). However, the Super King Size bed type has a greater width than all of the previously mentioned types of beds. It is commonly used by adults, who want to have enough individual space in width.

16.4 PROCEDURES FOR DETERMINING DIMENSIONS OF THE BEDS

As it can be noticed, there are many types of beds with a significant variation in the dimensions between them. It may be noted that there is no type of bed that has the biggest measures according to the length as well as according to the width. It should be noted that apart from the mentioned conventional dimensions of beds, there is a greater number of non-standard types and sizes of beds, which do not have large-scale use in practice. Bearing all this in mind, the question is, in what way the dimensions of the surface of the beds for the laying position of users were determined?

Although the previous question seems to be relatively simple, the answer is not so simple and straightforward. Existing ergonomic literature on this question almost does not offer any answer. The only information that one can find in the ergonomic literature on this topic is the dimensions of only four types of beds: Single, Twin, 3/4 and Double (Woodson 1981). Woodson alleges that the length of all four types of beds should be the same, i.e., 198 cm. The widths of the beds of the specified models respectively are 91 cm, 99 cm, 122 cm and 137 cm. However, Woodson does not provide any specific recommendations for the design of the lying surface of these beds, i.e., recommendations for determining dimensions and selection of a certain type of bed. Except for the specifications of the four types of beds and their dimensions, other information related to this usability problem cannot be found.

However, the answer to this question is also difficult to find in the literature that relates to furniture design and industrial design. In other words, it is extremely difficult to find a textbook, professional or scientific journal, which offers recommendations for the dimensions of the bed. Having that in mind, the answer to this question may not be obtained in the domain of the existing literature. In order to provide answers to this question, in this case, usability research will be applied. In this regard, three procedures with the application of anthropometry will be considered, which can give an adequate answer regarding the selection of the most convenient dimensions of beds.

16.4.1 The First Procedure for Determination of the Dimensions of Beds

The first procedure for the evaluation and determination of dimensions of beds has more theoretical significance than its practical significance. It is based on the principle of maximum use of the bed surface, intended for sleeping in the lying position. This means that the length and width of the bed correspond to the length and width of the user, without additional free space in one or both directions (Figure 16.1). Thus, attention is focused on the accommodation of one or two persons, depending on the type of the bed, whereby it is not available additional interspace for the user. This situation is relatively rare in practice. However, this case should not be excluded from usability consideration, primarily because it is not impossible, especially if one takes into account the number of people on the planet, and if we take into account the population of people with greater body length and width than average.

For a particular type of bed (from Table 16.1), we cannot say for certain in what way their dimensions were determined since it is hard to find a written record about that. Given the existing large variations in the dimensions of beds, it is reasonable to assume that they are intended for people of different anthropometric characteristics. However, manufacturers and distributors of beds often do not specify precisely who

Dimensional Aspects of Bed Usability

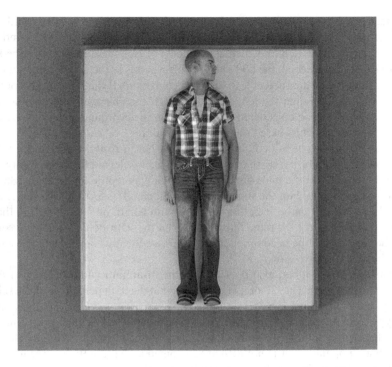

FIGURE 16.1. According to the criterion of the first procedure for the determination of dimensions of a bed, the length of a bed is not greater than the height of the user.

can use a particular type of bed. If at all any type of bed connects with a certain category of users, it is not uncommon that the descriptions of this kind are very superficial. It may be reduced to a description of the next type – younger people, children, adults, higher, lower, etc. When it comes to adult persons, almost without exception, the gender of users is not taken into account. In addition, it is difficult to find explicit information, which type of bed is recommended, for example, for people over 183 cm.

In order to evaluate the usability of existing types of beds in terms of anthropometric suitability, it is necessary to accurately determine to which percentile of people a particular type of bed is intended. This assessment will first be carried out in accordance with the above-mentioned properties, related to the first procedure. In this regard, for a particular type of bed, it is necessary to determine which are the maximum percentiles of the male and female population that can be accommodated in the available dimensions of a bed.

Let us consider first the cases of beds intended for one person. As an example, the beds that are intended for the North American market will be taken into consideration (these types of beds can also be found in other parts of the world, for example in Brazil). It is also necessary to select the anthropometric dimensions that will serve as a basis for usability evaluation. Heaving in mind the problem under consideration, it is necessary to adopt the highest body dimensions in terms of height and width. Accordingly, body height and shoulder width (bideltoid) are selected as anthropometric dimensions, based on which assessments of existing dimensions of beds will be carried out. In essence, the body height corresponds to the length of the body in a lying posture (when the subject is in the supine position). If these two anthropometric dimensions are less than the length and width of the observed bed, then a certain type of the bed can be considered as appropriate, in accordance with the settings of the assessment, which is based on the first procedure.

Given that the types of beds from the area of North America have been selected for the assessment, the population of USA residents is selected for the usability analysis. In this regard, for the male population, the 95th percentile for height is 1867 mm, and the 95th percentile for the shoulder width (bideltoid) is 535 mm (Kroemer 2001). From Table 16.1, it can be seen that the minimum length of a single bed in the area of North America is 1910 mm. This means that the 95th percentile of the selected male population can be accommodated lengthwise by any conventional type of bed from the area of North America.

From Table 16.1 it can also be seen that the minimum width of the bed for the area of North America is 990 mm. This means that the 95th percentile of the selected male population also can be accommodated across the width in any conventional bed type from the area of North America. Since that the anthropometric dimensions of the female population in terms of body height and shoulder width are less than of the male population, it is not difficult to conclude that the 95th percentile of the female population also can be accommodated in all major types of beds, which are represented in the US and Canada.

The calculated 99th percentile of the body height of the selected male population is 1912 mm. This means that the Single/Twin bed type has the boundary length, which can accommodate the 99th percentile. Moreover, in this type of bed female persons of the 99th percentile can be accommodated. The calculated 99th percentile of the shoulder width (bideltoid) is 552.6 mm (males), so without problems, male persons of this percentile can be accommodated in beds that are intended for one person. Females of the 99th percentile can also be accommodated in all types of single beds when the width of the bed is considered.

Now, let us consider the cases of beds that are intended for two people. The usability analysis that follows also relates to the types of beds from the area of North America, as well as to the same population of users. The minimum length of these beds according to Table 1.1 is 1900 mm. This means that all types of beds are able to accommodate both male and female persons of the 95th percentile. However, only the Double/Full bed type has a length slightly less than that required for the accommodation of male persons of the 99th percentile, according to the criteria of the first procedure. However, female persons of the 99th percentile can be accommodated lengthwise in the aforementioned type of bed. The minimum width of these beds is 137 cm. This means that this width is sufficient for the accommodation (one next to the other) of two male or two female persons of the 95th percentile, as well as of the 99th percentile.

Dimensional Aspects of Bed Usability

16.4.2 THE SECOND PROCEDURE FOR THE DETERMINATION OF THE DIMENSIONS OF BEDS

The second procedure for the evaluation and determination of dimensions of beds has more practical significance than it is in the case of the first procedure. This procedure is based on the principle of the existence of extra space along the length and width of the bed, in relation to the body height and width of a man in a lying position. According to the recommendation of one manufacturer of mattresses and beds (Dreams 2014), a mattress should be at least 10 cm longer than the tallest person who sleeps on it. According to the second recommendation (Schramm 2014), the length of a bed (mattress) is calculated by adding 20–30 cm to the body height. This additional space is primarily intended for performing minimum movements and partly for positioning pillows. So, in this way, the bed is designed to be greater than the user in length and the width, providing the additional free space in both directions (Figure 16.2). Attention is now focused on the accommodation of one or two persons, depending on the type of bed, with the little additional interspace that is available to the user. This situation exists most frequently in practice. However, this does not mean that this way of designing is optimal.

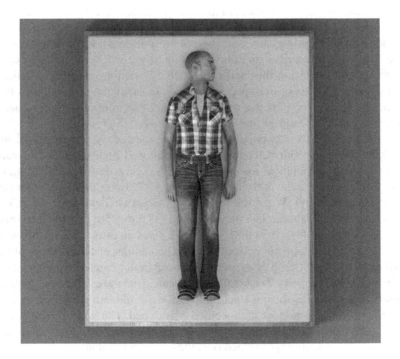

FIGURE 16.2. According to the criterion of the second procedure for the determination of dimensions of a bed, the length of a bed is greater to a certain extent than the height of the user.

In order to evaluate the existing types of beds in terms of usability, it is necessary to accurately determine which percentile of people can use a certain type of bed, if one takes into consideration that should be anticipated additional space (clearance) in length when designing. Bearing in mind the two previously mentioned recommendations for the size of the clearance along the length of the bed, in order to analyze, the free space of 20 cm for the length of a bed will be selected. In this regard, for a particular type of bed, it is necessary to determine which will be the maximum percentile of the male and female population that can be accommodated in the available dimensions of a bed.

In terms of the width of the bed, it should be noted that there are no recommendations in the literature that suggest that it is necessary to anticipate and provide additional space in width in relation to the physical dimensions of the user (even though this is the case in practice, what has been shown also by the analysis, which was conducted on the basis of the first procedure). Also, the literature does not indicate how much should amount an additional space in width, in relation to the body dimensions. Since the recommendations for a free space in relation to the width of a bed do not exist, the analysis within the second procedure will not include an assessment of the width of the additional space, in relation to the maximum width of a body (i.e., the width of shoulders).

Given that the width of the bed, in this case, is excluded from consideration, beds intended for one and two persons will be analyzed in parallel. Similar to the assessment done on the first procedure, beds that are intended for the North American market will be considered as an example. Body height will be again selected as the anthropometric dimension that will serve as the basis for assessment. If the body height with the addition of free space of 20 cm is less than the length of the observed bed, then a certain type of bed can be considered as appropriate in accordance with the settings of the assessment that is based on the second procedure.

For the analysis, the US population is again selected. Assessment will include 99th, 95th, 75th and 50th percentile of the mentioned population. With regard to this, for the male population, the calculated 75th percentile for the standing height equals 1801 mm while the 50th percentile equals 1756 mm (99th and 95th percentiles are given already). When the female population is considered, the calculated 99th percentile is 1778 mm, 95th percentile is 1737 mm, 75th percentile is 1672 mm and the 50th percentile is 1629 mm. Table 16.2 gives an estimation of the possibilities for the accommodation of persons of listed percentiles in certain types of beds in accordance with the criteria of the second procedure for assessing and determining the dimensions of the beds. The symbol Y indicates that the established criteria have been satisfied, as well as there is a possibility of accommodating a person of a certain percentile in lying position for the specified type of bed. The symbol N indicates that it is not possible to provide the adequate accommodation of a person of a certain percentile in the selected type of bed (i.e., the bed is not usable).

From Table 16.2, it can be seen that only the California King bed type allows the accommodation of persons of 99th and 95th percentile in length. Persons belonging to these percentiles can perceive other types of beds as inadequate for use due to their inadequate length. The additional concern is the fact that males of average

TABLE 16.2.
Evaluation of the Possibility of Adequate Positioning of a Person of the Specific Percentile in the Lying Position, in the Types of Beds of Different Lengths (in Accordance with the Criteria of the Second Procedure)

Type of the bed	Percentile (men)				Percentile (women)			
	99th	95th	75th	50th	99th	95th	75th	50th
Single/Twin	N	N	N	N	N	N	Y	Y
Extra-long Twin	N	N	Y	Y	Y	Y	Y	Y
Double/Full	N	N	N	N	N	N	Y	Y
Queen	N	N	Y	Y	Y	Y	Y	Y
King	N	N	Y	Y	Y	Y	Y	Y
California King	Y	Y	Y	Y	Y	Y	Y	Y

height cannot be adequately accommodated in the two types of beds, Single/Twin and Double/Full. Additional analysis showed that these types of beds could accommodate only male adults of the 20th percentile (and lower). This means that these types of beds are not intended for male adults. In addition to children, these types of beds can be used by female adults of the 75th percentile and below.

16.4.3 The Third Procedure for Determination of the Dimensions of Beds

The third procedure for the evaluation and determination of the dimensions of beds has the highest practical importance. This procedure is based on the positions that a man occupies during sleep. According to this procedure, the dimensions of beds depend on the positions that a man occupies during sleep. It is necessary that dimensions of beds support postures that a man occupies during sleep. This is a novel approach, because there is no known research in which this approach was previously used for the determination of dimensions of beds.

There are three basic positions by which humans sleep: lateral, supine and prone. In Western society, most people (60%) prefer a lateral sleep position (Verhaert 2011). According to Haex (2004), 65% of people sleep in the lateral position, 30% in the supine position, and the remaining 5% in the prone position. However, most Asian people prefer sleeping supine. The sleeping position may also depend on age, but also on the quality of the bed on which they sleep. Certainly, during sleep, the position can be changed. A research (Verhaert 2011) showed that 43.7% of the tested subjects spent more than 60% of the time in their dominant sleep position, indicating that a significant amount of time is spent in non-dominant postures as well.

As already mentioned, the lateral position is the most common sleeping position. If the sleep system, including a pillow, is ergonomically designed, the spine can also take the straight position and the natural curves can be maintained (Haex 2004). The lateral position is shown in Figure 16.3. However, in practice, there are variations to this position, since the upper and lower limbs can be moved as needed. During sleep, both body sides (hips) can be equally used.

The main advantage of sleeping in the supine position is that contact with the bed is achieved over a large area of the body, thereby limiting the concentrated pressure on certain organs and soft tissues. However, although the muscles can be relaxed in this position, usually it is not the case with the spine (Haex 2004). The supine position is shown in Figure 16.4.

Although contact with the bed is achieved over a large area in the prone position (as with the previously described position), most people avoid that position. The reason is that in this body position, in most cases, it is not possible to maintain the correct position of the spine, even though the system for sleeping is adequately designed. The prone position is shown in Figure 16.5. It should be noted that in practice there are certain variations in the body positions during the sleep in relation to the previously described positions, because large variations of the positions of arms and legs are possible, including their rotation. Certain rotation of the torso is also possible. However, any such position can be classified into one of the three basic positions during sleep.

However, there are no data relating to the physical dimensions of the people in sleeping positions. To be able to determine usable dimensions of the beds based on the postures people occupy during sleep, it is necessary to conduct the proper research. In this regard, the research is presented in the coming sections and it was conducted with the aim of collecting data that can serve as a basis for the determination of dimensions of beds that are calculated based on the criteria of the third procedure.

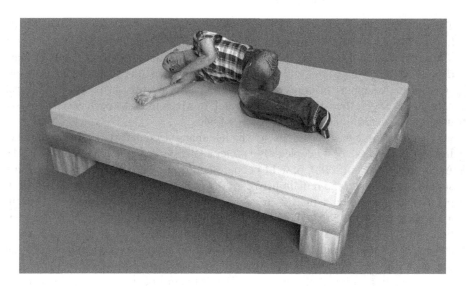

FIGURE 16.3. The lateral sleeping position.

Dimensional Aspects of Bed Usability

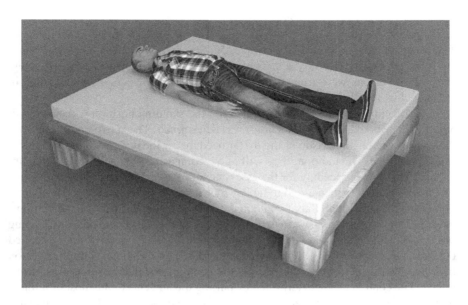

FIGURE 16.4. The supine sleeping position.

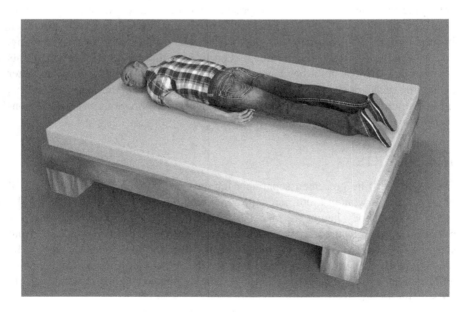

FIGURE 16.5. The prone sleeping position.

16.5 DATA COLLECTION PROCEDURE

The study included a total of 100 people, 50 men and 50 women. The age of subjects ranged from 18 to 68 years. The average age of the subjects was 30.9 years ($\sigma = 12.4$). Persons were from the territory of Serbia. All persons participated in testing voluntarily.

In the beginning, the height of the body in the standing position was measured, and then the shoulder width (bideltoid) was determined. These two anthropometric dimensions are characteristic of the first procedure for the determination of dimensions of the beds. In addition, they will allow comparison with the results that are related to the third procedure for the assessment of the dimensions of the beds.

The subjects were then asked to lie down on a flat surface for testing. The deflection of the area was such that it could not affect the results of measurements (minimum deflection). At first, the instruction was given to the subjects to take the sleeping position in which they spend most of their time during sleep. This position is called the usual sleeping position. After the accommodation of the subject into the position in which he/she most frequently sleeps, the contours for that position were determined. The contours were determined by pulling the tangents to the most prominent parts of the body in the horizontal and vertical directions. After that, the distance between the parallel tangents in the horizontal and the vertical direction was measured. In this way, the dimensions of the length and width that a person occupies in the usual sleeping position were obtained (Figure 16.6).

In many cases, the usual sleeping position is determined by the dimensions of the bed that is used. Often users are not aware that they could take a more comfortable lying position than they do it in practice when they might have the possibility of using a larger surface area for sleeping. With this in mind, the subjects were then instructed to lie in a position which they found to be most comfortable for sleeping, i.e., in which they would like to sleep when there would be no spatial constraints of any kind. This position has been named the preferred sleeping position. After occupying such a position, for each subject was determined another two dimensions, the maximum length and width, in the same way as in the previous case, which concerned the usual sleeping position.

16.6 RESULTS

The results will be presented separately for men and women. The mean value of body height for women (50th percentile) amounted to 167.1 cm ($\sigma = 6.1$ cm). The mean value of body height for men (50th percentile) amounted to 184.7 cm ($\sigma = 6.8$ cm). In order to present the distribution of the measured values for the usual sleeping position, the tabular presentation will be used for better understanding. The interval of 10 cm was used in Table 16.3 for the presentation of the frequency of female subjects, whose length of the usual sleeping position had certain values from the specified intervals.

Table 16.4 shows the number of female subjects (frequency), whose length of the preferred sleeping position had one of the values from the intervals.

Dimensional Aspects of Bed Usability

FIGURE 16.6. Example of determination of the length (L) and width (W) for the usual sleeping position.

TABLE 16.3.

Number of Female Subjects Whose Length of the Usual Sleeping Position Had Certain Value from the Specified Intervals

Length of the posture (cm)	100–110	110–120	120–130	130–140	140–150	150–160	160–170	170–180	180–190
Frequency	1	0	3	2	7	15	8	12	2

Table 16.5 shows the number of male subjects (frequency), whose length of the usual sleeping position had one of the values from the intervals.

Table 16.6 shows the number of male subjects (frequency), whose length of the preferred sleeping position had one of the values from the intervals.

Table 16.7 shows the number of female subjects (frequency), whose width of the usual sleeping position had one of the values from the intervals.

TABLE 16.4.
Number of Female Subjects Whose Length of the Preferred Sleeping Position Had Certain Value from the Specified Intervals

Length of the posture (cm)	150–160	160–170	170–180	180–190	190–200	200–210
Frequency	6	11	18	8	3	4

TABLE 16.5.
Number of Male Subjects Whose Length of the Usual Sleeping Position Had Certain Value from the Specified Intervals

Length of the posture (cm)	130–140	140–150	150–160	160–170	170–180	180–190	190–200	200–210	210–220
Frequency	1	1	2	3	12	20	7	3	1

TABLE 16.6.
Number of Male Subjects Whose Length of the Preferred Sleeping Position Had Certain Value from the Specified Intervals

Length of the posture (cm)	140–150	150–160	160–170	170–180	180–190	190–200	200-210	210–220	220–230
Frequency	1	1	0	1	19	20	5	2	1

TABLE 16.7.
Number of Female Subjects Whose Width of the Usual Sleeping Position Had Certain Value from the Specified Intervals

Width of the posture (cm)	30–40	40–50	50–60	60–70	70–80	80–90	90–100	100–110	110–120	120–130
Frequency	3	6	17	16	5	1	1	0	0	1

Table 16.8 shows the number of female subjects (frequency), whose width of the preferred sleeping position had one of the values from the intervals.

Table 16.9 shows the number of male subjects (frequency), whose width of the usual sleeping position had one of the values from the intervals.

Table 16.10 shows the number of male subjects (frequency), whose width of the preferred sleeping position had one of the values from the intervals.

TABLE 16.8.
Number of Female Subjects Whose Width of the Preferred Sleeping Position Had Certain Value from the Specified Intervals

Width of the posture (cm)	40–50	50–60	60–70	70–80	80–90	90–100	100–110	110–120	120–130	130–140
Frequency	4	9	15	10	5	6	0	0	0	1

16.6.1 ANALYSIS OF THE RESULTS

In order to make proper conclusions regarding the usability of certain types of beds, it is important to conduct the analysis that follows. Therefore, it is necessary to consider whether the length of the usual sleeping position is greater than the height of the body. If this is the case, it means that a bed whose length for sleeping is equal to the height of a subject would be practically small for him/her. In this case, preference should be given to a third procedure for determining the dimensions of beds in relation to the first procedure. This analysis will be performed separately for women and men.

In the beginning, it is necessary to determine the statistical significance of the difference between body height and the length of the usual sleeping position for women. For this analysis, the paired t-test will be used. This test includes all realized individual differences between the aforementioned two dimensions. The calculated value of the test statistics is $t = 4.012$. Since this value is greater than the tabular (1.67) we can conclude that the standing body height is greater than the length of the usual sleeping position for women.

Now, it is necessary to determine the statistical significance of the difference between body height and the length of the usual sleeping position for men. For this analysis also, paired t-test will be used. The calculated value of the test statistics is $t = 2.179$. Since this value is greater than the tabular (1.67), we conclude that the height of the body is greater than the length of the usual sleeping position for men.

Now it is necessary to determine the statistical significance of the difference between the length of the preferred sleeping position for women and their body height. For this analysis, again paired t-test will be used. The calculated value of the test statistics is $t = 4.714$. Since this value is greater than the tabular (1.67), we conclude that the length of the preferred sleeping position is greater than the body height for women.

It is also necessary to determine the statistical significance of the difference between the length of the preferred sleeping position for men and their body height. For this analysis, again paired t-test is used. The calculated value of the test statistics is $t = 3.551$. This value is bigger than the tabular (1.67) and so it can be concluded that the length of the preferred sleeping position is greater than the standing height of the body of males. However, in this case, unlike the previous ones, it is not a fulfilled precondition that the difference between the observed sizes should have a normal distribution. For this reason, the validity will be confirmed by using the Wilcoxon

TABLE 16.9.
Number of Male Subjects Whose Width of the Usual Sleeping Position Had Certain Value from the Specified Intervals

Width of the posture (cm)	50–60	60–70	70–80	80–90	90–100	100–110	110–120	120–130	130–140	140–150	150–160	160–170	170–180
Frequency	10	16	11	10	1	1	0	0	0	0	0	0	1

TABLE 16.10.
Number of Male Subjects Whose Width of the Preferred Sleeping Position Had Certain Value from the Specified Intervals

Width of the posture (cm)	50–60	60–70	70–80	80–90	90–100	100–110	110–120	120–130	130–140
Frequency	5	7	12	10	6	3	3	1	3

Signed Rank Test. Statistical parameters of the test are W = 770.0, T + = 1022.5, T = −252.5. Since the Z-Statistic (based on positive ranks) = 3.721, it is confirmed that the length of the preferred sleeping position is greater than the standing height of the body for men ($P \leq 0.001$).

Furthermore, it is necessary to determine whether the length of the preferred sleeping position is greater than the length of the usual sleeping position for women. For this analysis, the paired t-test will be used. The calculated value of the test statistics is $t = 6.54$. Since this value is greater than the tabular (1.67), it can be concluded that the length of the preferred position for sleeping is longer than the usual sleeping position for women. However, as in the previous case, the prerequisite is not met, that is, the difference between the observed sizes must have a normal distribution. For this reason, the validity will be confirmed by using the Wilcoxon Signed Rank Test. Statistical parameters for this test are W = −861.0, T + = 0.0 and T = −861.0. Since the Z-Statistic (based on positive ranks) = −5579, it has been confirmed that the length of the preferred sleeping position is greater than the length of the usual sleeping position for women ($P \leq 0.001$).

Also, it is necessary to determine whether the length of the preferred sleeping position is greater than the length of the usual sleeping position for men. For this analysis, the same paired t-test will be used. The calculated value of the test statistics is $t = 5.804$. Since this value is greater than the tabular (1.67), we can conclude that the length of the preferred position for sleeping is longer than the usual position for sleeping for men. However, again the precondition that the difference between the observed sizes must have the normal distribution is not met. For this reason, the validity will be confirmed by using the Wilcoxon Signed Rank Test. Statistical parameters of this test are W = −595.0, T + = 0.0 and T = −595.0. Since the Z-Statistic (based on positive ranks) = −5089, it has been confirmed that the length of the preferred sleeping position is greater than the length of the usual sleeping position for men ($P \leq 0.001$).

Now it is necessary to perform the analogous analysis, taking into account the relevant dimensions of width. In this regard, in the beginning, it is necessary to determine the statistical significance of the difference between the width of the usual sleeping position for women and the static maximum width of the body (bideltoid). For this analysis, the paired t-test will be used. The calculated value of the test statistics is $t = 8.196$. Since this value is greater than the tabular (1.67), it can be concluded that the width of the preferred sleeping position is greater than the width of the body (bideltoid) for women. However, again the precondition that the difference between the observed sizes should have the normal distribution is not met. For this reason, the validity will be confirmed by using the Wilcoxon Signed Rank Test. Statistical parameters of this test are W = 1054.0, 1067.5 = T+ and T = −13.5. Since the Z-Statistic (based on positive ranks) = 5.759, it is confirmed that the width of the usual sleeping position is greater than the width of the body (bideltoid) for women ($P \leq 0.001$).

It is also necessary to determine the statistical significance of the difference between the width of the usual sleeping position and the width of the body (bideltoid) for men. For this analysis, the paired t-test will be used. The calculated value of the test statistics is $t = 8.894$. Since this value is greater than the tabular (1.67), it can be concluded that the width of the preferred sleeping position is greater than the width of the body (bideltoid) for men. However, the precondition that the difference between the observed sizes should have a normal distribution is not met. For this reason, its validity will be confirmed by using the Wilcoxon Signed Rank Test. Statistical parameters for this test are W = 1225.0, T+ =1225.0, T− = 0.0. Since the Z-Statistic (based on positive ranks) = 6.094, it was confirmed that the width of the usual sleeping position is greater than the width of the body (bideltoid) for men ($P \leq 0.001$).

Let us consider now the statistical significance of the difference between the width of the preferred sleeping position and the width of the body (bideltoid) for women, and the paired t-test will be used for this purpose. The calculated value of the test statistics is $t = 10.886$. Since this value is greater than the tabular (1.67), it can be concluded that the width of the preferred sleeping position is greater than the width of the body (bideltoid) for women.

It is also necessary to determine the statistical significance of the difference between the width of the preferred sleeping position for men and their body width (bideltoid). For this analysis, again the paired t-test will be used. The calculated value of the test statistics is $t = 12.045$. This value is higher than the tabular (1.67), so it can be concluded that the width of the preferred sleeping position is greater than the width of the body (bideltoid) for men.

Furthermore, it is necessary to determine whether the width of the preferred sleeping position is greater than the width of the usual sleeping position for women. For this analysis, the same paired t-test will be used. The calculated value of the test statistics is $t = 3.341$. Since this value is greater than the tabular (1.67), it can be concluded that the width of the preferred sleeping position is greater than the width of the usual sleeping position for women. However, in this case, it is not a fulfilled precondition that the difference between the observed sizes must have a normal distribution. For this reason, the validity will be confirmed by using the Wilcoxon Signed Rank Test. Statistical parameters of this test are W = −696.0, T + = 192.5, T− = −888.5. Since the Z-Statistic (based on positive ranks) = −3804, it has been confirmed that the width of the preferred sleeping position is larger than the width of the usual sleeping position for women ($P \leq 0.001$).

Finally, it is necessary to determine whether the width of the preferred sleeping position is greater than the width of the usual sleeping position for men. For this analysis, the paired *t*-test will be used. The calculated value of the test statistics is *t* = 3.648. Since this value is greater than the tabular (1.67), it is concluded that the width of the preferred sleeping position is greater than the width of the usual sleeping position for men. However, in this case, also the precondition that the difference between the observed sizes must have the normal distribution is not met. For this reason, the validity will be confirmed by using the Wilcoxon Signed Rank Test. Statistical parameters of this test are W = −512.0, T+ = 59.0, T = −571.0. Since the Z-Statistic (based on positive ranks) = −4194, it has been confirmed that the width of the preferred sleeping position is larger than the width of the usual sleeping position for men ($P \leq 0.001$).

16.7 CONCLUSION

Now we are able to make adequate conclusions regarding the types of beds that are usable for certain percentiles of users. From Table 16.1, it can be seen that the smallest conventional dimension of the bed lengthwise amounts to 188 cm regardless of a region of the world. This means that persons of the 95th percentile (or less) from the region of North America can be accommodated in terms of length in any bed (from the front to the back edge of the bed – on the basis of criteria that are linked to the first procedure for determining of dimensions of beds). This applies to the beds intended for one person, as well as to beds designed for two people. However, a person of the 99th percentile cannot be accommodated lengthwise in all types of beds. There are a number of types of beds intended for one or two people, which, depending on the region of the world where they are used, do not allow the accommodation of persons of the 99th percentile in length from the front to the back edge of the bed. Such lengths of the beds cannot be considered usable for persons of the 99th percentile.

In accordance with the first procedure for determining the dimensions of the bed and in order to be able to accommodate the male persons of the 99th percentile, the minimum length of a bed should be greater than 192 cm. Designing of beds that are intended for sleeping of one or more persons with a length of over 192 cm will enable that even 99% of certain population are in possibility to be accommodated in any type of a bed (based on the criteria described in the first procedure for determining the dimensions of the beds).

From Table 16.1, it can be seen that the smallest conventional dimension of a bed that is intended for one person, regardless of the regions of the world, equals 75 cm in width, which is sufficient to accommodate the persons of the 99th percentile in the lying position. This means that existing types of beds are wide enough that a further increase in the width is not necessary for them (if the beds are designed based on the criteria of the first procedure for determining the dimensions of beds). From Table 16.1, it can also be noticed that the minimum width of the bed that is designed for two people (regardless of the region of the world) equals 120 cm. This width allows accommodation, one beside the other, even two males of the 99th percentile. For this reason, further increasing of the width is not necessary for the beds that are intended for the accommodation of two people, if the bed is designed on the principles of the first procedure for determining the dimensions of the beds.

Based on Table 16.1 and based on the application of criteria for assessing the adequacy of the dimensions of beds, which is based on the second procedure (the existence of clearance of 20 cm), it can be noted that from the aspect of usability only California King bed type provides accommodation of 99th and 95th percentile of the male population from the territory of the US. Other conventional types of beds around the world do not provide adequate accommodation (into the lying position) of the previously mentioned male population of the specified percentiles. In addition, none of the conventional types of beds in the UK and South Africa will allow the positioning of male persons of 75th percentile from the territory of the US (in accordance with the criteria that originate from the second procedure for the determination of dimensions of beds).

In accordance with the second procedure for determining the dimensions of the beds and in order to provide opportunities for accommodating male persons of the 99th percentile, the minimum length of the bed should be greater than 212 cm. Designing of beds intended for sleeping of one or more persons with a length of over 212 cm will ensure that as much as 99% of the population has the possibility of getting accommodated in any type of bed (based on the criteria described in the second procedure for determining of dimensions of beds). In a case where usability designers want to provide the accommodation of male persons up to the 95th percentile, the recommended length of the bed then equals 197 cm. It should be noted that these recommendations are based on the data of the male population from the territory of the US. If any population of residents has a higher body height in relation to the previously mentioned, then the adequate correction that is needed for increasing the length of the bed should be made. In addition, the clearance of 30 cm is allowed according to the second procedure for the determination of dimensions of the bed. If that is the case, the length of the bed calculated under both previous recommendations should be increased by 10 cm.

Conceptually, the third procedure for determining the dimensions of the beds that are linked to the lying position of users is most relevant from the aspect of usability, because it is based on the space that is necessary for a man for sleeping, depending on the position that he/she occupies during sleep. In this regard, it is necessary to distinguish the usual sleeping position from the preferred sleeping position. The usual sleeping position is the position in which a person spends most of his time during sleep. The preferred sleeping position is the one that would be adopted for sleeping when there would be no restrictions, which are primarily imposed by the existing dimensions of the beds. These restrictions may also be due to the presence of another person, with which he/she shares the surface of the bed. The results of the subjective responses of subjects in this research have shown that 28% of women consider that their bed is less than what it should be. In addition, 62% of men believe that their beds could be larger to achieve greater comfort. So, the users' needs related to the size of a bed cannot be accurately determined only based on the knowledge of their body sizes, or formulae that do not involve the aspect of the body position.

Analysis of the results that was conducted based on the third procedure has shown that the length of the usual sleeping position is statistically less than the height of the body of men and women. However, the width of the usual sleeping position is statistically greater than the width of the body (bideltoid). This indicates the possibility of the appearance of errors if boundary dimensions of beds are determined based on the maximum width of the body. On the other hand, the length and width of the preferred sleeping position are statistically significantly greater than the maximum height and width of the body. In addition, the analysis showed that both dimensions of the preferred sleeping position are larger in comparison with the dimensions of the usual sleeping position. All these should be taken into account in order to design beds that are usable for most of the population.

If the length of the bed is determined based on the usual sleeping position, manufacturers of beds do not have a reason for great concern. However, this does not mean that users are satisfied with such dimensions of beds. In order to achieve greater comfort and usability of beds in general, it is recommended that the sizes of beds be determined based on the dimensions related to the preferred sleeping positions.

REFERENCES

Dreams. 2014. Mattresses. www.dreams.co.uk

Haex, B. 2004. *Back and Bed*. Boca Raton: CRC Press/Taylor & Francis.

ISO 9098-2. 1994. Bunk beds for domestic use - Safety requirements and tests - Part 2: Test methods. International Organization for Standardization, Geneva, Switzerland. Standard.

Kroemer, K. H. E. 2001. Body sizes of US Americans. In *International Encyclopedia of Ergonomics and Human Factors*, ed. W. Karwowski, 292–296. London: Taylor &Francis.

Rosenblatt, P. C. 2006. *Two in a Bed: The Social System of Couple Bed Sharing*. New York: State University of New York Press.

Schramm. 2014. Beds. http://www.schrammwerkstaetten.de/en/company

Verhaert, V. 2011. *Ergonomic Analysis of Integrated Bed Measurements: Towards Smart Sleep Systems*. Leuven: Katholieke Universiteit Leuven – Faculty of Engineering.

WHO Europe. 2009. Night noise guidelines for Europe. https://www.euro.who.int/__data/assets/pdf_file/0017/43316/E92845.pdf

Woodson, W. E. 1981. *Human Factors Design Handbook*. New York: McGraw-Hill, Inc.

17 Usability of the Back Seat of Wagon Cars – Recommendations for Design

Aleksandar Zunjic and Vladimir Lesnikov

CONTENTS

17.1 Introduction 319
    17.1.1 Goal of the Research 320
17.2 Method 321
    17.2.1 Selection of Wagons 321
    17.2.2 Selection of the Anthropometric Dimensions for the Usability Study 322
    17.2.3 Selection of Subjects and Percentiles 323
    17.2.4 Selection of the Measurement Points 323
17.3 Results 323
    17.3.1 Evaluation of the Hip Room for Equal Percentiles 324
    17.3.2 Evaluation of the Shoulder Room for Equal Percentiles 324
    17.3.3 Evaluation of the Hip Room for Unequal Percentiles 325
    17.3.4 Evaluation of the Shoulder Room for Unequal Percentiles 326
17.4 Discussion 328
17.5 Conclusion 333
References 333

17.1 INTRODUCTION

A relatively large number of papers have been published in connection with designing and evaluation of anthropometric convenience of the driver's seat of passenger cars. However, in the scientific literature, it is very hard to find an article that deals with evaluation of usability of the rear seat of passenger cars. This segment of researching of the passenger cars has so far been neglected. The reason for this is probably due to the fact that it is of crucial importance that the driver's seat be adequately designed, since the safety of driving and the fulfillment of prerequisites for long-lasting driving (especially if one takes into account the population of professional drivers) in a great extent depend on the comfort and positioning of the drivers.

Wagon car (station wagon or estate) is defined and described in different ways. The wagon is an automobile with a full-height body all the way to the rear, with the load carrying space, which is accessed via a rear door or doors (Jazar 2008). It has the roofline extended to the rear of the body to enlarge its internal capacity (Hillier and Coombs 2004). The rear door can be opened in various ways, depending on the model. Folding the rear seats down provides a large floor area for the luggage or goods. The wagon is a type of vehicle that is typically intended to transport five passengers, including the driver. However, some models have space for up to nine passengers (Erjavec 2006).

From the standpoint of usability, the basic idea when designing wagons is to provide greater luggage space in relation to the sedan type of vehicle. This is achieved by means of changes in the construction of the chassis, particularly in the rear of the vehicle. In most cases, the station wagons have the four pillars, A, B, C and D. The aforementioned changes are mainly related to the shape and dimensions of the cargo space but to some extent (depending on model), these changes are reflected in the space that is intended for the passengers on the rear seat. This fact reflects on the usability of the rear seat from the standpoint of people who could be accommodated at the rear seat of wagon cars.

Primarily, wagons are used as a family vehicle, but they can be additionally used for the transport of goods in the case of realization of the small business. When buying a wagon, a future private owner of the vehicle will rarely take into account the factor of accommodation of all potential users of the new vehicle. A similar or even worse situation usually occurs when the wagon is purchased as a vehicle for the purpose of business. Owner or person in a firm who is responsible for the purchase of a new vehicle, most often, when buying does not take into account anthropometric dimensions of the persons who will be transported by the wagon, which results in affecting the usability of this kind of vehicle from the aspect of passengers. As a result of these circumstances, it is often the case that the passengers complain about the lack of comfort in the vehicle, when at the same time are transported all passengers. Due to that, it is particularly important to design this type of vehicle in such a way that allows comfortable transport of all passengers. If the width of the rear seat of the wagons is not adequately determined, the usability of the vehicle, including the comfort of passengers, will be understandably reduced.

Determining the width of the rear seat, basically, relates to the determination of the required width of the area in the rear of the car, at a location that is designed for the transport of passengers in a sitting position. This includes the planning of sufficient room for accommodation of all parts of the body in the coronal plane. However, a usability study that was previously conducted in connection with the evaluation and recommendations for determining the width of the space required for passengers in the back seat of the wagons is not known.

17.1.1 Goal of the Research

Nowadays, car designers are primarily trying to make a car look good, with the lowest possible coefficient of air resistance. However, with such an approach, they often neglect certain important aspects of the usability of the interior space for the accommodation of passengers. In other words, they do not give proper attention to the need of the car users in relation to comfortable transport. As a consequence of this approach, many concept vehicles that had an attractive look never left certain design bureaus or factories and found their place on roads.

Generally, wagons are considered vehicles of larger dimensions than sedans, which provide more space to passengers. The usability study that will be presented has the aim to examine the extent to which the existing design solutions of wagons really provide sufficient space for passengers on the rear seat. In this connection, the attention is focused on assessing the width of the available space for passengers on the rear seats of wagons. The width of this area directly affects the comfort of the people who are transported on the rear seats of wagons. As mentioned, it is not known a usability study that previously considered this aspect of the comfort of this type of vehicle.

Given the above, the primary aim of this chapter is to consider the usability of the interior space behind the driver in terms of the possibilities for positioning of the passengers with different anthropometric dimensions on the back seats of wagons. With that in mind, it is necessary to perform the assessment of the width of space at the rear seat for the accommodation of passengers, taking into account as many different models of wagons. In this way, insight into the existing state of this problem in the automotive industry can be obtained.

For this purpose, it should be taken into account the wide range of percentiles of persons for the selected anthropometric dimensions. For the reason that passengers on the rear seat of wagons can be males, females and children, the usability assessment should involve all these three categories of users. Such an approach can lead to the formation of the diagrams of comfortable accommodation of passengers.

17.2 METHOD

17.2.1 Selection of Wagons

In accordance with the goal of the research, this study included the majority of vehicle brands and the majority of types of wagons that are produced by different companies for the US market in 2014 and some are from 2015. A comprehensive list of the wagons for the US market has been provided by The Auto Channel TACH (2014). This list contains a large number of models and types of wagons. However, not all vehicles from this list have been included in the analysis. The reason for this is described below.

As previously mentioned, the wagon is the type of vehicle that has certain structural characteristics that distinguish it from other types of vehicles. However, the wagons have a lot of similarities with the hatchback type of vehicle. The definitions of these two types of vehicles are very similar. Except for the similarities, some authors have tried to present certain differences between these two types of vehicles. It certainly was possible with earlier models of wagons and hatchbacks, which had more pronounced mutual differences. However, in recent times, this difference in practice becomes increasingly blurred. Many wagons are made in the shape that possesses the most or all the characteristics of hatchbacks. Examples are the Porsche Panamera-Wagon and Aston Martin Rapide Wagon. In addition, some models are classified and ranked by quality within both of these categories of vehicles (see, for example, US News, 2014, http://usnews.rankingsandreviews.com/cars-trucks/).

Bearing in mind the aforementioned problem, we strived to form the list that contains the models that are without doubt wagons or rather wagons than any other types of vehicles. For the purposes of classification, an additional criterion was adopted. This criterion is the existence of a glass surface behind the c-pillar. If a particular type of vehicle does not possess the glass surface behind the c-pillar, with high reliability can be said that such a vehicle does not belong to the group of wagons. These vehicles, as a rule, have a smaller trunk and obviously they are not designed with the idea of providing greater cargo space. In addition, from the initial list of the TACH, models that obviously did not have the size of the trunk that characterizes wagons have been excluded. All this has led to a considerable reduction in the number of models that have remained from the aforementioned list.

In order to increase the number of models that will be included in the analysis, the option of expanding the list with the models for which the manufacturers themselves provide the necessary data has been accepted. In addition, when it was not possible to find the relevant data for individual models of wagons in both these ways, the list was extended with the models on which performing direct measurements was possible. From the TACH list, variants of particular models are not omitted because it was assumed that such information might be of interest to the existing customers or potential customers of these vehicles. In this way, a total of 49 different wagons was selected for the usability study in terms of the convenience of passenger accommodation on the back seat.

17.2.2 Selection of the Anthropometric Dimensions for the Usability Study

In the beginning, it is important to perform a selection of adequate anthropometric dimensions, which will serve as a basis for the usability study of wagons, in terms of ensuring the optimal width on the back seat for the accommodation of passengers. Most of the textbooks, as well as other literature, that take into account the problem of designing the chairs indicate that the most important human dimension for designing the width of chairs is the hip breadth. However, given the specific circumstance that on the back seat of a wagon should be positioned three people side by side, the previously mentioned recommendation can be considered justified only in the case when looking at whether is it possible in any way to accommodate three people on the back seat. In the mentioned case, passengers do not have the possibility for comfortable transportation, because it is not foreseen the room for adequate accommodation of upper extremities of all passengers. Such a situation would lead to a distortion of the torso of passengers and the appearance of other unpleasant postures, which are inconsistent with the principles of correct ergonomic seating. For this reason, in addition to the aforementioned anthropometric dimension as the starting point for the consideration of passenger accommodation, the shoulder breadth in a sitting position has been adopted as the anthropometric dimension, which adequately foresees the possibility for the comfortable accommodation of passengers on the back seat of wagons.

17.2.3 SELECTION OF SUBJECTS AND PERCENTILES

Given that the models of wagons that are mostly produced for the US market were selected, in order to assess the adequacy of available space in the position of the rear seat of wagons, the population of US residents has been selected. For the purpose of analysis, the data for adult females and males (18 to 79 years) that relate to the hip width, sitting (Woodson 1981) will be used. In order to estimate the area in the shoulder zone on the rear seat, the relevant data regarding the shoulder width (bideltoid) will be used for adult females and males, in the range of 17 to 51 years (Gordon et al. 1988). In addition to these data for adults, additional data for the hip width and shoulder width for children at the age of 4, 7 and 10 years (Weber et al. 1985) will be used. The analysis will cover 1st, 2.5th, 5th, 10th, 25th, 50th, 75th, 90th, 95th, 97.5th and 99th percentiles of both anthropometric dimensions, for all three categories of subjects. Therefore, different combinations of 55 human dimensions and the possibility of their positioning in the total of 49 types of wagons will be analyzed.

17.2.4 SELECTION OF THE MEASUREMENT POINTS

In the automotive industry, the hip width and shoulder width are dimensions that are also accepted as important for designing the area on the back seat. In connection with this, by designers it is foreseen particular space on the back seat for the accommodation of hips, as well as shoulders. The hip room is determined by the width of the rear seat cushion. Basically, determining the shoulder space on the back seat involves measurement of the really available width for sitting in the area of the shoulders from one door to another, inside the car. According to the recommendations (Macey and Wardle 2008), the shoulder room was measured in the interval of 200 mm, from the starting point, which is 254 mm above the H-point. According to the same recommendations, the hip room was measured at the height that is in the range of 25 mm below and 75 mm above the H-point (the recommended interval of 100 mm). For the calculation, as previously mentioned, the data for hip room and shoulder room provided by TACH (2014) and certain manufacturers of the wagons will be included in the analysis. In this regard, it should be mentioned that for all 49 models the data relating to the width on the back seat in the shoulder area have been provided. However, the data that refer to the region of hips have been provided for 41 models. The reason is that some manufacturers give only the information regarding the width of the shoulders.

17.3 RESULTS

To gain insight into the usability of the rear seats of selected models of wagons for accommodations of passengers, four cases will be tested. The first case relates to the assessment of the available space in the area of the hips, for the situation in which on the back seat three persons with identical percentiles (hips) are accommodated. The second case relates to the assessment of the available space in the shoulder area, in which three persons with identical percentiles (shoulders) are accommodated on the back seat. The third and fourth cases involve a more detailed analysis, which includes the placement of individuals with different percentiles on the back seat. In this regard, the third case refers to the assessment of the available space in the area of the hips, for the case in which three persons of different percentiles (hips) are placed on the back seat. The fourth case refers to the assessment of the available space in the shoulder area, in which on the back seat three persons with different percentiles (shoulders) are placed.

17.3.1 Evaluation of the Hip Room for Equal Percentiles

Taking into account the values of percentiles for hips of male and female persons, as well as the dimensions of the available space in the hip area for models of wagons that are considered, the results of the usability assessment indicate the following:

- 2.44% of models enable accommodation of maximum three females of 99th percentile
- 29.27% of models enable accommodation of maximum three females of 97.5th percentile
- 36.59% of models enable accommodation of maximum three females of 95th percentile
- 17.07% of models enable accommodation of maximum three males of 99th percentile
- 12.19% of models enable accommodation of maximum three females of 90th percentile
- 2.44% of models enable accommodation of maximum three males of 90th percentile.

17.3.2 Evaluation of the Shoulder Room for Equal Percentiles

Taking into account the values of percentiles for shoulders of male and female persons, as well as the dimensions of the available space in the shoulder area for models of wagons that are considered, the results of the usability assessment indicate the following:

- 4.08% of models enable accommodation of maximum three females of 99th percentile
- 2.04% of models enable accommodation of maximum three females of 97.5th percentile
- 12.24% of models enable accommodation of maximum three males of 25th percentile
- 2.04% of models enable accommodation of maximum three females of 95th percentile

- 48.97% of models enable accommodation of maximum three females of 90th percentile
- 8.16% of models enable accommodation of maximum three males of 10th percentile
- 6.12% of models enable accommodation of maximum three males of 5th percentile
- 6.12% of models enable accommodation of maximum three females of 75th percentile
- 2.04% of models enable accommodation of maximum three males of 2.5th percentile
- 6.12% of models enable accommodation of maximum three males of 1st percentile
- 2.04% of models enable accommodation of maximum three females of 1st percentile.

17.3.3 Evaluation of the Hip Room for Unequal Percentiles

The analyses that relate to the two previous cases enable acquiring the general insight into the availability of free space in wagons in the areas of the hips and shoulders. However, in practice, it is very rare that on the back seat three persons of identical percentile are positioned. It is understandable that there is a very large number of possible combinations of three persons with mutually different percentiles. For this reason, in order to assess the available space on the back seat of the existing models of wagons, the criterion of accommodation of people of different percentiles has been adopted. In that respect, first, selection of the person of a certain percentile is performed, whom we want to place on the back seat. After that, the selection of the second person with the maximum possible percentile is performed, but that there is a third person of any percentile who can still be accommodated in the available width on the back seat.

The female person of 95th percentile here will be chosen as the first person. The following list in addition to the name of models for the evaluation of usability contains the marks of percentiles of the second and the third passenger, who may be accommodated in accordance with the adopted criterion. Mark m refers to men, mark w refers to women, while mark c refers to children. The number beside the mark c refers to the years of age (4, 7 or 10). The third person is the one with the highest possible percentile, who can be accommodated next to the first and the second person.

Model	Percentiles
2015 TOYOTA Venza	$P_{99(w)}, P_{99(w)}$
2015 VOLVO XC70 (2015.5) 3.2 AWD	$P_{99(w)}, P_{99(w)}$
2015 VOLVO XC70 3.2 AWD	$P_{99(w)}, P_{99(w)}$
2015 VOLVO XC70 T5 Drive-E Premier Plus FWD	$P_{99(w)}, P_{99(w)}$
2015 VOLVO XC70 T6 Platinum AWD	$P_{99(w)}, P_{99(w)}$
2015 SUBARU Outback	$P_{99(w)}, P_{99(w)}$
2014 HYUNDAI i40 Tourer	$P_{99(w)}, P_{99(w)}$
2014 ACURA TSX Sport Wagon 5-Spd AT	$P_{99(w)}, P_{99(w)}$

2014 HONDA Civic Tourer	$P_{99\,(W)}, P_{99\,(W)}$
2014 TOYOTA Prius V Five	$P_{99\,(W)}, P_{97.5\,(W)}$
2014 TOYOTA Prius V Three	$P_{99\,(W)}, P_{97.5\,(W)}$
2014 TOYOTA Prius V Two	$P_{99\,(W)}, P_{97.5\,(W)}$
2014 SKODA Octavia Combi	$P_{99\,(W)}, P_{97.5\,(W)}$
2015 VOLVO V60 (2015.5) T5 Platinum AWD	$P_{99\,(W)}, P_{95\,(W)}$
2015 VOLVO V60 (2015.5) T6 R-Design Platinum AWD	$P_{99\,(W)}, P_{95\,(W)}$
2015 VOLVO V60 T5 Drive-E FWD	$P_{99\,(W)}, P_{95\,(W)}$
2015 VOLVO V60 T5 Platinum AWD	$P_{99\,(W)}, P_{95\,(W)}$
2015 VOLVO V60 T6 R-Design Platinum AWD	$P_{99\,(W)}, P_{95\,(W)}$
2014 CADILLAC CTS Sport Wagon Performance AWD	$P_{99\,(W)}, P_{95\,(W)}$
2014 CADILLAC CTS-V Sport Wagon RWD	$P_{99\,(W)}, P_{95\,(W)}$
2014 FIAT 500L Easy	$P_{99\,(W)}, P_{95\,(W)}$
2014 FIAT 500L Lounge	$P_{99\,(W)}, P_{95\,(W)}$
2014 FIAT 500L Pop	$P_{99\,(W)}, P_{95\,(W)}$
2014 FIAT 500L Trekking	$P_{99\,(W)}, P_{95\,(W)}$
2014 OPEL Insignia Sports Tourer	$P_{99\,(W)}, P_{95\,(W)}$
2014 FORD Focus Wagon	$P_{99\,(W)}, P_{99\,(M)}$
2014 FORD C-Max Energi SEL	$P_{99\,(W)}, P_{97.5\,(M)}$
2015 FORD C-Max Hybrid SEL	$P_{99\,(W)}, P_{97.5\,(M)}$
2014 RENAULT Megane Estate	$P_{99\,(W)}, P_{90\,(M)}$
2014 HYUNDAI i30 Tourer	$P_{99\,(W)}, P_{90\,(M)}$
2014 KIA cee'd sportswagon	$P_{99\,(W)}, P_{75\,(W)}$
2014 VOLKSWAGEN Golf Variant	$P_{99\,(W)}, P_{75\,(W)}$
2014 DACIA Logan MPV	$P_{99\,(W)}, P_{75\,(M)}$
2014 TOYOTA Auris Touring Sports	$P_{99\,(W)}, P_{75\,(M)}$
2014 SKODA Fabia Combi	$P_{99\,(W)}, P_{50\,(W)}$
2014 RENAULT Clio Estate	$P_{99\,(W)}, P_{25\,(W)}$
2014 KIA Soul Red Zone Special Edition	$P_{99\,(W)}, P_{25\,(W)}$
2015 KIA Soul !	$P_{99\,(W)}, P_{25\,(W)}$
2015 KIA Soul +	$P_{99\,(W)}, P_{25\,(W)}$
2015 KIA Soul Base	$P_{99\,(W)}, P_{25\,(W)}$
2014 NISSAN cube 1.8 SL	$P_{99\,(W)}, P_{2.5\,(W)}$

17.3.4 Evaluation of the Shoulder Room for Unequal Percentiles

The male person of 95th percentile here will be chosen as the first person. Usability evaluation of models given in the list below also has been performed in accordance with the criterion that is described in the assessment of the hip area.

2015 TOYOTA Venza	$P_{99(M)}, P_{10(W)}$
2014 AUDI A6 Avant	$P_{99(M)}, P_{10(W)}$
2015 SUBARU Outback	$P_{99(M)}, P_{90\,(C,10)}$
2015 MERCEDES-Benz E-Class E63 AMG S-Model 4MATIC Wagon	$P_{99(M)}, P_{75\,(C,10)}$

Usability of the Back Seat of Wagon Cars

2015 VOLVO XC70 (2015.5) 3.2 AWD	$P_{99(M)}, P_{75(C,10)}$
2015 VOLVO XC70 3.2 AWD	$P_{99(M)}, P_{75(C,10)}$
2015 VOLVO XC70 T5 Drive-E Premier Plus FWD	$P_{99(M)}, P_{75(C,10)}$
2015 VOLVO XC70 T6 Platinum AWD	$P_{99(M)}, P_{75(C,10)}$
2014 ACURA TSX Sport Wagon 5-Spd AT	$P_{99(M)}, P_{99(C,7)}$
2014 HYUNDAI i40 Tourer	$P_{99(M)}, P_{97.5(C,7)}$
2014 MAZDA 6 Wagon	$P_{99(M)}, P_{90(C,7)}$
2014 FORD C-Max Energi SEL	$P_{99(M)}, P_{25(C,10)}$
2014 TOYOTA Prius V Five	$P_{99(M)}, P_{25(C,10)}$
2014 TOYOTA Prius V Three	$P_{99(M)}, P_{25(C,10)}$
2014 TOYOTA Prius V Two	$P_{99(M)}, P_{25(C,10)}$
2015 FORD C-Max Hybrid SEL	$P_{99(M)}, P_{25(C,10)}$
2015 VOLVO V60 (2015.5) T5 Platinum AWD	$P_{99(M)}, P_{25(C,10)}$
2015 VOLVO V60 (2015.5) T6 R-Design Platinum AWD	$P_{99(M)}, P_{25(C,10)}$
2015 VOLVO V60 T5 Drive-E FWD	$P_{99(M)}, P_{25(C,10)}$
2015 VOLVO V60 T5 Platinum AWD	$P_{99(M)}, P_{25(C,10)}$
2015 VOLVO V60 T6 R-Design Platinum AWD	$P_{99(M)}, P_{25(C,10)}$
2015 BMW 3 Series Sports Wagon 328d xDrive	$P_{99(M)}, P_{25(C,10)}$
2014 HYUNDAI i30 Tourer	$P_{99(M)}, P_{75(C,7)}$
2014 KIA cee'd Sportswagon	$P_{99(M)}, P_{75(C,7)}$
2014 CADILLAC CTS Sport Wagon Performance AWD	$P_{99(M)}, P_{10(C,10)}$
2014 CADILLAC CTS-V Sport Wagon RWD	$P_{99(M)}, P_{10(C,10)}$
2014 FIAT 500L Easy	$P_{99(M)}, P_{10(C,10)}$
2014 FIAT 500L Lounge	$P_{99(M)}, P_{10(C,10)}$
2014 FIAT 500L Pop	$P_{99(M)}, P_{10(C,10)}$
2014 FIAT 500L Trekking	$P_{99(M)}, P_{10(C,10)}$
2014 KIA Soul Red Zone Special Edition	$P_{99(M)}, P_{10(C,10)}$
2015 KIA Soul !	$P_{99(M)}, P_{10(C,10)}$
2015 KIA Soul +	$P_{99(M)}, P_{10(C,10)}$
2015 KIA Soul Base	$P_{99(M)}, P_{10(C,10)}$
2014 AUDI A4 Avant	$P_{99(M)}, P_{50(C,7)}$
2015 AUDI allroad 2.0T Premium quattro Tiptronic	$P_{99(M)}, P_{50(C,7)}$
2014 RENAULT Megane Estate	$P_{99(M)}, P_{50(C,7)}$
2014 SKODA Octavia Combi	$P_{99(M)}, P_{50(C,7)}$
2014 TOYOTA Auris Touring Sports	$P_{99(M)}, P_{95(C,4)}$
2014 DACIA Logan MPV	$P_{99(M)}, P_{1(C,10)}$
2014 OPEL Insignia Sports Tourer	$P_{99(M)}, P_{75(C,4)}$
2014 FORD Focus Wagon	$P_{99(M)}, P_{5(C,7)}$
2014 VOLKSWAGEN Jetta SportWagen TDI w/Sunroof & Nav	$P_{99(M)}, P_{5(C,7)}$
2014 VOLKSWAGEN Golf Variant	$P_{99(M)}, P_{2.5(C,7)}$
2014 HONDA Civic Tourer	$P_{99(M)}, P_{25(C,4)}$
2014 NISSAN cube 1.8 SL	$P_{99(M)}, P_{1(C,7)}$
2014 RENAULT Clio Estate	$P_{99(M)}, P_{1(C,4)}$
2014 SKODA Fabia Combi	$P_{97,5(M)}, P_{1(C,4)}$
2014 MINI Cooper Clubman John Cooper Works	$P_{10(W)}, P_{1(C,4)}$

17.4 DISCUSSION

Based on the results of the usability assessment that refers to the third analyzed case (the hip room for unequal percentiles), it can be seen that all models of wagons enable the accommodation of three adults of different percentiles on the back seat. However, the fourth analyzed case reveals a somewhat dramatic situation, when considering the free space on the back seat of wagons in the field of the shoulder. Therefore, 95.91% of the considered models do not enable accommodation of three adults on the back seat, when the first person is $P_{95(M)}$, and when the second person is $P_{99(M)}$ (shoulders).

In the literature cannot be found a recommendation which precisely determines the width of the space on the rear seat. The data that can be found relate to the global recommendations for designers such as how much should amount approximately the width of the back seat (Macey and Wardle 2008) in the region of the hips and shoulders. With just these two numerical data, the designers certainly cannot plan accurately the number of people that can be comfortably accommodated on the back seat of a wagon. It is necessary to establish more precise and comprehensive recommendations that will help usability designers to plan the space for the accommodation of passengers on the back seats.

Bearing in mind the foregoing, in order to be in the ability to plan the usability of the cabin space in the area of the back seat of wagons, in the continuation, the diagrams intended for the comfortable accommodation of passengers will be formed and presented. If the dimensions of one of the three persons that should be placed on the back seat of a wagon are known, these diagrams allow determination of the maximum dimensions of two other persons who may be accommodated on the back seat. In addition, if the design solution specifies certain percentiles of persons who should be accommodated on the back seat, based on these diagrams it is possible in a quick way to determine the required optimum width of the back seat, in the area of the hips and shoulders.

Figure 17.1 shows the diagram for determination of the comfortable accommodation of passengers in the area of the hips, when it is known that the fixed person who we wish to transport on the back seat is the 5th percentile male.

Figure 17.2 shows the diagram for determination of the comfortable accommodation of passengers in the area of the hips, when it is known that the fixed person who we wish to transport on the back seat is the 95th percentile male.

Figure 17.3 shows the diagram for determination of the comfortable accommodation of passengers in the area of the shoulders, when it is known that the fixed person who we wish to transport on the back seat is the 5th percentile female.

Figure 17.4 shows the diagram for determination of the comfortable accommodation of passengers in the area of the shoulders, when it is known that the fixed person who we wish to transport on the back seat is the 95th percentile female.

The diagrams of comfortable accommodation of passengers can be formed for each percentile (e.g., 50th). The use of these diagrams that have been formed for the purpose of planning the usability of the cabin space in the area of the back seats of wagons is relatively easy. In connection with Figure 17.4, an example is given for

Usability of the Back Seat of Wagon Cars

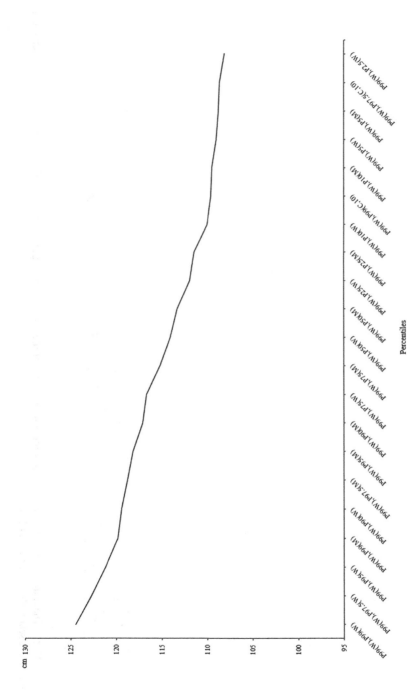

FIGURE 17.1. Diagram of comfortable accommodation for the area of the hips, for three passengers on the back seat of a wagon, when the first person on the seat is the 5th percentile man.

330 Handbook of Usability and User Experience

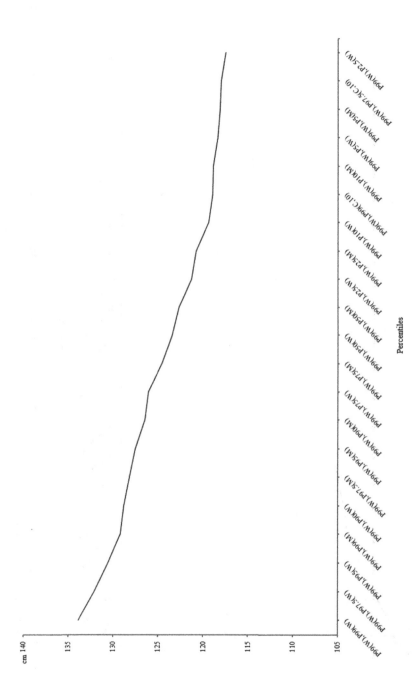

FIGURE 17.2. Diagram of comfortable accommodation for the area of the hips, for three passengers on the back seat of a wagon, when the first person on the seat is the 95th percentile man.

Usability of the Back Seat of Wagon Cars

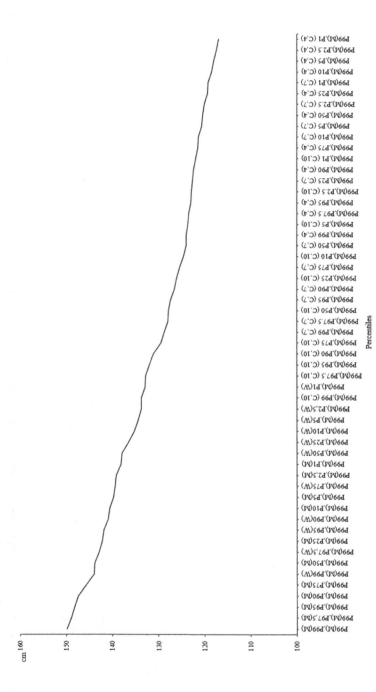

FIGURE 17.3. Diagram of comfortable accommodation for the area of the shoulders, for three passengers on the back seat of a wagon, when the first person on the seat is the 5th percentile woman.

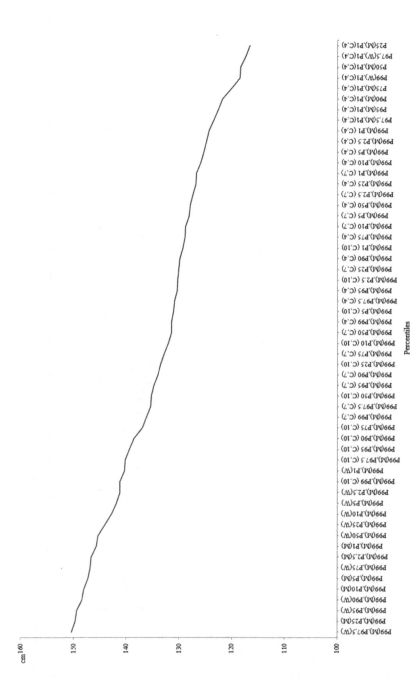

FIGURE 17.4. Diagram of comfortable accommodation for the area of the shoulders, for three passengers on the back seat of a wagon, when the first person on the seat is the 95th percentile woman.

determining the maximum percentile of people that can be accommodated on the back seat of a wagon. Let the planned width on the back seat in the shoulder area amounts to 130 cm, and let it be known that the person who will certainly be transported is the 95th percentile female person (shoulder). From Figure 17.4 it can be seen that the percentiles of the other two persons are $P_{99\,(M)}$ and $P_{25\,(C,\,7)}$. So, for the planned width of 130 cm, the maximum percentiles of persons who can be comfortably accommodated in addition to the first $P_{95\,(W)}$ passenger are $P_{99\,(M)}$ and $P_{25\,(C,\,7)}$. This means that all subjects of smaller percentile than the above mentioned can be comfortably transported together.

17.5 CONCLUSION

The results of this usability study show that almost all models of wagons have enough space on the back seat in the area of the hips. In this regard, based on the performed analysis, it should be noted that even the wagon model with minimum width allows comfortable accommodation of three males of the 90th percentile (in the area of the hips). But, in the area of the shoulders, the situation is different. There is no model of a wagon that can accommodate comfortably three males of 50th percentile. If only male passengers are considered, there were no models of a wagon that could accommodate more than 3 males of the 25th percentile. Generally speaking, based on the comparisons with the results related to sedans (Zunjic and Lesnikov 2014), it can be concluded that the existing construction solutions of wagons do not provide more space on the back seat in the regions of shoulders and hips.

So, the basic recommendation for usability designers is to pay more attention to the provision of sufficient width of space on the back seat in the area of the shoulders. It can be presumed that the lack of published researches on this topic is one of the reasons for the mentioned omission. Diagrams that relate to the comfortable accommodation of the passengers that have been created within this study give the opportunity that usability designers can determine in an easy and quick manner the required width of a wagon in the area of the hips and shoulders, in accordance with the percentiles of people that should be accommodated on the back seat. Implementation of this approach facilitates planning the usability of cabin space of the vehicle and at the same time provides the conditions for comfortable transport, from the aspect of providing adequate width of space for all passengers on the back seat of wagons.

REFERENCES

Erjavec, J. 2006. *Automotive Technology: A Systems Approach* (2nd edition). New York: Delmar Cengage Learning.

Gordon, C. C., Bradtmiller, B., Churchill, T., Clauser, C. E., McConille, J. T., Tebbetts, I., and Walker, R. A. 1988. *Anthropometric Survey of U.S. Army Personnel: Methods and Summary Statistics.* Technical Report NATICK/TR-89/044. Natick: U.S. Army Natick RD&E Center.

Hillier, V. A. W., and Coombs, P. 2004. *Hillier's Fundamentals of Motor Vehicle Technology* (book 1). Cheltenham: Nelson Thornes.

Jazar, R. N. 2008. *Vehicle Dynamics: Theory and Application.* New York: Springer.

Macey, S., and Wardle, G. 2008. *H-POINT, The Fundamentals of Car Design & Packaging.* Culver City, LA: Design Studio Press.
TACH. 2014. Rank cars by specs. http://www.theautochannel.com/
Weber, K., Lehman, R. J., and Schneider, L. W. 1985. *Child Anthropometry for Restraint System Design, UMTRI-85-23.* Ann Arbor: The University of Michigan, Transportation Research Institute.
Woodson, W. E. 1981. *Human Factors Design Handbook.* New York: McGraw-Hill, Inc.
Zunjic, A., and Lesnikov, V. 2014. Evaluation of an optimal width of a rear seat of sedans. In *Proceedings of the 5th International Conference on Applied Human Factors and Ergonomics AHFE 2014.* Krakow: AHFE.

18 System Perspective in Usability and UX Design
A Case Study of an Indian Cooking Spatula

Somnath Gangopadhyay and Sourav Banerjee

CONTENTS

18.1 Introduction	335
18.2 SystemEnd User and Product: An Understanding between User and Product	336
18.2.1 Case Study on Redesigning Cooking Spatula	336
18.2.1.1 Background Literature about Cooking Spatula and Users' Problems during Cooking	336
18.3 Results	341
18.3.1 Cognitive Walkthrough for the New Design	343
18.3.2 Kansei Questionnaire for the New Design	344
18.4 Conclusion	346
References	346

18.1 INTRODUCTION

A new design becomes user-friendly when the end users can interact with that design efficiently. A perfect design can significantly affect the system and will increase the interactions between different components of the system, which means it will increase the interaction between the user and the product. If the product is difficult to use then it will fail to meet the user's needs and this condition will become the reason for the rejection of a product. A design should support the user's need for improvement of interactions.

User-experience (UX) design is the process of inclusion of user behavior and perspective in a product. The process will proceed through usability, accessibility and desirability of users with a product. Magnification of the interaction with a product is the outcome of UX design.

When all aspects of the new design are perceived by the users then the product will become user-friendly. User's behavior and thought about that design play a significant role in the success of the design. The main purpose of the UX design is to solve the end user's problems in using a product. This will provide a delightful experience in using a product. The designer, in this stage, will identify the perfect relationship between the developer and the user.

18.2 SYSTEMEND USER AND PRODUCT: AN UNDERSTANDING BETWEEN USER AND PRODUCT

Take an example of a kitchen tool. In the kitchen, the chef makes a system with tools. He or she makes continuous interaction with tools during cooking. This interaction will depend on the understanding of using the tool during a specific activity. Try to elaborate this condition in the following way: cooking spatula, a product, is used frequently in cooking activities. Chefs are end users of this cooking spatula. They may face wrist pain and other upper extremities disorders during the long use of this cooking spatula. The design needs to be changed for the betterment of the wrist posture and to give much comfort to the end users of the system.

18.2.1 CASE STUDY ON REDESIGNING COOKING SPATULA

A study was done in the Ergonomics Laboratory, University of Calcutta, on redesigning cooking spatula, which may be taken as an example of the development of the better product through enhancement of end user–product relationship.

18.2.1.1 Background Literature about Cooking Spatula and Users' Problems during Cooking

Wu et al. (2015) concluded in their study on using of spatula to mix cooked food. They observed that the operation involved dorsiflexion, palmary flexion, and radial and ulnar deviations. These movements of the wrist cause disorders in the upper extremities and, in particular, in the carpal tunnel.

The cooking spatula has seldom been investigated; poorly designed culinary spatula will be ergonomically inefficient and cause injury to the hand and wrist.

Wu and Hsieh (2002) revealed that a spatula with 20cm handle length and 25°lift angle was the best; however, in their opinion, a 25cm long handle and 15°lift angle were second in the row in acceptability. However, to prevent users from touching the edge of the hot pan, the spatula with 25cm handle length and 25°lift angle was suggested.

Lewis and Narayan (1993) showed that the ergonomically designed handle allows higher working efficiency than the existing handles. During the use of the existing cooking spatula, the wrist of the users undergoes repeated bending and twisting; a new intervention design of cooking spatula is required for the correction of wrist posture that gives less bending to the wrists of the users.

At first, the prevalence of wrist pain was assessed by questionnaires prepared for this purpose, and Nordic Musculoskeletal Questionnaire (Kuorinka et al., 1987) was used. The questionnaire was modified accordingly to suit the context of this study. The developed questionnaire was tested for internal consistency by the Cronbach's alpha test (Cronbach, 1951).

System Perspective in Usability & UX Design

A prototype of a cooking spatula was designed using the 3D modeling software SketchUp Pro 2015 version. For designing a bent-handled cooking spatula, some dimensions of the hand were measured using the caliper. The dimensions are hand-breadth with the thumb, index breadth, thumb breadth, index depth and grip inside diameter. Figures 18.1–18.3 illustrate that three different prototypes were designed according to the ideas and the data collected from anthropometry, and finally the developed intervention was given to the users and the response was taken to find which prototype is much better and comfortable to use.

The 5th and 95th percentile value of each anthropometric dimension was calculated by the IBM-SPSS 25.0 statistical software. Table 18.1 shows the result of anthropometric statistical data.

The working posture of the users while mixing the food with the existing spatula and the newly designed bent-handled spatula were analyzed by Rapid Entire Body Assessment (REBA) (Hignett & McAtamney, 2000) and Individual Risk Assessment (ERIN) (Rodriguez, Vina & Montero, 2013) methods. Figures 18.4 and 18.5 illustrate the average score of posture analysis of the subjects after and before intervention.

The strain index (SI) score (Moore 1995) was calculated and identified the risk of occurring musculoskeletal disorder while using the existing cooking spatula and designed bent-handled spatula. The repetitiveness of the work while using the

FIGURE 18.1 Prototype A (20 cm, 25°) and 45° hand grip

FIGURE 18.2 Prototype B (20 cm, 15°) and 35° hand grip

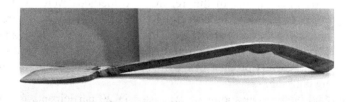

FIGURE 18.3 Prototype C (20 cm, 20°) and 35° hand grip (final prototype)

TABLE 18.1.
Percentile Statistics for All Anthropometric Dimensions of the Design

		Hand breadth with thumb (meter)	Index breadth (meter)	Thumb breadth (meter)	Index depth (meter)	Grip inside diameter (meter)
N	Valid	20	20	20	20	20
	Missing	0	0	0	0	0
Mean		0.0939	0.012767	0.014633	0.010300	0.037767
Std. Deviation		0.00288	0.0012507	0.0011592	0.0016640	0.0064095
Percentiles	5	0.0890	0.011000	0.013000	0.00800	0.027000
	50	0.0945	0.013000	0.015000	0.010000	0.039000
	95	0.0980	0.015000	0.016450	0.013000	0.045450

FIGURE 18.4 Posture analysis by REBA with existing and designed cooking spatula

existing and designed spatula was analyzed by the Assessment of Repetitive Tasks (ART) method. Figures 18.6 and 18.7 illustrate ART analysis for the assessment of repetitiveness of work while using existing and designed spatula respectively.

The post-intervention study mainly included the response data after the implementation of the intervention. In this regard, two methods were used: Cognitive walkthrough (CWT), Kansei Engineering (KE). These two methods are basically used for the assessment of user-experiences (UX) about the new design and identifying the desirability, accessibility and usability of the product. This assessment will help to choose the correct product that will solve the end user's problems of wrist bending during cooking.

The modified Nordic Questionnaire was used in 20 participants for analysis of their wrist comfort during the use of the existing cooking spatula. The designed and existing spatulas were given to 20 participants while cooking and the observation

System Perspective in Usability & UX Design

FIGURE 18.5 Posture analysis by ERIN with existing and designed cooking spatula

Score sheet

Enter the colour band and numerical score for each risk factor in the table below.
Follow the instructions on page 10 to determine the task score and exposure score.

Risk factors		Left arm		Right arm	
		Colour	Score	Colour	Score
A1	Arm movements		0		3
A2	Repetition		0		3
B	Force		0		12
C1	Head/neck posture		0		0
C2	Back posture		0		0
C3	Arm posture		0		4
C4	Wrist posture		0		2
C5	Hand/finger grip		0		2
D1	Breaks		2		2
D2	Work pace		1		1
D3	Other factors		1		1
	Task score		4		30
D4	Duration multiplier	x	0.5	x	0.5
	Exposure score		2		15
D5	Psychosocial factors	1. High levels of attention and concentration 2. Excessive work demand			

FIGURE 18.6 The score sheet for assessment of repetitiveness, during the use of existing cooking spatula

Score sheet

Enter the colour band and numerical score for each risk factor in the table below.
Follow the instructions on page 10 to determine the task score and exposure score.

Risk factors	Left arm Colour	Left arm Score	Right arm Colour	Right arm Score
A1 Arm movements		0		3
A2 Repetition		0		3
B Force		0		4
C1 Head/neck posture		0		0
C2 Back posture		0		0
C3 Arm posture		0		4
C4 Wrist posture		0		0
C5 Hand/finger grip		0		0
D1 Breaks		2		2
D2 Work pace		1		0
D3 Other factors		1		0
Task score		4		16
D4 Duration multiplier	x	0.5	x	0.5
Exposure score		2		8
D5 Psychosocial factors				

FIGURE 18.7 The score sheet for assessment of repetitiveness, during the use of designed cooking spatula

time was 30 minutes. Three different prototypes were given (0.2 m × 20°, 0.2 m ×15°, 0.2 m× 25°) in random order. These prototypes have different angles for handgrip (35°, 45°). The angle of this handgrip was given for reducing the wrist bend while cooking and because of this handgrip wrist did not deviate much from the midline.

The most comfortable prototype was identified by the participants and used for further analysis.

The cognitive walkthrough is a way of interpreting people's thoughts and actions when they use an interface for the first time. Four cognitive questions are asked to each participant and the thought about this new design is identified.

a) Will the user try to achieve the effect that the subtask has?
b) Will the user notice that the correct action is available?
c) Will the user understand that the wanted subtask can be achieved by the action?
d) Does the user get appropriate feedback?

System Perspective in Usability & UX Design   341

Kansei Engineering (Nagamachi 1995) is a product development methodology, which translates customer's impressions, feelings and demands on existing products or concepts to design solutions and concrete design parameters. Twenty-one psychological questions are created with Kansei terms and used for the problem analysis of existing spatula. These problems are corrected in the new design. The psychological feeling about this new design is compared with that of the existing spatula by a line diagram plot using the Kansei Questionnaire.

18.3 RESULTS

Figures 18.8–18.11 show Nordic Questionnaire results and identified wrist pain of the user; they felt maximum pain during the repetitive use of existing cooking spatula.

The posture analysis study, strain index sheet and ART score with existing spatula showed that the use of existing cooking spatula was hazardous for the wrist. For

Pain felt during

sleep	40.00%
rest	15%
cooking	45.00%

0.00% 10.00% 20.00% 30.00% 40.00% 50.00%

FIGURE 18.8 Percentage of response about wrist pain during cooking, rest and sleeping hour

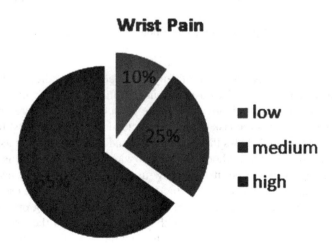

FIGURE 18.9 Grade of wrist pain and percentage of responses

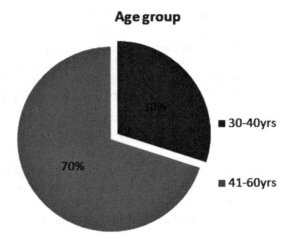

FIGURE 18.10 Age group for intervention study

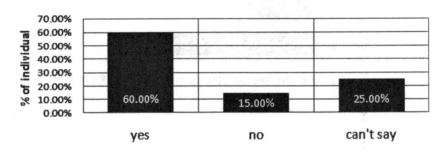

FIGURE 18.11 Percentage of individual responses about the need of modified cooking spatula

reducing this wrist stress, an ergonomically designed bent-handled cooking spatula was designed and considered as a health intervention. After the intervention, the posture, strain index and repetitiveness of the wrist of the users were analyzed and the analysis showed that the risk of musculoskeletal disorders (MSDs) for the wrist was reduced. Table 18.2 shows the result of pre- and post-intervention assessment.

A usability inspection method was used to identify the usability issues in this design, focusing on how easy it is for new users to accomplish the task with this design. The cognitive walkthrough method was used to identify the user's response about this design, and this design is much better than the existing one because this bend in the design could be correlated with the natural bend of the wrist. Therefore, anyone can use this spatula to accomplish the cooking task without the further bending of the wrist. Using the Kansei Engineering questionnaire, the problems associated with existing spatula were reduced in this newly designed product.

System Perspective in Usability & UX Design

TABLE 18.2.
Pre- and Post-intervention Score of REBA, ERIN, SI, ART Assessments

Tests	Pre-intervention	Post-intervention
REBA (average) score	4.8	2.1
ERIN (average) score	26.3	13.8
SI score (right hand)	27	1.5
ART score (right hand)	15	8

18.3.1 Cognitive Walkthrough for the New Design

A cognitive walkthrough is a process, by which one can identify the efficacy of a new design. At first, the tasks related to the product will be identified and the users are asked to perform those tasks. All the feedback and responses given by the users during performing the tasks are considered as subjective responses regarding the usability of the newly designed product. In this case study, selected participants are the end users of the system (cooking in the kitchen). The newly designed products are given to those users to assess the cognitive acceptance and feeling about the design. The product was given to each of the users in their kitchen setup for mixing food.

The end user's thoughts and actions were assessed by asking some cognitive questions and the responses were collected.

Typically, four questions were asked

1. *Will the user try to achieve the effect that the subtask has?* E.g., does the user understand that this subtask is needed to reach the user's goal? The subtask is to use the spatula and the goal is to mix cooked materials.
   Ans: The bent handle of the spatula is for the handgrip of the user during cooking.
2. *Will the user notice that the correct action is available?* E.g., is the wrist bending of the user during the usage of the cooking spatula is reduced?
   Ans: Yes, wrist bending is reduced and feeling much comfortable.
3. *Will the user understand that the wanted subtask can be achieved by the action?* E.g., the subtask is to mix the cooked material using the bent-handled spatula with the hand (action).
   Ans: Yes, the subtask is achieved using the designed spatula.
4. *Does the user get appropriate feedback?* E.g., will the user know that they have done the right thing after performing the action?
   Ans: Yes, while using the designed (prototype) cooking spatula they feel that they have less wrist bending and they can achieve their goal of the task with the least strength on the wrist and they have a proper handgrip for using the designed spatula.

So, using the cognitive walkthrough method, the cognition of the users about this new design was recognized and the responses about this new prototype were noted down. The performances of the users during the usage of cooking spatula improved.

18.3.2 KANSEI QUESTIONNAIRE FOR THE NEW DESIGN

During the testing phase of my study, the participants were invited to interact with both existing spatula and designed bent-handled prototype of spatula and during a given task of cooking, allowing them to acquire information about both spatulas. In the meantime, the participants could evaluate the rate of comfort, strength and wrist bend difference between the two sets of cooking spatula. The rating of the response is noted through the questionnaire with Kansei terms. Other factors about the design of the two cooking spatulas obtained from the questionnaire were proper finger rest, affordability and grip support.

After getting all the responses from the participants, the internal consistency of the Kansei questionnaire was tested by Cronbach's alpha test. Then the average comfort score for both existing spatula and the designed prototype was plotted in a line diagram against each Kansei term, then the two line diagrams were compared (one for existing spatula and another for designed one) and the problems with the existing spatula were identified and corrected in case of the newly designed spatula. The whole data is collected based on user response and psychological feelings about the tool (Naddeo et al. 2015). From the response, this study is suggesting as to which one of the spatulas is much comfortable and less stressful for the wrist of the homemakers during cooking.

	1. Comfortable			12. Exciting	
Not at all	1 2 3 4 5 6 7	Very Much	Not at all	1 2 3 4 5 6 7	Very much
	2. Elegant			13. Realistic	
Not at all	1 2 3 4 5 6 7	Very Much	Not at all	1 2 3 4 5 6 7	Very Much
	3. Pleasant			14. easy to operate	
Not at all	1 2 3 4 5 6 7	Very Much	Not at all	1 2 3 4 5 6 7	Very Much
	4. Relaxing			15. affordable	
Not at all	1 2 3 4 5 6 7	Very Much	Not at all	1 2 3 4 5 6 7	Very Much
	5. Grip support			16. acceptibility	
Not at all	1 2 3 4 5 6 7	Very Much	Not at all	1 2 3 4 5 6 7	Very Much
	6. Wrist bent			17. useful	
Not at all	1 2 3 4 5 6 7	Very Much	Not at all	1 2 3 4 5 6 7	Very Much
	7. Strength on wrist			18. powerful Material	
Not at all	1 2 3 4 5 6 7	Very Much	Not at all	1 2 3 4 5 6 7	Very Much
	8. Innovative			19. Proper finger rest	
Not at all	1 2 3 4 5 6 7	Very Much	Not at all	1 2 3 4 5 6 7	Very Much
	9. Robust			20. satisfaction	
Not at all	1 2 3 4 5 6 7	Very Much	Not at all	1 2 3 4 5 6 7	Very Much
	10. Good design			21. user friendly	
Not at all	1 2 3 4 5 6 7	Very Much	Not at all	1 2 3 4 5 6 7	Very Much
	11. I will continue to like				
Not at all	1 2 3 4 5 6 7	Very Much			

Figures 18.12 and 18.13 show the line diagram of twenty-one Kansei questions that were plotted after and before the implementation of the final prototype.

System Perspective in Usability & UX Design 345

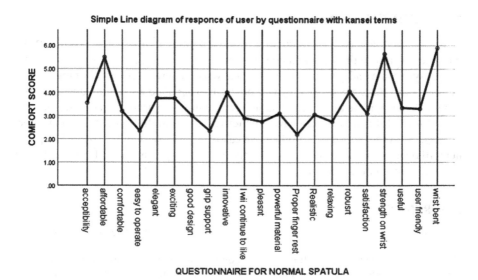

FIGURE 18.12 Line diagram of response of users while using existing cooking spatula

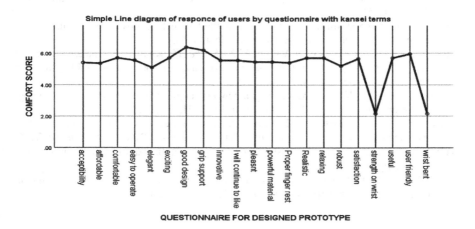

FIGURE 18.13 Line diagram of response of users while using designed bent handled spatula

The questions were asked to the users about the existing cooking spatula while cooking and the problems were identified and rectified in the new design. The major problems while using the existing spatula were the high wrist bending, improper handgrip and huge mechanical stress developed on the wrist. These problems were reduced in the newly designed cooking spatula. Modified spatula has proper handgrip, requires less wrist bending during cooking and exerts less strength on the wrist. These psychological feelings were identified through the Kansei Engineering method.

18.4 CONCLUSION

This intervention acts in the reduction of wrist pain by giving better posture of the wrist, less strain on the wrist and powerful grip for the users. People who have carpal tunnel syndrome and ganglion/mass in their wrist also will feel comfortable during the use of this design because they will not have to bend the wrist more during activities and hence will be more beneficial to their health. The usability, accessibility and desirability of this new design have a better review from the end user's perspective. All aspects of the design of a product can easily be perceived by the users. The product will become more user-friendly as importance was given from the perspective of the users during modification.

REFERENCES

Cronbach LJ. 1951. Coefficient alpha and the internal structure of tests. *Psychometrika*. 16:297–334.

Hignett S, McAtamney L. 2000. Rapid entire body assessment (REBA). *Appl Ergon*. 31:201–5.

Kuorinka I, Jonsson B, Kilbom A, Vinterberg H, Biering-Sørensen F, Andersson G, Jørgensen K, 1987. Standardised nordic questionnaires for the analysis of musculoskeletal symptoms. *Appl Ergon*. 18:233–7.

Lewis WG, Narayan CV. 1993. Design and sizing of ergonomic handles for hand tools. *Appl Ergon*. 24:351–6.

Moore JS. 1995. The Strain Index: a proposed method to analyze jobs for risk of distal upper extremity disorders. *Am Ind Hyg Assoc J*. 56:443–58.

Naddeo A, Califano R, Cappetti N, Vallone M 2015. *The effect of external and environmental factors on perceived comfort: the car-seat experience*. Dept. of Industrial Engineering. University of Salerno Italy, 291–308.

Nagamachi M. 1995. Kansei Engineering: a new ergonomic consumer-oriented technology for product development. *Int J Ind Ergon*. 15:63–74.

Rodriguez Y, Vina S, Montero R. 2013. A method for non-experts in assessing exposure to risk factors for work-related musculoskeletal disorders—ERIN. *Ind Health*. 51:622–6.

Wu SP, Hsieh CS. 2002. Ergonomics study on the handle length and lift angle for the culinary spatula. *Appl Ergon*. 33:493–501.

Wu SP, Yang C, Lin CH, Pai PK. 2015. Ergonomic design of Bent-Handled culinary spatula for female cook's stir-frying task. *J Food Process Technol*. 6:476.

Index

AAA games, 136
Accessibility features, crossing of, 149
ACC system, *see* Adaptive cruise control system
AD, *see* Autonomous driving
Adaptive cruise control (ACC) system, 63
ADS, *see* Automated driving system
Affective computing, 38
Affective disposition theory, 160
Agile startup environment, user testing in, 221
    agile software engineering (ASE), user-experience and HCD, 223
    Fanbot Places (case study), 223
        product, 224
        UX team at Fanbot, 224–225
    Lean Startup and Human-Centered Design (HCD), 222–223
    startup way, 222
    user testing iterations, 225–239
AI, *see* Artificial intelligence
Anthropometrical study, 281
    breast arc length approach, 281–282
    folding line approach, 282
Antic Meet (1958), 176
AR, *see* Augmented reality
Arc lengths, 283–284
Armani/Silos Museum (2015), 174
Artificial devices, 35
Artificial intelligence (AI), 36–38, 247
Assessment of Repetitive Tasks (ART) method, 338
Assistive technology (AT), 259, 261
Augmented reality (AR) concept, 83, 86, 88, 90, 92–94, 96–111
Automated driving system (ADS), 61, 66
Automated vehicle (AV) technologies, 59
Automated vehicles' human–machine interface (HMI), 59
    method, 65–66
    results, 66–75
    usability recommendations for, 61–65
Automotive industry, 47
Autonomous driving (AD), 82; *see also* Manual and autonomous driving
Avatar (2009), 179
AV technologies, *see* Automated vehicle technologies

Band panel algorithm, 287
Beds, dimensional aspects of, 293
    data collection procedure, 308
    dimensions of beds, procedures for determining, 299–307
    existing dimensions of beds, 294–295
    results, 308–315
    types of beds, 295, 296
        California King bed type, 299
        double/full bed type, 298
        Extra-long twin bed type, 298
        Extra small single bed type, 297
        King bed, 299
        King single bed type, 298
        Large single bed type, 297
        Queen bed type, 298–299
        Single/twin bed type, 297
        Single/twin extra-long bed type, 297
        Small double bed type, 298
        small single-sized bed, 297
        Super King bed type, 299
        Super King Size bed type, 299
        Super single bed type, 298
Benchmark, 204–209
Bioco, 184, 185
Bluetooth, 250, 251
Body Meets Dress and Dress Meets Body (1997), 175
Brainstorming, 25
Bras
    design and development, 283
        arc lengths, 283–284
        band panel algorithm, 287
        center panel algorithm, 287
        cup coverage algorithm, 287
        generating personalized bra models, 287
        landmarks, 283
        left and right bra cups algorithm, 287
        original references and base bra model, 284
        rhino/grasshopper, mechanism of, 285
        self-measuring approach, 283–284
    design exploration, 281
        anthropometrical study, 281–282
        bra design, 282
        parametric design method, 282–283
    future study, 289
        fit tests, 291
        optimization of parametric algorithm, 289–291
        personalization era for women's bras, 291
        prototyping and manufacturing, 291

motivation and goal, 280–281
traditional bra sizing method, 280
user test and evaluations, 287
   discussion, 288–289
   evaluation results, 288
   procedures, 288
   purpose of user test, 288
Breast arc length approach, 281–282

Calçada-Mar/Coblestone-Sea (2009), 180, 181
California King bed type, 299
Center panel algorithm, 287
Centro Technologic Automation of Galicia (CTAG), 84
Chanel Mobile Art Container (2008), 174
Chatbot, 224
Cinderella (2018), 182, 185–188
Cloth Dance (2012), 181, 182
Coat Stand (1929), 176
Cognitive walkthrough (CWT), 338, 343
Computer–user relationship, 34–36
Conceptual fashion, 175
Conflicts, resolving, 24–25
Consumer decision-making process, 194
Consumer Protection Code, 195
Consumer socialization, 192–194
Consumers' rights, 196, 197
Conversational speech accommodation, 118
Cooking spatula redesigning (case study), 336–341
Corporate social responsibility (CSR), 192
Costume, 173–174, 176
   and embodied experience in immersion, 177–179
Costume design
   for *Cinderella* (2018), 185–188
   for *Garb'urlesco* (2019), 183–185
   and scope of fashion as art, 174–177
Council of Supply Chain Management Professionals (CSCMP), 199
CSR, *see* Corporate social responsibility
CTAG, *see* Centro Technologic Automation of Galicia
Cup coverage algorithm, 287
Customer experience (CX), 10
CWT, *see* Cognitive walkthrough
CX, *see* Customer experience

Dark Souls game franchise, 138
Data collection and analysis, methods for, 266–270
Data flow diagrams (DFDs), 23
Deaf players, digital games accessibility for, *see* Digital games accessibility for deaf players
Decision-making process, 194

Dematerialization of tangible objects, 28–31
Design concept emergence, 25
   brainstorming, 25
DFDs, *see* Data flow diagrams
Digital assistant in student house; *see also* Voice assistants, survey on the use of; Voice-based digital home assistant
   aim, 125–126
   method, 126
   results, 126
Digital games accessibility for deaf players, 135
   analyzing games, 146
      accessibility features, crossing of, 149
      God of War (2018), 148
      The Last of Us Part 2 (2020), 148–149, 153
      Marvel's Spider-Man (2018), 147–148
   development, 137
      exclusion of deaf people in digital games, 141–142
      Universal Design, 137–139
      usability, 139–141
      user-experience, 142–143
   methodology, 143
      analyzed games, choice of, 144–146
   results evaluation, 153
Digital inclusion, 260–262
Digital interface, 3, 5, 7
Documenting processes, 19
Double/full bed type, 298
Dressed body, 176–177

Earcons, 66
Easy returns, 195–196
E-commerce logistics *vs.* traditional logistics, 200
E-commerce user-experience, 191
   analysis, 200–202
   consumer socialization, 192–194
   corporate social responsibility (CSR), 192
   market and user context
      in Brazil, 202–203
      in Europe, 203
      in Portugal, 203
   project, 204
      benchmark, 204–209
      ideation, 209–212
   selling and buying of clothes online, 194
      easy returns, 195–196
      reverse logistics system, impact on, 199–200
      serial returners, 194–195
      size inconsistencies, 196–199
Electroencephalography, 9
Elements of Nature, 184
Eliciting requirements, 18, 19
   documenting processes, 19–20
   generating requirements, 20

Index

development of personas, 20–21
stakeholder meeting, 20
Embodiment, 36
Encounters and difficulties, 259
   case studies, 265
      data collection and analysis, methods for, 266–270
      description, 265–266
   conceptual context, 260
      digital inclusion, 260–262
      human–computer interaction (HCI), human work in, 262–263
      user-experience, 263–264
   results and discussion, 270–276
Enterprise Resource Planning (ERP) resource, 242
Equal percentiles
   hip room evaluation, 324
   shoulder room evaluation, 324–325
Equitable Use, 139, 153
Ergonomics, 48
ERP resource, *see* Enterprise Resource Planning resource
ESTAL (*Escola Superior de Tecnologias e Artes de Lisboa*), 181
Exoskeleton systems, usability evaluation of, 47
   approach to assess exoskeleton usability level, 51–52
   features, 49
   operators participation in exoskeleton test, 51, 52
   usability score and arithmetic mean, 53
   workstation, 50
Experience design methods' application, user requirements analysis, 17
   analyzing requirements, 19, 23
      clarity, uniqueness and traceability, 24
      conflicts, resolving, 24–25
      design concept emergence, 25
   eliciting requirements, 18, 19
      documenting processes, 19–20
      generating requirements, 20–21
      stakeholder meeting, 20
   recording requirements, 19, 21
      data flow diagrams (DFDs), 23
      use cases, 21–22
      user stories, 22–23
   user requirements analysis activities, 18
Experience Society, 4
Expropriating bodies, 173
   costume design and scope of fashion as art, 174–177
   costumes and embodied experience in immersion, 177–179
   personal and social construction, 179–182

   practice-led research, immateriality and emotion in, 182
   Cinderella (2018), costume design for, 185–188
   Garb'urlesco (2019), costume design for, 183–185
Extra-long twin bed type, 298
Extra small single bed type, 297

Fanbot Places (case study), 223, 226
   product, 224
   UX team at Fanbot, 224–225
FCW, *see* Forward collision warning
Finger Gloves (1972), 178, 181
Flexibility in Use, 139, 153, 154
FMCode, 224, 227
Folding line approach, 282
Fondazione Prada (1993), 174
Forward collision warning (FCW), 90
Four elements of UX, 267–268
Functional requirements, 21

Garb'urlesco (2019), 182–185
Gender differences, 162
Generating requirements, 20–21
God of War (2018), 146, 148
God of War 2 (2007), 141
Google, 201
Google Play Games, 145

Haptic feedback, 67
HCD, *see* Human-Centered Design
HCI, *see* Human–computer interaction
Head-up display (HUD), 61, 67, 72, 83
Health/Cosmetics, 203
Health Safety and Environmental (HSE) Management System, 241–247
Hedonic system
   arousal and affect theories, 159–160
   examining sports media experience, 159
HFE, *see* Human factors and ergonomics professional
Hip room evaluation
   for equal percentiles, 324
   for unequal percentiles, 325–326
HMD (360-degree) video *vs.* 2D (monitor), 164
HMI, *see* Human–machine interaction; Human–machine interface
Homo sapiens, 37
Horizon Zero Dawn‡ (2017), 146
Horse gallop, 182
HSE Management System, *see* Health Safety and Environmental Management System
HTC Vive, 158
HUD, *see* Head-up display

Human agents and computational agents, equity between, 40
Human-Centered Design (HCD), 33, 222–223
Human–computer interaction (HCI), 5–7, 28, 32, 262–263
Human factors and ergonomics professional (HFE), 51–52
Humanity, 37
Human–machine interaction (HMI), 86, 87, 92, 94
Human–machine interface (HMI), 60, 64; *see also* Automated vehicles' human–machine interface (HMI)

IC, *see* Instrument cluster
Ideation, 209–212
Implied movement, 176
INCoDE.2030, 261
Inhabitant, 184
Instrument cluster (IC) concept, 83, 86, 88, 90, 92, 93, 94, 96–111
Intelligent life, 38
Interaction Design (IxD), 32, 33
Interaction designers, 31
Interaction experience (IxX), 40–41
Interactor, 39
    designing, 27
        affective computing, 38
        artificial intelligence (AI), 36–38
        computer–user relationship, 34–36
        dematerialization of tangible objects, 28–31
        equity between human agents and computational agents, 40
        Interaction Design (IxD), 32
        interaction experience (IxX), 40–41
        user *vs.* interactor, 38–40
Interface design, 36
International Organization for Standardization (ISO), 199
iPhones, 247
ISO, *see* International Organization for Standardization
ITC Sense of Presence Inventory, 165
iTMI, 201
IxD, *see* Interaction Design
IxX, *see* Interaction experience

Just Before Dawn (1967), 181

Kansei Engineering (KE), 338, 341, 342
Kansei questionnaire for new design, 344–345
KE, *see* Kansei Engineering
King bed, 299
King single bed type, 298

Lamentation (1930), 181
Large single bed type, 297
The Last of Us (2015), 136
The Last of Us Part 2 (2020), 136, 137, 146, 148–149, 153
Lateral sleep position, 305–306
LDUF, *see* Little Design Up Front
Lean Startup and Human-Centered Design (HCD), 222–223
Left and right bra cups algorithm, 287
Le Musée Yves Saint Laurent Paris (2017), 174
Little Design Up Front (LDUF), 222

Machine learning, 247
Major League Baseball (MLB), 158
Manual and autonomous driving
    activation, 97–100
    apparatus, 84–86
    experimental design, 90–94
    manual mode, 94–97
    participants, 84
    procedure, 86–90
    rear-end collision, 92, 93, 104–105
    stimuli, 86
    takeover request, 100–104
Market and user context
    in Brazil, 202–203
    in Europe, 203
    in Portugal, 203
Marvel's Spider-Man (2018), 146, 147–148, 149
Masked Feelings (2019), 180
Minimum viable product (MVP), 222
MLB, *see* Major League Baseball
Mobile app development, HSE management, 245–247
Mobile computing devices, 247
Mobile Web applications, 247
MSDs, *see* Musculoskeletal disorders
MUDA, 261
Muros de Abrigo, 180
Musculoskeletal disorders (MSDs), 342
MVP, *see* Minimum viable product

Native applications, 247
NDRTs, *see* Non-driving-related tasks
New Human Factors, 7
NextVR, 158
The Nielsen Company report, 202
Non-digital attributes, 194
Non-driving-related tasks (NDRTs), 67, 69
Non-functional requirements, 21
NPD Group (2020), 146

OA, *see* Operational Area
Oculus Rift, 158
Operational Area (OA), 106

Index

Paired *t*-test, 311
Parametric design method, 282–283
Parametric personalized bra algorithm
   band panel algorithm, 287
   center panel algorithm, 287
   cup coverage algorithm, 287
   generating personalized bra models, 287
   left and right bra cups algorithm, 287
   original references and base bra model, 284
   rhino/grasshopper, mechanism of, 285
PDA, *see* Personal digital assistant
Perceptible Information, 153, 154
Perfume and Cosmetics category, 202
Perfumery, 203
Personal digital assistant (PDA), 243
PlayStation 4, 145
PlayStation console, 141
Pong (1972), 141
Poor Theater, 177
Practice-led research, immateriality and emotion in, 182
   Cinderella (2018), costume design for, 185–188
   Garb'urlesco (2019), costume design for, 183–185
Prada Transformer (2009), 174
Product presentation, 194
Professional sports organizations, 158
Project, 204
   benchmark, 204–209
   ideation, 209–212
Prone sleeping position, 305–307
Pronomes/Pronouns (2001), 180
Prototyping, 23
PrüfExpress software, 245, 248
   architecture of system, 251–255
   concept, technology and programming languages, 250–251
   legal documentation, 249
   state-of-the-art mobile application, 248–249
   testing and documentation process, 250

Queen bed type, 298–299

Radio Frequency Identification (RFID) sensors, 242
Rapid Entire Body Assessment (REBA), 337
Reaction time, 93
Ready-to-wear garments, 198
Rear-end collision, 92, 93, 104–105
REBA, *see* Rapid Entire Body Assessment
Recording requirements, 19, 21
   data flow diagrams (DFDs), 23
   use cases, 21–22
   user stories, 22–23
Reverse logistics system, impact on, 199–200

RFID sensors, *see* Radio Frequency Identification sensors
RFID tags, 250–251
Rhino and Grasshopper program, 284, 285

SAE International, *see* Society of Automotive Engineers
Scenography, 179
Self-discovery, 180
Self-measuring approach, 283
   arc lengths, 283–284
   landmarks, 283
Selling and buying of clothes online, 194
   easy returns, 195–196
   reverse logistics system, impact on, 199–200
   serial returners, 194–195
   size inconsistencies, 196–199
Serial returners, 194–195
Shaping Things (Sterling), 30
Short-term memory load, minimizing, 118
Shoulder room evaluation
   for equal percentiles, 324–325
   for unequal percentiles, 326–327
Single/twin bed type, 297
Single/twin extra-long bed type, 297
SI score, *see* Strain index score
Size inconsistencies, 196–199
Small double bed type, 298
Small single-sized bed, 297
Smartphone application, 241
   final considerations, 255–257
   mobile app development, HSE management, 245–247
   PrüfExpress System case study, 248
   architecture of system, 251–255
   concept, technology and programming languages, 250–251
   legal documentation, 249
   state-of-the-art mobile application, 248–249
   testing and documentation process, 250
   theoretical framework, 242–245
Smartphones, 246
Society of Automotive Engineers (SAE International), 82
Software psychology, 6
Somatic costumes, 178
Sony's PlayStation 4, 145
Sound Suits (since 1992), 179
Spimes, 30
Sports fan, 162
Sports media, virtual reality in, 157
   examining sports media experience, hedonic system, 159
   expanding sports media viewing beyond 2D, 163–164

hedonic systems, arousal and affect theories, 159–160
pilot study, 164–166
technology manipulations to enhance arousal, 160–161
user characteristics, understanding impact of, 161–162
viewing context, 162–163
Sportswear and clothing, 203
SQLite database, 251
Stakeholder meeting, 20
Steam, 145
Strain index (SI) score, 337
Super King bed type, 299
Super King Size bed type, 299
Super single bed type, 298
Supine sleeping position, 305–307
SUS, *see* System Usability Scale
Suspended vehicle, 50
Sustainable Development, defined, 191–192
System perspective, 335
  cognitive walkthrough for new design, 343
  Kansei questionnaire for new design, 344–345
  system end user and product, 336
    cooking spatula redesigning, 336–341
System Usability Scale (SUS), 51–52

Take-over request (TOR), 60, 63, 64, 66, 67, 69–70, 71
Takeover/transition of control, 59
Tangible objects, dematerialization of, 28–31
Textile Resistance, 176
Time-to-collision (TTC) threshold, 90
TOR, *see* Take-over request
Traditional logistics *vs.* e-commerce logistics, 200
Transition of control, 59
Triadic Ballet (1927), 176
T-shirt-styled bra 3D model, 284
TTC threshold, *see* Time-to-collision threshold

UCD, *see* User-centered design
UI, *see* User interface
Uncanny valley, 132
Understanding Media: The Extensions of Man (McLuhan), 35
Unequal percentiles
  hip room evaluation, 325–326
  shoulder room evaluation, 326–327
Unicorn (1972), 178
Unique Forms of Continuity in Space (1913), 176
Universal Design, 137–139
Usability, defined, 6, 34, 139, 242
Usability to UX, 5–7
Use analysis, context of, 22

Use case modeling, 21–22
User-centered design (UCD), 33
User characteristics, understanding impact of, 161–162
User experience (UX); *see also Individual entries*
  concept, 33–34
  defined, 34
  design process, 10–11
  interviews and observations, 20
User-experience goals, 22
User interface (UI), 36, 224
User journey map, 19
User perception
  broad system interaction, 11–12
  digital product/usability of digital product interaction, 8–9
  specific product/service interaction, 9–10
User requirements analysis activities, 18
User versus interactor, 38–40

Viewing context, 162–163
Virtual reality (VR) technology, 158
Virtual reality in sports media, *see* Sports media, virtual reality in
Voice assistants; *see also* Digital assistant in student house
  acceptance of voice-based systems, 119
  functions considered available but not used, 120–121
  usefulness of voice-based assistant, 121–125
  use of voice systems and purposes of use, 119–120
Voice-based digital home assistant, 126
  aim, 126
  design guidelines to improve user growth, 132–133
  discussion, 131–132
  method, 127
  results, 127
    from experienced users, 129–131
    from inexperienced users, 128–129
Voice-based functions, encouraging learning of, 133
Voice interaction technology, 118
VR technology, *see* Virtual reality technology

Wagon cars, usability of back seat of, 319
  discussion, 328–333
  goal of the research, 320
  method
    anthropometric dimensions, selection of, 322–323
    measurement points, selection of, 323
    subjects and percentiles, selection of, 323
    wagons, selection of, 321–322

results, 323
    hip room evaluation, equal percentiles, 324
    hip room evaluation, unequal percentiles, 325–326
    shoulder room evaluation, equal percentiles, 324–325
    shoulder room evaluation, unequal percentiles, 326–327
Wear To Move (2014), 178
Webshopper report, 202

Wilcoxon analyses, 99, 102, 104
Wilcoxon Signed Rank Test, 313, 314
Work-related musculoskeletal disorders, 48
Workstation, 50
    at assembly sector, 51

Xamarin tool, 251

Z-Statistic, 313–315
Zumero, 251

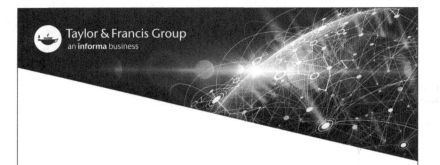

9780367357719